IET ENERGY ENGINEERING 194

# Offshore Wind Power

**Other volumes in this series:**

Volume 1 **Power Circuit Breaker Theory and Design** C.H. Flurscheim (Editor)
Volume 4 **Industrial Microwave Heating** A.C. Metaxas and R.J. Meredith
Volume 7 **Insulators for High Voltages** J.S.T. Looms
Volume 8 **Variable Frequency AC Motor Drive Systems** D. Finney
Volume 10 **$SF_6$ Switchgear** H.M. Ryan and G.R. Jones
Volume 11 **Conduction and Induction Heating** E.J. Davies
Volume 13 **Statistical Techniques for High Voltage Engineering** W. Hauschild and W. Mosch
Volume 14 **Uninterruptible Power Supplies** J. Platts and J.D. St Aubyn (Editors)
Volume 15 **Digital Protection for Power Systems** A.T. Johns and S.K. Salman
Volume 16 **Electricity Economics and Planning** T.W. Berrie
Volume 18 **Vacuum Switchgear** A. Greenwood
Volume 19 **Electrical Safety: A guide to causes and prevention of hazards** J. Maxwell Adams
Volume 21 **Electricity Distribution Network Design, 2nd Edition** E. Lakervi and E.J. Holmes
Volume 22 **Artificial Intelligence Techniques in Power Systems** K. Warwick, A.O. Ekwue and R. Aggarwal (Editors)
Volume 24 **Power System Commissioning and Maintenance Practice** K. Harker
Volume 25 **Engineers' Handbook of Industrial Microwave Heating** R.J. Meredith
Volume 26 **Small Electric Motors** H. Moczala *et al.*
Volume 27 **AC-DC Power System Analysis** J. Arrillaga and B.C. Smith
Volume 29 **High Voltage Direct Current Transmission, 2nd Edition** J. Arrillaga
Volume 30 **Flexible AC Transmission Systems (FACTS)** Y-H. Song (Editor)
Volume 31 **Embedded Generation** N. Jenkins *et al.*
Volume 32 **High Voltage Engineering and Testing, 2nd Edition** H.M. Ryan (Editor)
Volume 33 **Overvoltage Protection of Low-Voltage Systems, Revised Edition** P. Hasse
Volume 36 **Voltage Quality in Electrical Power Systems** J. Schlabbach *et al.*
Volume 37 **Electrical Steels for Rotating Machines** P. Beckley
Volume 38 **The Electric Car: Development and future of battery, hybrid and fuel-cell cars** M. Westbrook
Volume 39 **Power Systems Electromagnetic Transients Simulation** J. Arrillaga and N. Watson
Volume 40 **Advances in High Voltage Engineering** M. Haddad and D. Warne
Volume 41 **Electrical Operation of Electrostatic Precipitators** K. Parker
Volume 43 **Thermal Power Plant Simulation and Control** D. Flynn
Volume 44 **Economic Evaluation of Projects in the Electricity Supply Industry** H. Khatib
Volume 45 **Propulsion Systems for Hybrid Vehicles** J. Miller
Volume 46 **Distribution Switchgear** S. Stewart
Volume 47 **Protection of Electricity Distribution Networks, 2nd Edition** J. Gers and E. Holmes
Volume 48 **Wood Pole Overhead Lines** B. Wareing
Volume 49 **Electric Fuses, 3rd Edition** A. Wright and G. Newbery
Volume 50 **Wind Power Integration: Connection and system operational aspects** B. Fox *et al.*
Volume 51 **Short Circuit Currents** J. Schlabbach
Volume 52 **Nuclear Power** J. Wood
Volume 53 **Condition Assessment of High Voltage Insulation in Power System Equipment** R.E. James and Q. Su
Volume 55 **Local Energy: Distributed generation of heat and power** J. Wood
Volume 56 **Condition Monitoring of Rotating Electrical Machines** P. Tavner, L. Ran, J. Penman and H. Sedding
Volume 57 **The Control Techniques Drives and Controls Handbook, 2nd Edition** B. Drury
Volume 58 **Lightning Protection** V. Cooray (Editor)
Volume 59 **Ultracapacitor Applications** J.M. Miller
Volume 62 **Lightning Electromagnetics** V. Cooray
Volume 63 **Energy Storage for Power Systems, 2nd Edition** A. Ter-Gazarian
Volume 65 **Protection of Electricity Distribution Networks, 3rd Edition** J. Gers
Volume 66 **High Voltage Engineering Testing, 3rd Edition** H. Ryan (Editor)
Volume 67 **Multicore Simulation of Power System Transients** F.M. Uriate
Volume 68 **Distribution System Analysis and Automation** J. Gers
Volume 69 **The Lightening Flash, 2nd Edition** V. Cooray (Editor)

Volume 70 **Economic Evaluation of Projects in the Electricity Supply Industry, 3rd Edition** H. Khatib
Volume 72 **Control Circuits in Power Electronics: Practical issues in design and implementation** M. Castilla (Editor)
Volume 73 **Wide Area Monitoring, Protection and Control Systems: The enabler for smarter grids** A. Vaccaro and A. Zobaa (Editors)
Volume 74 **Power Electronic Converters and Systems: Frontiers and applications** A.M. Trzynadlowski (Editor)
Volume 75 **Power Distribution Automation** B. Das (Editor)
Volume 76 **Power System Stability: Modelling, analysis and control** A.A. Sallam and B. Om P. Malik
Volume 78 **Numerical Analysis of Power System Transients and Dynamics** A. Ametani (Editor)
Volume 79 **Vehicle-to-Grid: Linking electric vehicles to the smart grid** J. Lu and J. Hossain (Editors)
Volume 81 **Cyber-Physical-Social Systems and Constructs in Electric Power Engineering** S. Suryanarayanan, R. Roche and T.M. Hansen (Editors)
Volume 82 **Periodic Control of Power Electronic Converters** F. Blaabjerg, K. Zhou, D. Wang and Y. Yang
Volume 86 **Advances in Power System Modelling, Control and Stability Analysis** F. Milano (Editor)
Volume 87 **Cogeneration: Technologies, optimisation and implementation** C.A. Frangopoulos (Editor)
Volume 88 **Smarter Energy: From smart metering to the smart grid** H. Sun, N. Hatziargyriou, H.V. Poor, L. Carpanini and M.A. Sánchez Fornié (Editors)
Volume 89 **Hydrogen Production, Separation and Purification for Energy** A. Basile, F. Dalena, J. Tong and T.N. Veziroğlu (Editors)
Volume 90 **Clean Energy Microgrids** S. Obara and J. Morel (Editors)
Volume 91 **Fuzzy Logic Control in Energy Systems with Design Applications in MATLAB®/Simulink®** İ.H. Altaş
Volume 92 **Power Quality in Future Electrical Power Systems** A.F. Zobaa and S.H.E.A. Aleem (Editors)
Volume 93 **Cogeneration and District Energy Systems: Modelling, analysis and optimization** M.A. Rosen and S. Koohi-Fayegh
Volume 94 **Introduction to the Smart Grid: Concepts, technologies and evolution** S.K. Salman
Volume 95 **Communication, Control and Security Challenges for the Smart Grid** S.M. Muyeen and S. Rahman (Editors)
Volume 96 **Industrial Power Systems with Distributed and Embedded Generation** R Belu
Volume 97 **Synchronized Phasor Measurements for Smart Grids** M.J.B. Reddy and D.K. Mohanta (Editors)
Volume 98 **Large Scale Grid Integration of Renewable Energy Sources** A. Moreno-Munoz (Editor)
Volume 100 **Modeling and Dynamic Behaviour of Hydropower Plants** N. Kishor and J. Fraile-Ardanuy (Editors)
Volume 101 **Methane and Hydrogen for Energy Storage** R. Carriveau and D.S-K. Ting
Volume 104 **Power Transformer Condition Monitoring and Diagnosis** A. Abu-Siada (Editor)
Volume 106 **Surface Passivation of Industrial Crystalline Silicon Solar Cells** J. John (Editor)
Volume 107 **Bifacial Photovoltaics: Technology, applications and economics** J. Libal and R. Kopecek (Editors)
Volume 108 **Fault Diagnosis of Induction Motors** J. Faiz, V. Ghorbanian and G. Joksimović
Volume 110 **High Voltage Power Network Construction** K. Harker
Volume 111 **Energy Storage at Different Voltage Levels: Technology, integration, and market aspects** A.F. Zobaa, P.F. Ribeiro, S.H.A. Aleem and S.N. Afifi (Editors)
Volume 112 **Wireless Power Transfer: Theory, technology and application** N. Shinohara
Volume 114 **Lightning-Induced Effects in Electrical and Telecommunication Systems** Y. Baba and V.A. Rakov
Volume 115 **DC Distribution Systems and Microgrids** T. Dragičević, F. Blaabjerg and P. Wheeler
Volume 116 **Modelling and Simulation of HVDC Transmission** M. Han (Editor)
Volume 117 **Structural Control and Fault Detection of Wind Turbine Systems** H.R. Karimi

Volume 119 **Thermal Power Plant Control and Instrumentation: The control of boilers and HRSGs, 2nd Edition** D. Lindsley, J. Grist and D. Parker
Volume 120 **Fault Diagnosis for Robust Inverter Power Drives** A. Ginart (Editor)
Volume 121 **Monitoring and Control Using Synchrophasors in Power Systems with Renewables** I. Kamwa and C. Lu (Editors)
Volume 123 **Power Systems Electromagnetic Transients Simulation, 2nd Edition** N. Watson and J. Arrillaga
Volume 124 **Power Market Transformation** B. Murray
Volume 125 **Wind Energy Modeling and Simulation, Volume 1: Atmosphere and plant** P. Veers (Editor)
Volume 126 **Diagnosis and Fault Tolerance of Electrical Machines, Power Electronics and Drives** A.J.M. Cardoso
Volume 128 **Characterization of Wide Bandgap Power Semiconductor Devices** F. Wang, Z. Zhang and E.A. Jones
Volume 129 **Renewable Energy From the Oceans: From wave, tidal and gradient systems to offshore wind and solar** D. Coiro and T. Sant (Editors)
Volume 130 **Wind and Solar Based Energy Systems for Communities** R. Carriveau and D.S-K. Ting (Editors)
Volume 131 **Metaheuristic Optimization in Power Engineering** J. Radosavljević
Volume 132 **Power Line Communication Systems for Smart Grids** I.R.S. Casella and A. Anpalagan
Volume 134 **Hydrogen Passivation and Laser Doping for Silicon Solar Cells** Brett Hallam and Catherine Chan (Editors)
Volume 139 **Variability, Scalability and Stability of Microgrids** S.M. Muyeen, S.M. Islam and F. Blaabjerg (Editors)
Volume 145 **Condition Monitoring of Rotating Electrical Machines** P. Tavner, L. Ran, C. Crabtree
Volume 146 **Energy Storage for Power Systems, 3rd Edition** A.G. Ter-Gazarian
Volume 147 **Distribution Systems Analysis and Automation, 2nd Edition** J. Gers
Volume 151 **SiC Power Module Design: Performance, robustness and reliability** Alberto Castellazzi and Andrea Irace (Editors)
Volume 152 **Power Electronic Devices: Applications, failure mechanisms and reliability** F. Iannuzzo (Editor)
Volume 153 **Signal Processing for Fault Detection and Diagnosis in Electric Machines and Systems** Mohamed Benbouzid (Editor)
Volume 155 **Energy Generation and Efficiency Technologies for Green Residential Buildings** D. Ting and R. Carriveau (Editors)
Volume 156 **Lithium-Ion Batteries Enabled by Silicon Anodes** Chunmei Ban and Kang Xu (Editors)
Volume 157 **Electrical Steels, 2 Volumes** A. Moses, K. Jenkins, Philip Anderson and H. Stanbury
Volume 158 **Advanced Dielectric Materials for Electrostatic Capacitors** Q Li (Editor)
Volume 159 **Transforming the Grid Towards Fully Renewable Energy** O. Probst, S. Castellanos and R. Palacios (Editors)
Volume 160 **Microgrids for Rural Areas: Research and case studies** R.K. Chauhan, K. Chauhan and S.N. Singh (Editors)
Volume 166 **Advanced Characterization of Thin Film Solar Cells** N. Haegel and M. Al-Jassim (Editors)
Volume 167 **Power Grids with Renewable Energy: Storage, integration and digitalization** A.A. Sallam and B. Om P. Malik
Volume 170 **Reliability of Power Electronics Converters for Solar Photovoltaic Applications** Frede Blaabjerg, Ahteshamul Haque, Huai Wang, Zainul Abdin Jaffery and Yongheng Yang (Editors)
Volume 171 **Utility-Scale Wind Turbines and Wind Farms** A. Vasel-Be-Hagh and D.S-K. Ting
Volume 172 **Lighting Interaction with Power Systems, 2 Volumes** A. Piantini (Editor)
Volume 174 **Silicon Solar Cell Metallization and Module Technology** T. Dullweber (Editor)
Volume 193 **Overhead Electric Power Lines: Theory and practice** S. Chattopadhyay and A. Das
Volume 198 **Battery Management Systems and Inductive Balancing** A. Van den Bossche and A. Farzan Moghaddam
Volume 199 **Model Predictive Control for Microgrids: From power electronic converters to energy management** J. Hu, J.M. Guerrero and S. Islam
Volume 905 **Power System Protection, 4 Volumes** The Electricity Training Association (Editor)

# Offshore Wind Power

## Reliability, availability and maintenance
## 2nd edition

Peter Tavner

The Institution of Engineering and Technology

Published by The Institution of Engineering and Technology, London, United Kingdom

The Institution of Engineering and Technology is registered as a Charity in England & Wales (no. 211014) and Scotland (no. SC038698).



First published 2012

Second edition 2021



The Institution of Engineering and Technology
Michael Faraday House
Six Hills Way, Stevenage
Herts, SG1 2AY, United Kingdom

www.theiet.org



**British Library Cataloguing in Publication Data**
A catalogue record for this product is available from the British Library

**ISBN 978-1-83953-333-4 (hardback)**
**ISBN 978-1-83953-334-1 (PDF)**

Typeset in India by MPS Limited
Printed in the UK by CPI Group (UK) Ltd, Croydon

*This book is dedicated to*
*Charles and Daisy Tavner*
*Sarah, Felix, Rafa and Stella Pyne*

Behold, the sea itself
And on its limitless, heaving breast, the ships;
See where their white sails, bellying in the wind,
Speckle the green and blue sea.

*Walt Whitman, put to music in the Sea Symphony by*
*Ralph Vaughan Williams*

# Contents

# About the author

**Peter Tavner**, Eur Ing, CEng, FIET is an Emeritus Professor of Durham University, Department of Engineering, UK. He received an MA from Cambridge University in 1972, a PhD from Southampton University in 1978 and a DSc from Durham University in 2012.

He served as a Weapons Electrical Officer in the Royal Navy in General Service on HM Ships ASHANTI, BRISTOL and LONDON, subsequently working in an MRC laboratory, then teaching at the University of Benin, Benin City, Nigeria. He then held research and technical positions in the electrical supply and manufacturing industries, including Group Head of a former CEGB laboratory, Technical Director of both Laurence, Scott & Electromotors Ltd (LSE) and Brush Electrical Machines Ltd (BEM), two of the UK's large electrical machine manufacturers, and Group Technical Director of FKI Energy Technology, the owner of LSE and BEM, at that time an international business manufacturing wind turbines, electrical machines and drives in the United Kingdom, Holland, Italy, Germany and the Czech Republic.

In 2003 he was appointed a Professor of New and Renewable Energy at Durham University, leading several UK and European research initiatives to improve wind, wave and tidal reliability, being the Head of the Engineering Department at Durham from 2006 to 2010. More recently, he has provided consultancy in the United Kingdom with his former company LSE and in China to Dong Fang Electrical Machinery (DFEM) in Deyang, Szichuan, he has also taught at Chongqing University (CQU).

He retains an interest in practical electro-magnetic analysis, the application of condition monitoring to improve the reliability of power engineering electrical machines and converters and especially for the development of wind, tidal and wave power.

In 1988 he was winner with Richard Jackson of the IEE Institution Premium and in 2020 received the Scientific Award of the European Academy of Wind Energy for his pioneering role in wind farm reliability, availability and maintenance research.

He was formerly a non-Executive Director of UK's New and Renewable Energy Centre (NaREC), Blyth, now part of ORE Catapult, is an Honorary Member of the European Academy of Wind Energy and is or has been a Chair or Member of the Scientific Advisory Boards of these UKRI research consortia:

- Centre for Power Electronics led by Nottingham University;
- Holistic Operation and Maintenance for Energy from Off-shore Wind Farms (HOME) Project led by Manchester University;
- Prosperity Partnership for Off-shore Wind (NPOW) a collaboration between Sheffield, Durham and Hull Universities;
- Future Electrical Machine Manufacturing (FEMM) Hub a collaboration between Sheffield, Strathclyde and Newcastle Universities.

# Preface

This book is not about wind turbines, per se, but about off-shore wind farm power generation and collection systems, including the foundations, turbines, electrical connections to shore, sub-stations and converter stations, but particularly about their overall availability, reliability and maintenance.

This is a subject of paramount importance to the long-term survivability of off-shore wind power generation. There have been several modern, substantial and relevant wind power texts, including the following:

1. Hau, E. (2006) *Wind Turbines-Fundamentals, Technologies, Application and Economics*. 2nd Ed. Springer-Verlag, 783 pp.
2. Manwell, J. F., McGowan, J. G., Rogers, A. L. (2009) *Wind Energy Explained: Theory, Design and Application*, 2nd Ed. Wiley, Chichester, UK, 688 pp.
3. Twidell, J. W., Gaudiosi, G., Eds (2009) Off-shore Wind Power, Multi-science Publishing Co., UK, 300 pp.
4. Burton, T., Jenkins, N., Sharpe, D., Bossanyi, E. (2011) *Wind Energy Handbook*, 2nd Ed., John Wiley & Sons, Ltd, UK, 775 pp.
5. Gasch, R., Twele, J., Eds (2012) *Wind Power Plants-Fundamentals, Design, Construction and Operation*. 2nd Ed., Springer Verlag, Berlin, Germany, 548 pp.
6. Hogg, S., Crabtree, C. J., Eds (2017) *UK Wind Energy Technologies*. Routledge, Earthscan, Abingdon, UK, 290 pp.
7. Gipe, P. (2016) *Wind Energy for the Rest of Us: A Comprehensive Guide to Wind Power and How to Use It, Wind Works*. Bakersfield (CA) USA, 491 pp.
8. Ng, C., Ran, L., Eds (2016) *Off-shore Wind Farms-Technologies, Design and Operation*. Woodhead, Elsevier, Oxford, 634 pp.

However, these important works deal only peripherally with reliability, availability and maintenance, whereas this present book considers this aspect specifically but each of the previous texts will be quoted in this book, where appropriate.

The development of wind power has become an expanding and successful modern world energy activity in which the United Kingdom and other European, Asian and American countries have been taking a major part, driven by the need to find new electrical power sources avoiding the use fossil fuels. However, the deployment of off-shore wind farms has resulted even in greater interest and investment because of their improved performance in those locations.

In fact the extensive wind resource above the Irish and British seas, the prospect of long-term generation earnings and the knowledge that their environmental impact is likely to be low, has accelerated their development around our islands.

Major issues are being solved to reduce off-shore wind power costs and these revolve around the need to capture this energy at a cost per MWh competitive with other practicable sources, initially by reducing capital expenditure to secure the repayment of the initial investment with interest. But it also depends upon the reliability, availability and longevity of the wind turbines and their collection systems, making up these off-shore wind farms, that is their operational expenditure (OPEX). The cost-effectiveness of maintenance needed to achieve that availability and longevity is essential to improve off-shore wind life cycle costs and the industry's emerging future can only be achieved if OPEX is reduced. It is estimated that OPEX costs are currently between 25% and 40% of the lifetime Cost of Energy from an offshore wind farm.

This book goes beyond previous books to address the off-shore wind generation OPEX issue head-on, based on the Author's experience, to demonstrate clearly to manufacturers, developers and operators the facts and figures of off-shore wind turbine operation and maintenance in an inclement off-shore environment.

However, since the first edition of this book, it has become clear that the future success of off-shore wind depends not only on wind turbine performance but also on the reliability and operational performance of the very large arrays in which they are being deployed. This book will recommend how the maintenance of large off-shore wind farms could be done to achieve lower life cycle costs. The Author believes that this is not only important for the off-shore wind industry but also for the future of the other embryonic renewable energy technologies, such as wave and tidal, because there will be similar operational problems for all three emerging renewable energy industries.

The Author has been working on the wind industry reliability problem for 20 years but power equipment reliability had been the Author's technical interest for 50 years, initiated in the 1970s and 1980s during my PhD and in the UK power generation industry, when we were subjected in the United Kingdom to a series of difficult generator failures, on the introduction of our large 1–2-GW coal-, oil- and nuclear-fired power stations. It became rather like a technical detective story, and failure data were essential to resolve fundamental design, mechanical, electrical, quality and operational issues. The analysis of those data proved to be the only thorough engineering way to advance. Within a few years, most problems had been solved and those power stations with problematic reliability in 1970 were improved and went onto deliver more than 40–48 years of reliable service, no mean feat in a predominantly coal-fired era.

That experience, in the fossil- and nuclear-fired electricity supply industry, has provided many wind industry through-life cost lessons. However, fossil- and nuclear-fired power stations are in effect purpose-designed, well-housed, power factories, manned 24 h a day, 7 days a week, whose effectiveness has been improved and demonstrated.

The Author had an early naval training and knows from ship operations that good design, manufacture and maintenance is essential to keep ships operational on the high seas, assisted by the fact that they are also manned 24/7, 365 days a year. The efficacy of our maritime trade over the last 500 years demonstrates how this can be achieved.

The oil and gas industries have learnt how to install, maintain and operate effective off-shore engineering structures in production over 50 years, including important lessons about reducing operational offshore manning levels.

Off-shore wind technology has some similarities to all the previously mentioned ones but consists of unmanned, off-shore, robotic power units operating 24/7, 365 days a year, controlled from remote on-shore control rooms where manning levels must be kept very low. The engineering issues facing us as we build, maintain and keep these wind power stations at high degrees of operational readiness with low manning levels presents a fascinating challenge, assisted by former marine, power station and off-shore oil and gas experiences.

However, the off-shore wind industry also needs innovation, new technology, good manufacture and excellent management to become successful and that now applies to the industry's reliability, availability and maintenance.

Andrew Garrad, the co-founder of the UK wind consultancy Garrad Hassan, now part of DNV GL, said 'that for a long time the mantra of the wind turbine industry has been "bigger and bigger" but now it has moved to "better and better" and this marks a change in the areas of innovation', see Jamieson (2011), essential for future off-shore wind farm systems' lower CAPEX and, from this book's point of view, lower OPEX.

For this Second Edition I have revised the earlier edition to bring it up-to-date and expanded it to include power collection and transmission from off-shore wind farms, particularly including some materials on floating off-shore wind turbines and their support structures, where reliability and availability projections are available but unproven.

I have updated the references into a single list of key books and papers, in date order and Harvard format, including relevant doctoral theses, at the end of the book, based on subject bibliographies. I have also added a list of appropriate off-shore wind standards.

I have updated the appendices with a defect reporting form, methods of collecting reliability data, condition monitoring and SCADA review documents, and these have proved valuable for operators and maintainers who need to know precisely what information they should collect and collate. The appendices also include a summary of key equipment failure rates, taken from literature, which will enable researchers, using an FMEA approach, to develop their own view of off-shore wind farm reliability.

Hau (2006) said 'The utilisation of wind energy is not a new technology but draws on the rediscovery of a long tradition of wind power', therefore I have updated the historical timeline at the end of the book, to show the development of wind power over many centuries, but particularly the rapid development in the last

130 years and growth of off-shore wind power in the last 20 years. I have found this timeline valuable for younger engineers to put our present work in context.

I hope that this book, written from a UK perspective and based upon our own research at Durham, will help you to achieve our wind energy future.

Peter Tavner
Durham and Cambridge, 2021

# Acknowledgements

I am indebted to a number of colleagues whose work has contributed to this book, most particularly my former PhD students Christopher Crabtree, Tatiana Delorm, Fabio Spinato, Michael Wilkinson, Mahmoud Zaggout and Donatella Zappalá but also undergraduates, research students and postdoctoral research workers, including Hooman Arabian-Hoseynabadi, Wenjuan Blackwood, Lucy Collingwood, Nathaniel Cowton, Sinisa Djurovic, Yanhui Feng, Rosa Gindele, Peter Greaves, Andrew Higgins, Mark Knowles, Ting Lei, Luke Longley, Yingning Qiu, Paul Richardson, Sajjad Tohidi, Xiaoyan Wang, Matthew Whittle, Peter Wyllie, Jianping Xiang and Wenxian Yang.

I make a particular acknowledgement to Professor Percy Hammond, my PhD Supervisor whose inspiration taught me electromagnetism and to Professor Kurt Schwarz, my former Technical Director, who taught me the consequent logic of electrical machines and drives, on which my understanding of reliability has been based.

I am also indebted to other academic colleagues Janusz Białek, Jim Bumby, Will Coombs, Rob Dominy, Simon Hogg, Grant Ingram, Hui Long, Li Ran and Phil Taylor in Durham, Mike Barnes, Geoff Dutton, Sandy Smith and Simon Watson, members of the Supergen Wind and HOME Consortia, Stefan Faulstich, Berthold Hahn, Stefan Barth and other European Academy of Wind Energy colleagues who opened my eyes towards the knowledge developed in wind power throughout Europe and to Shawn Sheng of NREL, USA for their collaboration.

I would particularly like to mention five wind power exponents who I have met or worked with, whose work has been a foundation for me:

1. Professor Jurgen Schmid of Kassel University, founder of the European Academy of Wind Energy, who started wind turbine performance investigation with his book Schmid *et al.* (1991);
2. Professor Gerard van Bussel of Delft University, who wrote the first off-shore wind power reliability papers, drawing my attention to this important issue, see van Bussel *et al.* (1997);
3. Professor Bill Leithead, of Strathclyde University, an expert on wind turbine control, who had formerly worked for Wind Energy Group on the design of their huge LS1 wind turbine installed in Orkney, see Leithead (2007);
4. Professor Joachim Peinke, Carl von Ossietzky Universität Oldenburg, Germany, who taught me the importance of turbulence on wind turbine and farm performance, see Milan *et al.* (2013) and described results from the Growian, Große Windenergieanlage, which demonstrated these issues;

5. Peter Jamieson, formerly of Howdens, GL Garrad Hassan and subsequently Strathclyde University, with long wind industry knowledge and experience, see Jamieson (2011), having worked on Howden's HWP 55 wind turbine installed at Richborough, for which my own company supplied the induction generator, for his contribution to improving large wind turbine design;
6. Nick Hayward, formerly of Converteam, GE and subsequently Siemens, for his advice on the design and reliability of power electronics, particularly for use in wind turbines.

I also express appreciation for research funding which made the student, postgraduate and research contributions to this book possible:

- UKRI, Engineering and Physical Science Research Council:
  - Supergen Wind Phases 1, 2 and 3 plus Hub, lead initially by Durham University and subsequently by Strathclyde University;
  - EPSRC Centre for Power Electronics lead by Nottingham University;
  - HOME Off-shore, led by Manchester University;
  - NPOW New Prosperity Partnership in Off-shore Wind, led by Sheffield with Durham and Hull Universities.
- European Union for FP 7 ReliaWind involving Durham University.

Finally I thank colleagues in a number of industrial organisations who have assisted by providing data or photographs, including Alnmaritec, ABB Drives, Alstom Wind Power, Clipper Wind Marine, Converteam, GL Garrad Hassan, Flylogix, Hansen Transmissions, Off-shore Catapult, Oersted, Siemens-Gamesa Wind Power, Swire Blue Ocean & Windcats.

# Nomenclature

| Symbol | Explanation |
|---|---|
| $A$ | designates wind turbine category for higher turbulence characteristics |
| $B$ | designates wind turbine category for medium turbulence characteristics |
| $C$ | designates wind turbine category for lower turbulence characteristics |
| $A$ | availability, $A = MTBF/(MTBF+MTTR)$ |
| $Acc$ | acceleration factor for accelerated life testing |
| $AEP$ | annual energy production, MWh |
| $C$ | capacity factor, % |
| $CoE$ | cost of energy, £/MWh |
| $F$ or $F^{-1}$ | forward or backward fast Fourier transform |
| $FCR$ | fixed charge rate for interest, % |
| $\eta$ | efficiency |
| $H_s$ | wave height for sea-state |
| $ICC$ | initial capital cost, £ |
| $I$ | drive train inertia, kg m$^2$ |
| $I$ | turbulence intensity, $\sigma/u$, defined by IEC 61400-1 |
| $I_{char}$ | turbulence characteristic, defined by IEC 61400-1 |
| $I_{ref}$ | expected value of turbulence intensity at $u_{ref}$ 15 m/s |
| $k$ | constant in power balance equations |
| $ku_n$ | turbulence coefficient at a wind speed $u$ of $n$, m/s |
| $\lambda(t)$ | instantaneous hazard function for a sub-assembly or machine, failures/sub-assembly/year |
| $\lambda$ | failure rate of a sub-assembly or machine varying with time, failures/sub-assembly/year, $1/MTBF$ |
| $\mu$ | instantaneous repair rate, $1/MTTR$ |
| $N$ | speed of a machine rotor, rev/min |
| $n$ | number of years |
| $P$ | power, watt |
| $P_{det}$ | probability of detection of a fault |

| | |
|---|---|
| $p$ | integer number of pole pairs in a motor or generator |
| $Q$ | heat flow, watt/m$^2$ |
| $R$ | resistance, Ω |
| $R(t)$ | reliability or survivor function of a population of sub-assemblies as a function of time, failures/machine/year |
| $r$ | discount rate, % |
| $S$ | specific energy yield, MWh/m$^2$/year |
| $\sigma$ | wind speed standard deviation |
| $T$ | torque, N m |
| $T$ | temperature, °C |
| $\Delta T$ | temperature rise, K or deg C |
| $T$ | period of a wave, s |
| $u$ | wind speed, m/s |
| $u_{ref}$ | reference wind speed, m/s |
| $\theta$ | MTBF of an assembly or sub-assembly, $\theta = 1/\lambda$, h |
| $V_{ref}$ | mean wind speed at wind turbine hub height, m/s |
| $V$ | rms voltage, $V$ |
| $W$ | work done in a wind turbine drive train, MWh |
| $\omega$ | angular frequency, rad/s |

# Abbreviations

| Symbol | Explanation |
|---|---|
| AM | Asset management |
| AEP | Annualised energy production |
| AIP | Artemis Intelligent Power now part of Danfoss |
| ALT | Accelerated life testing |
| BDFIG | Brushless doubly red induction generator |
| BEM | Brush Electrical Machines, UK |
| BMS | Blade monitoring system |
| BOP | Balance of plant |
| CAPEX | Capital expenditure |
| CBM | Condition-based maintenance |
| CMS | Condition monitoring system |
| CMTR | Condition monitoring test rig |
| CoE | Cost of energy |
| DE | Drive end of generator or gearbox |
| DCS | Distributed control system |
| DDT | Digital drive technology (AIP) |
| DDPMG | Direct drive permanent magnet synchronous generator |
| DDWRSGE | Direct drive wound rotor synchronous generator with exciter |
| DFEM | Dong Fang Electrical Machinery, China |
| DFIG | Doubly fed induction generator |
| DP | Dynamic positioning |
| EAWE | European Academy of Wind Energy |
| EFC | Emergency feather control |
| EHV | Extra high voltage, 345–765 kV |
| EPRI | Electric Power Research Institute, USA |
| EWEA | European Wind Energy Association |
| FM | Field maintenance |
| FBG | Fibre Bragg grating |

| | |
|---|---|
| FCR | Fixed charge rate, interest rate on borrowed money |
| FFT | Fast Fourier transform |
| FMEA | Failure modes and effects analysis |
| FMECA | Failure modes, effects and criticality analysis |
| FSV | Oil field support vessel |
| GB | Ground benign |
| HAWT | Horizontal axis wind turbine |
| HM | Health monitoring |
| H&S | Health and safety |
| HV | High voltage, 100–345 kV |
| HPP | Homogeneous Poisson process |
| HSS | High-speed shaft, e.g. gear-wheel of a gearbox |
| IM | Information management |
| IMS or IS | Intermediate speed, e.g. gear-wheel of a gearbox |
| I/O | Input–output information, SCADA |
| IP | Intellectual property |
| ICS | Integrated control system |
| IEC | International Electrotechnical Commission |
| IET | Institution of Engineering and Technology (formerly IEE) |
| IMS | Intermediate shaft, gearbox |
| IEEE | Institute of Electrical and Electronic Engineers |
| LCC | Life cycle costing |
| LCOE | Lifetime cost of energy |
| LV | Low voltage, 100–1,000 V |
| LSE | Laurence, Scott & Electromotors Ltd, UK |
| LSS | Low-speed shaft, e.g. gear-wheel of a gearbox |
| LWK | Landwirtschaftskammer Schleswig-Holstein database from Germany |
| MM | Maintenance management |
| MM | Marelli Motori, Italy |
| MCA | Marine and Coastguard Agency |
| MEC | Marine energy converter, could be an off-shore wind or wave or tidal device |
| MIL-HDBK | US Reliability Military Handbook |
| MOB | Man over-board boat |
| MOTS | MOMAC off-shore transfer system |

| | |
|---|---|
| MTBF | Mean time between failures, h |
| MTTR | Mean time to repair, h |
| MV | Medium voltage, 1–100 kV |
| NAREC | National Renewable Energy Centre, Blyth, UK, now ORE Catapult |
| NDE | Non-drive end of generator or gearbox |
| NHPP | Normal homogeneous Poisson process |
| NPRD | Non-electronic parts reliability data |
| NREL | National Renewable Energy Laboratory, Denver, CO, USA |
| OAS | Off-shore access system, Ampelmann |
| OM | Operations management |
| O&M | Operations and maintenance |
| OEM | Original equipment manufacturer |
| OFGEM | Office of Gas and Electricity Markets, UK |
| OFTO | Off-shore Transmission Operator, UK |
| OPEX | Operational expense |
| ORE | Offshore Renewable Energy Catapult, UK |
| OREDA | Off-shore reliability data |
| OWT | Off-shore wind turbine |
| PLC | Programmable Logic Controller |
| PLP | Power law process |
| PMG1G | Permanent magnet synchronous generator with 1 stage gearbox |
| PMSG | Permanent magnet synchronous generator |
| PSD | Power spectral density |
| PTS | Personnel transfer system |
| RBD | Reliability block diagram |
| RCM | Reliability-centred maintenance |
| RMP | Reliability modelling and prediction |
| RPN | Risk priority number |
| SCADA | Supervisory control, or signal conditioning, and data acquisition |
| SCIG | Squirrel cage induction generator |
| SHM | Structural health monitoring |
| SLILAD | Sliding ladder |
| TBF | Time between failures |
| TTF | Time to failure |
| TTT | Total time on test |
| VAWT | Vertical axis wind turbine |

| | |
|---|---|
| VSD | Variable speed drive |
| WF | Wind farm |
| WT | Wind turbine |
| WMEP | Wissenschaftliche Mess-und Evaluierungsprogramm database from Germany |
| WRIG | Wound rotor induction generator |
| WRIGE | Wound rotor induction generator and exciter |
| WRSGE | Wound rotor synchronous generator with exciter |
| WSD | Windstats Database for Germany |
| WSDK | Windstats Database for Denmark |
| WTCMTR | Wind turbine condition monitoring test rig |

# Chapter 1
# Introduction to off-shore wind

## 1.1 Development of wind power

The human development of rotating machine wind power started more than 2,000 years ago at various locations around the globe but particularly in Iran and China, see Appendix F.

However, the technology of wind turbines for generating electricity dates to the end of the nineteenth century and to three particular historic wind turbines, respectively, as follows:

- a horizontal axis wind turbine (HAWT) in the USA in 1883, the Brush Turbine;
- a vertical axis wind turbine (VAWT) in Scotland in 1887, the Blyth Turbine;
- a HAWT in Denmark in 1887, the la Cour Turbine.

Larger electric power wind turbines from 100 kW to 1 MW were envisaged and built in Denmark, Germany, Russia, France and the USA in the 1930s and 1940s.

However, modern large wind turbine development dates to work in the USA and Europe, later stimulated by US Department of Energy and European Union (EU) experimental programmes in the 1970s–1980s, following the oil price rises after the 1973 Yom Kippur war among Egypt, Israel and Syria.

A detailed description of wind turbine development is given with photographs in Appendix F, but the key large wind turbine projects world-wide from 1931 to 2020 are listed in Table 1.1. Their evolution has been profoundly influenced by the la Cour wind turbine and by reliability and availability issues, for more details see Hau (2006), Burton *et al.* (2011) and Gasch *et al.* (2012) plus a valuable market study comparing the evolution of wind turbines heights, diameters and drive train technologies in Serrano-González *et al.* (2016).

This design evolution, with competing two or three blade, upwind or downwind HAWT, geared or direct drive configurations or VAWTs, has affected subsequent developments, which is interesting as the reliability of many of these early on-shore wind turbine prototypes was extremely poor.

The large wind turbines at Grandpa's Knob, the USA, Orkney, the United Kingdom (UK) and Growian, Germany only operated for some hundreds of hours, suffering catastrophic failures in the turbine hub or blades.

*Table 1.1 Large wind turbine development world-wide 1931–2020*

| Year | Location | Type | Power (MW) | Rotor diameter (m) | Tower height (m) | Blade no | Drive | Pitch | Speed | Comment |
|---|---|---|---|---|---|---|---|---|---|---|
| 1931 | WIMIE-3D, Yalta, USSR | Upwind HAWT | 0.1 | | 30 | 3 | Geared-drive | Adjustable blade-flaps | Variable | Connected to 6.3-kV distribution system, and 32% capacity factor<br>Post-mill with the whole structure rotate along track<br>Early large three-blade machine |
| 1941 | Grandpa's Knob, Vermont, USA | Downwind HAWT | 1.25 | 57 | 40 | 2 | Geared-drive | Pitch-controlled, stall-regulated | Fixed | Grid-connected |
| 1954 | John Brown Engineering, Orkney, UK | Upwind HAWT | 0.1 | 18 | | 3 | Geared-drive | Full-span pitch-regulated | Fixed | Grid-connected |
| 1956 | Station d'Etude de l'Energie du Vent, Nogent-le-Roi, France | Downwind HAWT | 0.8 | | | 3 | Geared-drive | Full-span pitch-regulated | Variable | Grid-connected |
| 1956 | Johannes Juul, Gedser, Denmark | Upwind HAWT | 0.2 | 24 | | 3 | Geared-drive | Fixed pitch, stall-regulated, aerodynamic tip brakes on rotor blades, released automatically in over-speed | Fixed | The so-called Gedser Mill, defining the Danish three-blade concept |
| 1979 | Nibe, Denmark | Upwind HAWT | 0.63 | | | 3 | Geared-drive | Fixed pitch, stall-regulated | Fixed | Danish Concept |
| 1981 | Boeing, MOD2, USA | Downwind HAWT | 2.50 | 91 | 60 | 2 | Geared-drive | Full-span pitch-regulated | Variable | |
| 1983 | Große Windenergieanlage (Growian), Germany | Downwind HAWT | 3.0 | 100 | 100 | 2 | Geared-drive | Full-span pitch-regulated | Variable | Grid-connected with a fully rated cyclo-converter |

| | | | | | | | | | | |
|---|---|---|---|---|---|---|---|---|---|---|
| 1985 | Wind Energy Group, LS1, Orkney, UK | Upwind HAWT | 3.0 | 60 | | 2 | Geared-drive | Adjustable tip-flap pitch-regulated | Variable | Grid-connected with a fully rated converter | |
| 2007 | Enercon E126, Cuxhaven, Germany | Upwind HAWT | 7.58 | 127 | 135 | 3 | Direct drive | Full-span pitch-regulated | Variable | Grid-connected with a fully rated converter | |
| 2018 | MHI Vestas Off-shore WindV174-9.5, Denmark and Japan | Upwind HAWT | 9.5 | 174 | | 3 | Geared-drive | Full-span pitch-regulated | Variable | Grid-connected with a fully rated converter | |
| 2019 | GE Haliade-X 12, France | Upwind HAWT | 12.0 | 200 | 260 | 3 | Direct drive | Full-span pitch-regulated | Variable | Grid-connected with a fully rated converter | Grid-connected with a fully rated converter |
| 2020 | DEC DEW-D7000-186, China | Upwind HAWT | 7.0 | 186 | | 3 | Direct drive | Full-span pitch-regulated | Variable | Grid-connected with a fully rated converter | |
| 2020 | Siemens Gamesa, SG 14-222 DD, Denmark and Germany | Upwind HAWT | 14.0 | 222 | | 3 | Direct drive | Full-span pitch-regulated | Variable | Grid-connected with a fully rated converter | |

But the three-blade Gedser HAWT in Denmark ran for 11 years without extensive maintenance, and this successful configuration, built upon as the Danish Concept, has come to dominate the development of modern wind turbines.

From these small beginnings, modern commercial wind electrical power generation has expanded rapidly to the present day, as represented by Figure 1.1(a) and (b), showing the world installed wind capacity, summarised by

- 2000 total installed wind power 17 GW, with off-shore contribution <1 GW;
- 2017 total installed wind power 539 GW, with off-shore contribution, 19 GW.

These figures and graphs demonstrate the tremendous expansion of wind power but particularly emphasise the rapid growth in wind power off-shore, 1–19 GW in 17 years, simply due to the enormous wind resource available off-shore relatively near to the coast and the falling capital cost of harnessing that power.

The recording of wind turbine reliability started in Europe in 1985 with the growth of the German and Danish wind industry, Schmid *et al.* (1991), and in the USA in 1987, following the California Wind Rush from 1973.

Various reports summarising wind turbine reliability have been published, including Tavner *et al.* (2006), Ribrant *et al.* (2007).

Work in the Netherlands in the 1990s, van Bussel *et al.* (1997), when off-shore wind farms were contemplated in the North Sea off the Dutch Coast, led to concerns about the influence of maintenance access to the wind turbines and a wider consideration not only of reliability but also of maintenance and the need to achieve a high wind turbine availability. It was predicted that this should lead to a low cost of energy (CoE) for off-shore wind power, so that it could compete against low-cost fossil fuels.

Energy production from on-shore turbines in the larger sizes >1 MW based on the Danish Concept has now been achieving operational availabilities of >98% and mean time between failures (MTBF) of >7,000 h that is a failure rate of just over 1 failure/turbine/year, where a failure could be described as a stoppage with a duration of 24 h. BVG Associates *et al.* (2019) give an excellent overview of the issues and costs surrounding off-shore wind.

The results of early recording of wind turbine failures are summarised in Figure 1.2. The details of these reliability developments will be dealt with in Chapter 2.

## 1.2 Reliability of on-shore wind turbines

Figure 1.2 taken from Tavner, Xiang *et al.* (2006) is a graph from various USA and European public domain sources, comparing wind turbine reliability with other grid-connected and distributed generation sources. This showed a steady improvement in on-shore wind turbine reliability from 1987 to 2005. However, reliability still needs further improvement to achieve that for electricity from large steam turbine generators, and this situation will be substantially affected by deployment off-shore.

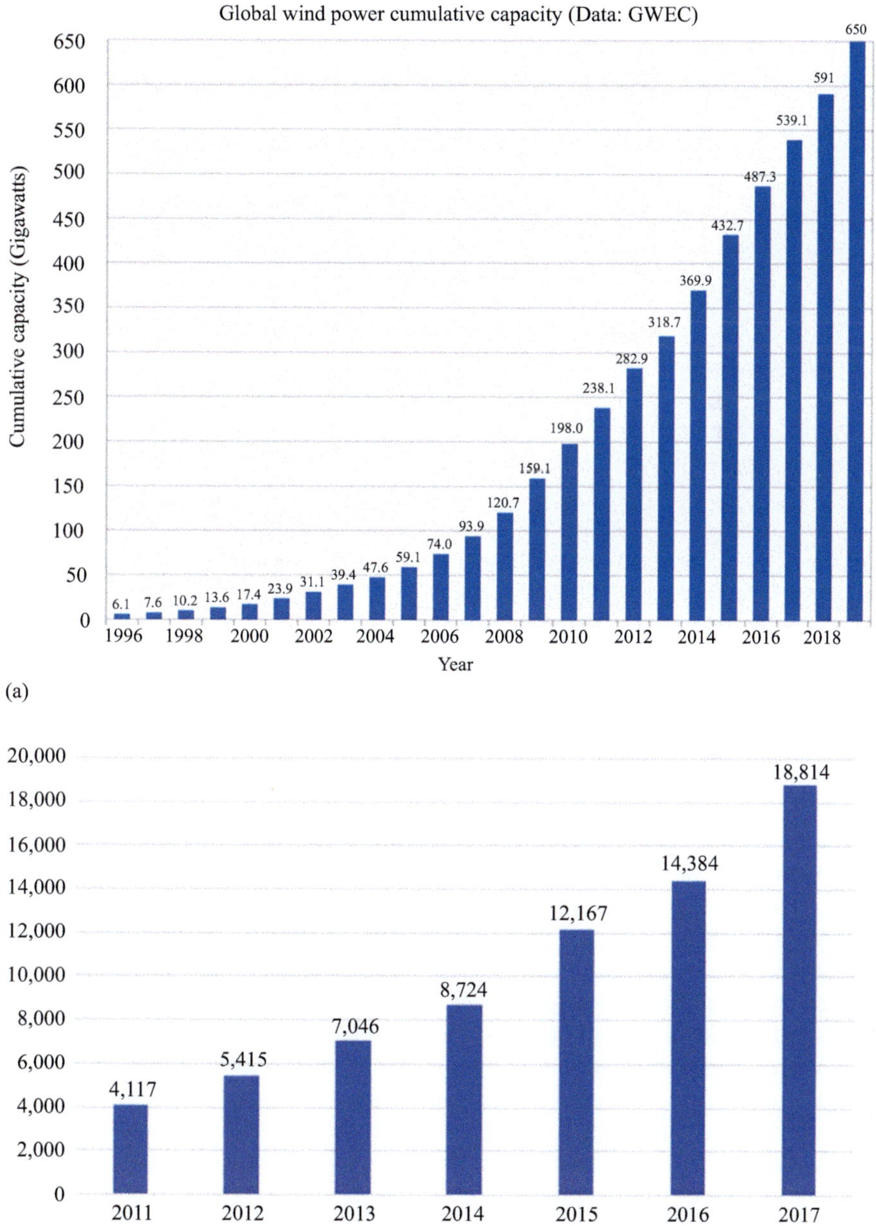


*Figure 1.1 Development of annually installed wind capacity world-wide 1996–2018: (a) total installed wind capacity, on-shore and off-shore, 1996–2018; (b) off-shore installed wind capacity, 2011–17*

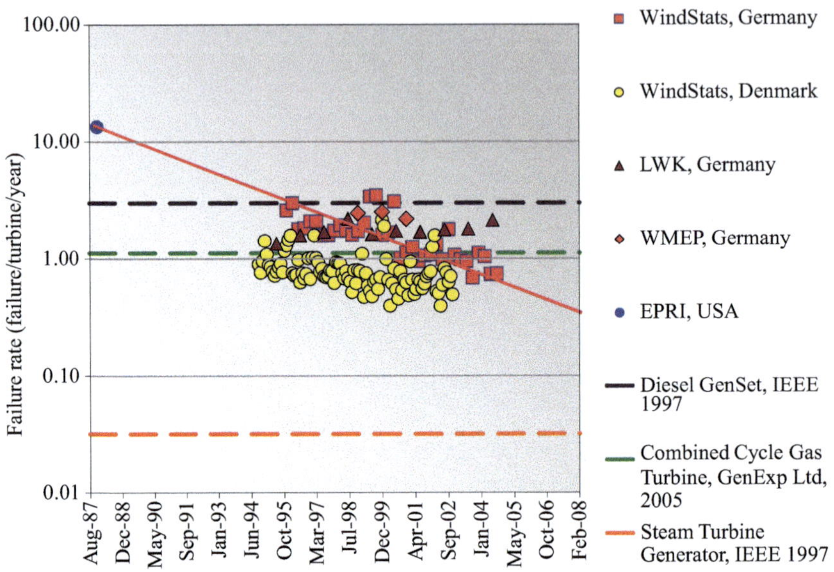


*Figure 1.2 Gross failure rate trends for on-shore wind turbines over the period 1987–2005, Tavner et al. (2006)*

## 1.3 Large wind farms

Deployment of wind turbines in larger wind farms has been a feature of modern electrical power since the 1980s as we try to harness the geographical extent of the distributed wind resource. The California wind farms built in the 1970s and 1980s, see Figure 1.3, were established with large numbers of relatively small wind turbines, $\leq$ 100 kW arranged in arrays of more than 100 wind turbines.

An advantage of an extensive wind farm is that the combined electrical resource will be substantial, justifying the cost of grid connection and considerable maintenance benefits accrue for a large wind farm, because personnel, tools, parts and facilities can be concentrated at or close to the wind turbine farm site.

It is currently not possible to tell whether the increasing reliability of wind turbines, shown in Figure 1.2, can be partially ascribed to their deployment in larger wind farms, although it is likely that this has been a contributory factor.

A disadvantage of large on-shore wind farms is their visual impact and this is particularly important in crowded countries, such as the UK, where citizens put space, amenity and visual impact high on the agenda during any wind farm approval process.

In general while large wind farms have been established in the USA, Spain and northern Germany, they are less common in the UK where the planning process has militated against the concentration process. Therefore, UK on-shore wind farms have generally ranged from only 1 to 30 wind turbines, although the largest on-shore wind farm currently operating in the UK, opened in 2010, is Whitelee, close to Glasgow, Figure 1.4, which has 140 Siemens 2.3 MW HAWTs.

Figure 1.3 Example of a large wind farm of >100 wind turbines in California in the early 1980s, Gipe (2016)

Figure 1.4 A large on-shore wind farm in the UK at Whitelee near Glasgow with 140 Siemens 2.3 MW HAWTs

## 1.4 First off-shore developments

The first off-shore wind farm was deployed in Denmark in 1991 at Vindeby with 11 wind turbines in sheltered, non-tidal Baltic waters close to Fyn island.

A small off-shore wind farm, two wind turbines, was installed in the tidal waters of the North Sea close inshore at Blyth, Northumberland, UK in 2001, see Figure 1.5.

The large capital investment required for off-shore installation has subsequently encouraged developers to increase the size of later off-shore wind farms to reduce overall unit costs.

*Figure 1.5 The first off-shore wind farm in the UK at Blyth, 2 Vestas V66 HAWTs*

The first substantial off-shore wind farm was installed at Middelgrunden near Copenhagen in Denmark in 2000, with 20 Bonus B76, 2 MW turbines, diameter 76 m, see Figure 1.7.

## 1.5 Off-shore wind in Northern Europe

### *1.5.1 Introduction*

A summary of planned and built off-shore wind farms in Northern Europe up to 2011 in Table 1.2 clearly shows the smaller earlier wind farms in Denmark and the UK with an expanding size as the years advance with further developments in Germany, the Netherlands and Sweden. The cumulative power generation capacity of the wind farms listed in Table 1.2 is 5.3 GW. Table 1.2, summarised in Figure 1.6, shows the increasing sizes of off-shore wind farms in Northern Europe. Research in the Netherlands on their off-shore programme has been reported in Beurksens (2011).

### *1.5.2 Baltic Sea, German, Swedish and Danish waters*

The Baltic Sea has non-tidal but windy conditions with potential ice and wave hazards.

The first large off-shore wind farm in the Baltic Sea was installed in 2000 at Middelgrunden (20 wind turbines) close to Copenhagen in Denmark, see Figure 1.7.

This process has accelerated rapidly since the Middelgrunden installation with a number of off-shore wind farms being installed, including Nysted, Denmark (72 wind turbines), Lillgrund, Sweden (48 wind turbines), Figure 1.8 and Rodsand, Denmark (90 wind turbines) see Lundberg (2006) for details on wind farm layouts.

### *1.5.3 North Sea, UK waters*

After the UK Blyth installation a process of licensing of UK off-shore wind farm sites was initiated from UK Crown Estate.

In Denmark after accelerating the process in more benign Baltic waters, off-shore wind farm size was dramatically increased at Horns Rev 1 (80 wind turbines) in the North Sea.

Operational problems in the first years of operation at Horns Rev, caused essentially by on-shore wind turbines being installed off-shore in a North Sea environment considerably less benign than in the Baltic Sea, caused a major rethink by wind turbine original equipment manufacturers (OEMs) and wind farm developers for future North Sea designs. This leads to a slowing down of development.

The UK Crown Estate's response to this North Sea experience was to organise a structured approach, organised in four separate leasing rounds:

- Round 1, 'Testing the Technology', adopted a cautious approach, with a model of 25 or 30 wind turbines per wind farm close inshore, 12 wind farms were to

*Table 1.2 European off-shore wind farms under construction up to 2011*

| Wind farm | Capacity (MW) | Country | Wind turbine no. | Maker | Type | Turbine rating (MW) | Commissioned |
|---|---|---|---|---|---|---|---|
| Vindeby | 4.95 | Denmark | 11 | Siemens | | 0.45 | 1991 |
| Blyth off-shore | 4 | UK, Round 1 | 2 | Vestas | V66 | 2.0 | 2000 |
| Middel-grunden | 40 | Denmark | 20 | Siemens | SWT-2.0-76 | 2.0 | 2000 |
| Horns Rev I | 160 | Denmark | 80 | Vestas | V80 | 2.0 | 2002 |
| Samsø | 23 | Denmark | 10 | Siemens | SWT-2.3-82 | 2.3 | 2002 |
| Rønland | 9.2 | Denmark | 4 | Siemens | SWT-2.3-93 | 2.3 | 2002 |
| Rødsand/ Nysted I | 166 | Denmark | 72 | Siemens | SWT-2.3-82 | 2.3 | 2003 |
| Frederik-shavn | 2.3 | Denmark | 1 | Siemens | SWT-2.3-82 | 2.3 | 2003 |
| North Hoyle | 60 | UK, Round 1 | 30 | Vestas | V80 | 2.0 | 2003 |
| Scroby Sands | 60 | UK, Round 1 | 30 | Vestas | V80 | 2.0 | 2004 |
| Kentish Flats | 90 | UK, Round 1 | 30 | Vestas | V90 | 3.0 | 2005 |
| Barrow | 90 | UK, Round 1 | 30 | Vestas | V90 | 3.0 | 2006 |
| Egmond aan Zee | 108 | Nether-lands | 36 | Vestas | V90 | 3.0 | 2007 |
| Lillgrund | 110 | Sweden | 48 | Siemens | SWT-2.3-93 | 2.3 | 2007 |
| Burbo Bank | 90 | UK, Round 1 | 25 | Siemens | SWT-3.6-107 | 3.6 | 2007 |
| Beatrice | 10 | UK | 2 | RePower | 5M | 5.0 | 2007 |
| Prinses Amalia | 120 | Nether-lands | 60 | Vestas | V80 | 2.0 | 2008 |
| Hywind | 2.3 | Norway | 1 | Siemens | SWT-2.3-82 | 2.3 | 2009 |
| Rhyl Flats | 90 | UK, Round 1 | 25 | Siemens | SWT-3.6-107 | 3.6 | 2009 |
| Horns Rev II | 209 | Denmark | 91 | Siemens | SWT-2.3-92 | 2.3 | 2009 |
| Lynn and Inner Dowsing | 194 | UK, Round 1 | 54 | Siemens | SWT-3.6-107 | 3.6 | 2009 |
| Alpha Ventus | 60 | Germany | 12 | RePower and Areva | 5M and M5000 | 5.0 | 2009 |
| Gunfleet Sands | 173 | UK, Round 1 | 48 | Siemens | SWT-3.6-107 | 3.6 | 2010 |
| Rødsand II | 207 | Denmark | 90 | Siemens | SWT-2.3-93 | 2.3 | 2010 |
| Thanet | 300 | UK, Round 2 | 100 | Vestas | V90 | 3.0 | 2010 |
| Walney I | 184 | UK, Round 2 | 51 | Siemens | SWT-3.6-107 | 3.6 | 2011 |
| Robin Rigg | 180 | UK, Round 2 | 60 | Vestas | V90 | 3.0 | 2010 |
| Baltic I | 48 | Denmark | 21 | Siemens | SWT-2.3-93 | 2.3 | 2010 |
| Bligh Bank, Belwind | 165 | Belgium | 55 | Vestas | V90 | 3.0 | 2010 |

*Table 1.2 (Continued)*

| Wind farm | Capacity (MW) | Country | Wind turbine no. | Maker | Type | Turbine rating (MW) | Commissioned |
|---|---|---|---|---|---|---|---|
| Greater Gabbard | 504 | UK, Round 2 | 140 | Siemens | SWT-3.6-107 | 3.6 | 2012 |
| London Ar-ray | 630 | UK, Round 2 | 175 | Siemens | SWT-3.6-120 | 3.6 | 2013 |
| Sheringham Shoal | 317 | UK, Round 2 | 88 | Siemens | SWT-3.6-107 | 3.6 | 2012 |
| Anholt | 400 | Denmark | 111 | Siemens | SWT-3.6-120 | 3.6 | 2013 |
| Pori | 2.3 | Finland | 1 | Siemens | SWT-2.3-101 | 2.3 | 2010 |
| Walney II | 183 | UK, Round 2 | 51 | Siemens | SWT-3.6-120 | 3.6 | 2012 |
| Teesside | | UK, Round 1 | 27 | Siemens | SWT-2.3–93 | 2.3 | 2013 |
| Borkum Riffgat | 108 | Denmark | 30 | Siemens | SWT-3.6-107 | 3.6 | 2014 |
| Baltic II | 288 | Denmark | 80 | Siemens | SWT-3.6-120 | 3.6 | 2015 |
| Dan Tysk | 288 | Denmark | 80 | Siemens | SWT-3.6-120 | 3.6 | 2014 |
| **Total** | **5,967** | – | **1,882** | – | – | – | – |

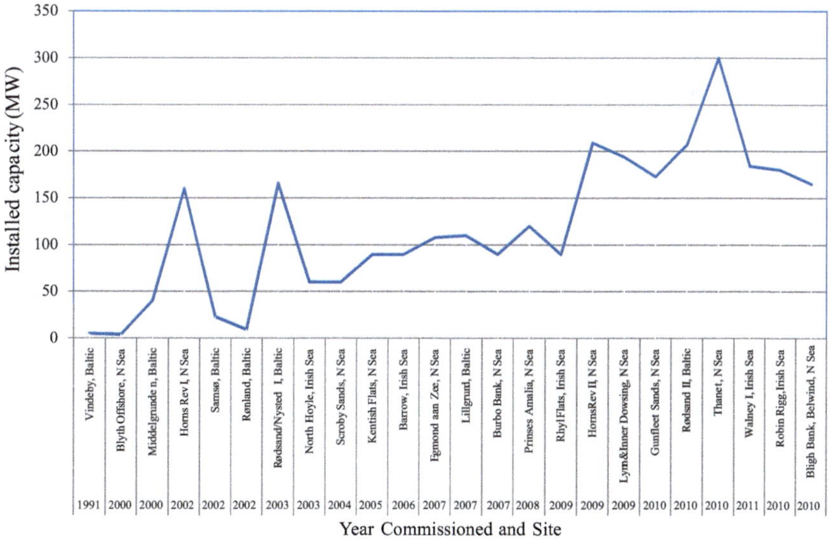


*Figure 1.6 North European growth in off-shore wind farm size, 1991–2010*

be built, intended to allow developers, installers and operators to gain experience, announced in 1999, to generate up to 1.2 GW.

- Round 2, 'First Projects at Commercial Scale', considered larger wind farms up to 175 turbines, announced in 2003, to generate up to 7.2 GW.

*Figure 1.7 First Baltic large off-shore wind farm, Middelgrunden, Copenhagen, 2000 20 Bonus B76, 2 MW turbines*

*Figure 1.8 A large Swedish off-shore wind farm at Lillgrund, 48 wind turbines*

- Round 1 and 2 extensions, announced in 2010, adding 2 GW more off-shore capacity.
- Round 3, 'Transformative Growth', large-scale project arrays of 50–300 wind turbines playing a role industrialising UK off-shore wind, announced in 2007, to generate up to 25 GW.

- Round 4, 'Core Energy Technology', building on significant off-shore wind cost reduction in large and far off-shore locations, including both fixed and floating arrangements, announced in 2019, to generate up to 7 GW.

This approach has proved a very successful model for the UK, the early caution of which can be clearly seen at the centre of the graph in Figure 1.6. However, this caution has paid off as the approach has encouraged many valuable developments, including risk sharing between turbine manufacturers, developers, improved contract management methods and the development of off-shore transmission network operators.

The early operational success with small UK Round 1 sites encouraged developers, where Horns Rev's severe problems were avoided, even though some sites used the same wind turbines as Horns Rev.

Therefore, the installation of Round 2 wind farms accelerated, with the first of these operational in 2011 at Thanet (100 wind turbines) transforming into a controlled roll-out of Rounds 2 and 3.

### *1.5.4 North Sea, German, Dutch, Belgian and Danish waters*

Meanwhile German, Dutch, Belgian and Danish developers similarly accelerated larger North Sea installations at BARD Off-shore 1 (80 wind turbines), Prinses Amalia (60 wind turbines), Belwind (55 wind turbines) and Horns Rev II (91 wind turbines).

## 1.6 Off-shore wind rest of the world

### *1.6.1 The USA*

No off-shore wind farms have yet been built in the USA, which reflects the recent political climate there.

However, USA's National Renewable Energy Laboratory has carried out considerable resource measurement, and development is underway to consider both fixed and floating off-shore wind farm sites, respectively, on the eastern and western seaboards, important references are Musial *et al.* (2018, 2019, 2020).

### *1.6.2 Asia*

China has been vigorous in the development of a wind industry and started to develop an off-shore capability but this development has been technically cautious.

China carried out extensive wind resource mapping throughout the country and identified very large on-shore wind resources in the NW and NE of the country. Therefore, early large wind farms were erected in Xinjiang, Gansu and Nei Monguo provinces.

However, resource mapping was also carried in the Bo Hai, Dong Hai and Nan Hai seas, which demonstrated that there were significant wind resources, relatively close to large electrical load centres in Tianjin, Shanghai and Guangzhou.

However, there were concerns about the security of large off-shore wind farms in seas where typhoon damage risks could be severe, so again a cautious approach has been taken.

Therefore, work was initiated with 1 wind turbine in Bohai Bay in 2007 and at an inter-tidal wind farm at Rudong (16 wind turbines). A larger wind farm has been constructed alongside the Dong Hai bridge, Shanghai (34 wind turbines), and Figure 1.9(a) shows one of these 3-MW turbines being installed. It is important to note that the installation assets being used for that operation were those used to build the nearby Dong Hai bridge. Figure 1.9(b) shows a large direct drive permanent magnet generator built by DFEM for an off-shore wind turbine in 2018.

China's installed off-shore wind capacity in 2016 was 1.9 GW and currently has 336 off-shore wind turbines but is planning to increase numbers to generate 52 GW by 2030 (Table 1.3).

(a)

(b)

*Figure 1.9 Off-shore wind turbine developments in China: (a) installation of a 3-MW Sinovel wind turbine at Dong Hai bridge near Shanghai; (b) construction of a 10-MW Dong Fang wind turbine generator for an off-shore wind turbine*

*Table 1.3 China early off-shore and intertidal wind farms*

| Wind farm | Type | Capacity (MW) | Province | Wind turbine no. | OEM and type | Commissioned |
|---|---|---|---|---|---|---|
| Bo Hai Bay | Off-shore, connected to off-shore oil platform | 1.5 | Liaoning | 1 | Goldwind 1.5 MW | 2007 |
| Rudong | Intertidal, grid-connected | 30 | Jiangsu | 16 | Various manufacturers | 2009 |
| Dong Hai bridge | Off-shore, grid-connected | 102 | Shanghai | 34 | Sinovel SL3000/90 | 2010 |
| Total installed off-shore wind turbines | All types | – | All | 336 | Various | 2019 |

The Fukushima nuclear power station accident in 2011, caused by the Tōhoku earthquake and tsunami, resulted in Japan re-appraising its energy generation policies and considering a significant place for wind power.

By 2019 Japan had 3.9-GW installed wind capacity, with 85-MW installed off-shore. The country's mountainous geography and their lack of a continental shelf, except in the Inland Sea, have also encouraged Japan to develop floating off-shore wind projects, Ushiyama *et al.* (2004) refer.

## 1.7 Off-shore wind power terminology and economics

### *1.7.1 Terminology*

The definition of availability for wind turbines needs to be clarified. Since 2007 an International Electrotechnical Commission working group has been working to produce a standard IEC 61400-26-1 and 2 to define wind turbine availability in terms of time and energy output. This is the internationally agreed definition of availability either in terms of time or energy; an extract of the information model is shown in Figure 1.10.

Two common availability definitions have been generally adopted in the UK in reports, see BVG Associates *et al.* (2019), summarised as follows:

- Technical availability, also known as system availability, is the percentage of time that an individual wind turbine or wind farm is available to generate electricity expressed as a percentage of the theoretical maximum.

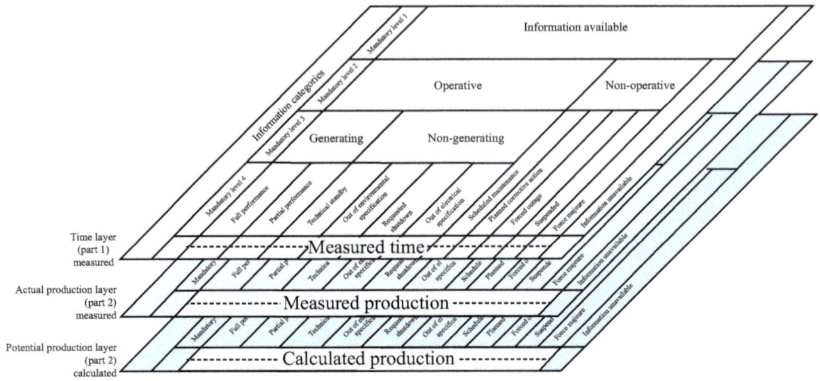


*Figure 1.10 Information model from IEC 61400-26-1 (2019)*

- Commercial availability, also known as turbine availability, is the focus of commercial contracts between wind farm owners and wind turbine OEMs to assess the operational performance of a wind farm project. Some commercial contracts may exclude downtime for agreed items, such as requested stops, scheduled repair time, grid faults and severe weather, when wind turbines cannot operate normally.

For the rest of the book, the term ‘availability’ refers to technical availability defined before, lending itself to comparison from project to project.

From the previous definitions, it follows that technical availability will always be lower than the commercial availability because there is more downtime alleviation in the former and an important issue off-shore is that availability, $A$, is affected by both time and wind speed, $u$, $A(u,t)$, see Faulstich, Lyding *et al.* (2011).

In respect of reliability, the following expressions are useful:

| | |
|---|---|
| Mean time to failure | $MTTF$ |
| Mean time to repair | $MTTR$ |
| Logistic delay time | $LDT$ |

$$\text{Downtime,} \quad MTTR + LDT \tag{1.1}$$

$$\text{Mean time between failures,} \quad MTBF \approx MTTF \tag{1.2}$$

$$MTBF \approx MTTF + MTTR = 1/\lambda + 1/\mu \tag{1.3}$$

$$MTBF = MTTF + MTTR + LDT \tag{1.4}$$

$$\text{Failure rate, } \lambda \quad \lambda = 1/MTBF \tag{1.5}$$

$$\text{Repair rate, } \mu \quad \mu = 1/MTTR \tag{1.6}$$

$$\text{Commercial availability,} \quad A = (MTBF - MTTR)/MTBF = 1 - (\lambda/\mu) \tag{1.7}$$

$$\text{Technical availability,} \quad A = MTTF/MTBF < 1 - (\lambda/\mu) \tag{1.8}$$

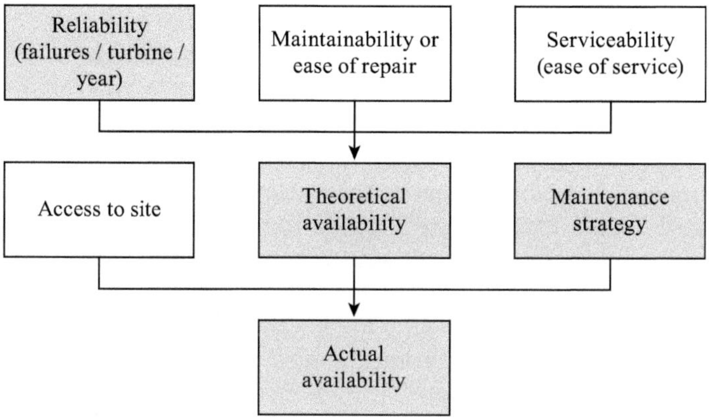


*Figure 1.11 Availability as a function of machine properties, access to site accessibility and maintenance strategy, van Bussel et al. (2001) and van Bussel, Henderson et al. (2001)*

Note that these all are expressed in terms of the variable time but availability can be expressed in terms of energy production and this may ultimately be more valuable for the operator, Figure 1.11.

Capacity factor and specific energy yield are two commonly used terms describing the productivity of a wind turbine or wind farm. Capacity factor, $C$, is defined as the percentage of the actual annual energy production $E$ (MWh) over the rated annual energy production, $AEP$, from a wind turbine or wind farm of rated power output $P$:

$$C = AEP \times 100/(P \times 8,760)\% \tag{1.9}$$

Specific energy yield, $S$ (MWh/m$^2$/year), is defined as the $AEP$ of a wind turbine normalised to its swept rotor area $A$ (m$^2$):

$$S = \frac{AEP}{A} \tag{1.10}$$

The ratio $R_S$ of rated power, $P$, over the swept rotor area, $A$, is a fixed value for a specific wind turbine type:

$$R_S = \frac{P}{A} \tag{1.11}$$

Or:

$$R_S = \frac{S}{(C \times 8,760)} \tag{1.12}$$

For a specific type of wind turbine, the specific energy yield is proportional to the capacity factor:

$$S = R_S \times C \times \ 8,760 \tag{1.13}$$

Therefore, the operational performance of a wind turbine or wind farm can be defined as the measured specific energy yield, $S$ over the predicted $S$.

### *1.7.2 Cost of installation*

Off-shore wind power uses large wind turbines whose capital cost is currently estimated at around £1.2 M/MW, compared to on-shore wind turbines at £0.65 M/MW. Off-shore wind turbine structures are large; the wind turbine hub for a 3.5-MW off-shore machine will be 90 m above the sea surface; the rotor diameter will be of the order of 100 m. Initially the structures were installed in relatively shallow water depth, 5–20 m, and the weight of each structure will be relatively low, ≈400 t, depending on rating. So, in contrast to a typical oil and gas off-shore structure, the applied vertical load to the foundation is relatively small compared to the wind and wave overturning moments. The foundation or flotation costs can be compared as follows:

- For the foundation and structure of a fixed off-shore wind turbine accounts for up to 35% of the whole installed structure cost.
- For the flotation structure and mooring of a floating off-shore wind turbine, albeit of much larger rating, may account for 75% of the whole installed structure cost, see Musial *et al.* (2004).

Therefore, off-shore wind turbine unit capital costs are large and will increase as larger wind turbines and more extensive wind farms are placed into deeper water but the cost per MW will reduce as size increases.

However, a single off-shore wind turbine design can be mass-produced for use over a whole wind farm or many wind farms, rather than each structure/foundation being individually engineered, as it has been in the oil and gas industry.

So off-shore wind farm capital costs will progressively fall with subsequent projects at later dates and this has been noted in Belgian, Danish, Swedish, UK, German and Dutch off-shore projects.

An interesting comparison has been made between the capital cost for off-shore wind in China at Dong Hai Da Qiao compared with UK late Round 1 projects as shown in Figure 1.12.

The capital cost for that off-shore wind development in China at £2.15 M/MW was greater than in the UK at £1.25 M/MW because China was at the very start of its off-shore development, whereas the UK has already learnt some of the lessons. At Dong Hai Da Qiao the facilities used to erect the wind turbines were those that had been deployed to erect the bridge sponsons and had already been paid for by that endeavour. So costs in China will fall as capacity increases.

Further details on costs were given in European Wind Energy Association (EWEA, 2009); however, more up-to-date figures are available in BVG Associates *et al.* (2019).

### *1.7.3 Cost of energy*

CoE is commonly used to evaluate the economic performance of different wind farms. This methodology was adopted in a joint report IEA, NEA Report (2005). It

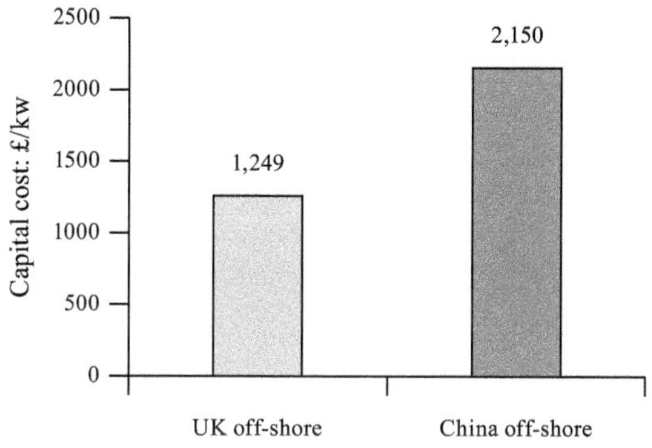


*Figure 1.12 Comparison of off-shore wind capital cost between the UK and China*

compared the cost of different electricity production options. A simplified calculation equation was adopted in the USA to calculate the *CoE* (£/MWh) for a wind turbine system, see Walford (2006):

$$CoE = (ICC + FCR\ O\&M)/AEP \tag{1.14}$$

where *ICC* is initial capital cost (£); *FCR* is annual fixed charge rate (%); *AEP* is annual energy production (MWh); *O&M* is annual operations and maintenance (O&M) cost (£).

The result of this approach is the same as that of levelised electricity generation cost used in Walford (2006), where the parameter FCR is a function of the discount rate $r$ used, as follows:

$$FCR = \frac{r}{(1 - (1 + r)^n)} \tag{1.15}$$

where $r \neq 0$. The discount rate $r$ is the sum of inflation and real interest rates. If inflation is ignored, the discount rate equals the interest rate. For the special case of a discount rate $r = 0$, unlikely in the real world, *FCR* will be *ICC* divided by the economic lifetime of the wind farm in years, currently estimated at $n = 20$ years.

A preliminary estimation of the *CoE* from off-shore wind was carried out by Feng *et al.* (2010) based on the early UK Round 1 sites. This showed that at that stage the CoE for off-shore wind in the UK was about 1.5× that on-shore, see Figure 1.13. It is probable that improvements in $\lambda$ and $\mu$ will have improved these figures.

The UK-subsidised CoE for off-shore wind is therefore estimated from Round 1 at about £69/MWh against £47/MWh for on-shore.

An interesting comparison, Figure 1.14, can be made with the CoE for off-shore wind from the Shanghai Donghai Bridge project in China of ¥980/MWh (i.e. ~£91/MWh), on a project installation cost of ¥23,000/kW (i.e. ~£2,150/kW) from Chinese sources.

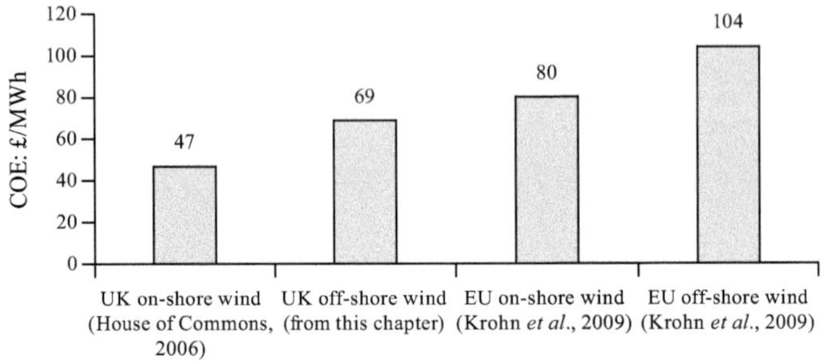


*Figure 1.13 Relative CoE for off-shore wind UK and Europe, Feng et al. (2010)*

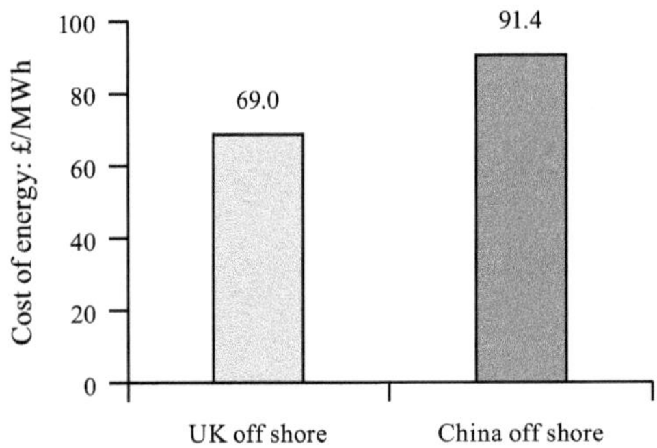


*Figure 1.14 Comparison of off-shore wind power cost of energy between the UK and China*

Again it should be expected that these CoE will fall as experience is gained, the O&M costs fall and the risks associated with the capital investment reduce.

These calculations were made on the basis of the subsidised CoE, and recent work has stripped away those benefits showing the true CoE for off-shore wind in 2012 around the UK coast to be closer to £120/MWh. Again this will fall as experience is gained and capital costs fall and life is extended, the latter being heavily influenced by the O&M regime surrounding the wind farm.

Early studies show clearly that operators who impose a higher quality O&M regime achieve higher availability, lower through-life costs and a lower CoE but incur maintenance costs. The relationship between CoE and the design and operations of the wind turbine have been presented in Jamieson (2011) and are shown in Figure 1.15, as the focus of this book is on the highlighted areas of the diagram.

### 1.7.4 Cost of O&M

The estimated cost of off-shore wind energy varies dependant on the site and project, but this chapter shows that off-shore wind projects are significantly more

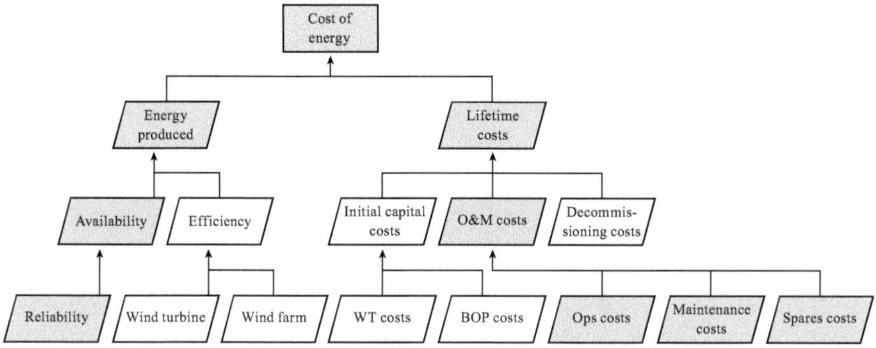


*Figure 1.15 Structure of cost of energy, showing highlighted in grey areas of interest for this book, Schmid et al. (1991) and Jamieson (2011)*

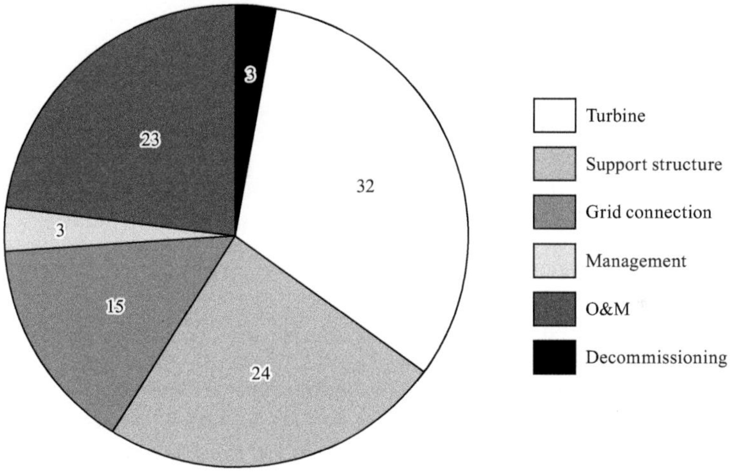


*Figure 1.16 Typical cost breakdown for an off-shore wind farm in shallow water*

costly than on-shore van Bussel *et al.* (1997). As wind turbine designs become adapted to off-shore conditions, the achievement of a favourable economic solution depends upon installing larger turbines on fewer foundations and controlling the wind farm system full life cycle cost. Figure 1.16 illustrates a breakdown of typical total system costs for an off-shore wind farm in shallow water, Byrne *et al.* (2003). Much of the price premium now being paid for electricity from off-shore wind can be attributed to wind turbine, foundations, grid connection capital expenditure costs associated with high operation and maintenance (operational expense costs).

O&M for off-shore wind farms is more complex than on-shore. As a consequence, O&M percentage costs for some European off-shore wind farms vary from 18% to 23%, much higher than the measured 12% for on-shore projects van Bussel *et al.* (2001) and van Bussel, Henderson *et al.* (2001).

Off-shore conditions require more onerous erection and commissioning operations, meanwhile accessibility for off-shore routine servicing and maintenance is a major issue. During winter, a whole wind farm may be inaccessible for many days due to harsh sea, wind or visibility conditions.

Even given favourable weather, O&M tasks are more costly than on-shore, being influenced by distance off-shore, site exposure, wind farm size, wind turbine reliability and maintenance strategy. Off-shore conditions require special lifting equipment to install and change out major sub-assemblies, which may not be available at short notice or be locally sourced.

Therefore, advanced techniques are needed to plan maintenance, using data from the Supervisory Control, or Signal Conditioning, and Data Acquisition (SCADA) and Condition Monitoring Systems fitted to wind turbines, requiring a thorough knowledge of off-shore conditions, qualitative physics theory and other design tools to predict failure modes in less conventional ways than has hitherto been done.

Off-shore remote monitoring and visual inspection, using boats, drones and possibly helicopters, will become much more important to maintain appropriate wind turbine availability and capacity factor levels.

### *1.7.5 Effect of reliability, availability and maintenance on cost of energy*

Equation (1.13) for *CoE* can be expressed as a function of $\lambda$ and $\mu$ allowing us to see the effect of reliability and maintenance on *A* and *CoE* as follows:

$$CoE = (ICC \times FCR + O\&M(\lambda, 1/\mu))/AEP(A(1/\lambda, \mu)) \tag{1.16}$$

Reductions in failure rate, $\lambda$, will improve reliability *MTBF*, $1/\lambda$, and availability, *A*, therefore reducing O&M costs.

Reductions in downtime, *MTTR*, will improve maintainability, $\mu$, and availability, *A*, therefore also reducing O&M. As a consequence, *CoE* will also reduce as $\lambda$ and $\mu$ improve.

### *1.7.6 Previous work*

Schmid *et al.* (1991) published the first data on European wind turbine performance from the EUROWIN project.

The EU FP7 ReliaWind project prepared a report on the previous literature on wind turbine reliability and a summary of results, see ReliaWind (2011) and Wilkinson *et al.* (2011).

## 1.8 Roles

### *1.8.1 General*

There are a number of roles in the off-shore wind industry and these are clearly set out in this chapter; a valuable guide to off-shore wind turbines is also given in BVG Associates *et al.* (2019), which gives more information on these roles.

Off-shore wind farms are emerging technologies that extract energy from an intermittent renewable resource, which are exposed to that resource 24/7 of 365 days/year, and they may not operate at full power all the time. They must, therefore, be designed to withstand all the loads the resource imposes under all these intermittent operational conditions. Being permanently installed at selected sites, they endure harsh climatic, wave and current load conditions, which their complex mechanical, electrical and control systems must withstand. Therefore, their reliability and survivability are an engineering challenge that has now largely been overcome.

These issues can be mitigated by selecting benign off-shore wind sites, although 30 years' experience of off-shore wind is encouraging the developers to plan for operations further from shore in more energetic and remote locations.

This chapter is outlining the resources available to an off-shore wind farm, which are substantial. However, these locations can impose difficult challenges.

There are many stakeholders within the task of innovating, developing, building and operating off-shore wind farms, whose actions define and shape our ability to achieve the objectives of that farm and their interrelationship with cost is shown in Figure 1.16. Those objectives are to generate electricity reliably from the wind's renewable source at competitive prices and provide an acceptable return to each of the stakeholders.

This book concerns the operation of the wind farm, once built, and the vital task of ensuring that the planned wind farm returns are extracted in an efficient and predictable way.

The following describes the role of each of these major stakeholders, so that the reader can understand their influence upon innovation, development, building, operation and the planned process. A graphical representation of these roles is given in Figure 1.17, from Hardisty (2009).

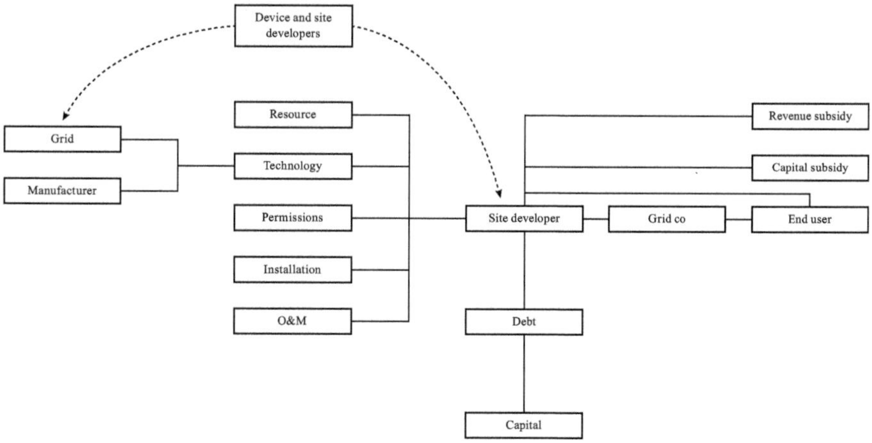


*Figure 1.17 Relationship between roles and capital for an off-shore renewable energy development, Hardisty (2009)*

### *1.8.2 Innovators*

Some off-shore wind technologies are still in their infancy, and there is still a need for innovation, whether in the wind turbines themselves or their foundations and off-shore sub-station infrastructure, cables or maintenance technologies. Innovators are, therefore, very present in the marine energy field and their input is essential to the healthy development of reliable plants. These innovators come from universities, entrepreneurs, specialist companies, main electricity suppliers and governments, and their contribution over the last 50 years has made off-shore wind a reality. There is currently a wide range of off-shore wind devices under development, as shown in the previous chapters, and it would be impossible to identify the reliability characteristics of all of them. However, some principles are beginning to emerge.

### *1.8.3 Governments*

Governments set national Energy Policies, which have played a large part in stimulating the renewable energy industry. The motivations of these policies are to secure long-term energy resources for the country; stimulate the economy; reduce greenhouse gas emissions and answer democratic calls for more sustainable energy production.

These energy policies are empowered by subsidies for certain energy sources, which are highly contentious in democratic countries, where citizens demand to see personal benefits in energy costs from these subsidies.

The issues are complex and vary from country to country but include capital subsidies to offset the cost of construction and revenue subsidies as feed-in payments. The former ones stimulate the off-shore wind farm construction process, whereas the latter ones provide long-term security for the off-shore wind farm operators.

However, the EU has sustained some degree of cohesion between member countries with respect to subsidising renewable energy sources. This has led to EU countries becoming world leaders in their development, manufacture and deployment.

It is, however, a truism that all electrical generation sources, since their inception in the 1880s, have received some sort of subsidy to sustain early technological development. This was especially true for large-scale deployment of new technologies, including the use of particular fuels, such as coal and nuclear or the application of high-speed steam or gas turbines. This has been so whether the subsidies came from governments, developers or the consumers themselves, via elevated electricity costs.

Our current array of reliable and profitable electrical energy supply sources, developed over the last 130 years, is a tribute to the efficacy of that process. In fact the UK is a good example of the early development of a national electricity network, from 1925 under the auspices of the Central Electricity Board, creating a National Grid, then in 1958 the creation of a nationalised Central Electricity Generating Board, incorporating large coal-, oil- and nuclear-fired steam turbine generators and the National Grid, entirely capitalised by the government and subsidised from elevated electricity charges, which ultimately fell to acceptable levels.

Most recently in the late 1980s, UK's Central Electricity Generating Board was broken up into a number of privatised generation and transmission entities. However, the EHV electrical grid was incorporated into National Grid plc in 1990 and subsequently merged in 2005 with Great Britain's gas distribution network.

### 1.8.4 Test facilities

Reliable devices for off-shore wind are now being developed with substantial investment in prototype testing, both of drive trains, turbines, blades, converters and their components. To facilitate this, there has been considerable development of National and University testing facilities, in the UK, Denmark, Germany, France, Spain, the USA and China. These test facilities prove the viability of technology on scaled or full-size prototypes. They are essential to the future reliable development of off-shore wind energy, particularly in remoter and more energetic locations and those in the UK are described in more detail in a later chapter.

### 1.8.5 Regulators

The price of electricity from on-shore or off-shore wind farms is important to their future financial liability, and off-shore wind, with its long transmission distances to shore and costly off-shore assets, requires special regulatory consideration. In the UK the Regulator, the Office of the Gas and Electricity Markets, sets the market landscape for off-shore wind. A particularly important aspect of this has been the development of the role for Off-shore Transmission Operators (OFTO) ensuring that off-shore wind farms will have secure and flexible connection assets to transport power into the on-shore transmission grid. The long-term availability of the OFTO's connection asset and its reliability will be essential to the achievement of off-shore wind farms objectives, but its technical reliability will be outside the scope of this book.

### 1.8.6 Investors

Investors in off-shore wind include banks, energy supply and energy manufacturing companies, and landowners, including, in the UK, the Crown Estate and Crown Estate (Scotland), who have licensed the off-shore areas for development. The issues of reliability and availability of off-shore assets are of importance to long-term investors, and the energy supply and manufacturing companies, since this is the means by which their investment can be reliably and predictably repaid, with the required return. The difficulty for investors, in emerging technologies as off-shore wind farms increase in size, is to understand the technical issues involved, so that the right parameters can be defined for their investment. It is also true that Investors are keen to benefit from short-term government subsidies for renewable energy, rather than for the profits available from the long-term operation of reliable renewable energy sources. By explaining the technical issues of off-shore wind farm reliability and availability, this book will enable those parameters to be more precisely defined.

### *1.8.7 Certifiers and insurers*

Certifiers, such as Det Norske Veritas-GL, will be responsible for ensuring that off-shore wind farm designs and their associated marine structures meet IEC standards and are therefore insurable. Project insurers are also important participants as they determine the premiums for large off-shore projects, which affect their invest-ability. An important aspect of these processes is imposing the necessary health and safety (H&S) regime on the installation and operational phases of the project, to ensure that the human risks are acceptable and that the losses due to such risks are minimised. These processes were developed for the on-shore wind industry and have successfully ensured that machines and structures are sound and safe invest-ments. They are even more important off-shore, where the environment is more challenging. However, this has meant that, so far, off-shore equipment designs may have focused more on meeting safety and certification requirements than on pro-duction demands. That will change as investors demand a return on their invest-ment during operation.

### *1.8.8 Developers*

Developers of off-shore wind farms are emerging as consortia of investors, energy companies, wind turbine, off-shore sub-station and converter and cable manu-facturers and operators. Their objective is to gain a return on the development of wind farm generation assets which are subsequently sold onto long-term operators such as the main electricity generating companies.

Because of the scale and complexity of the off-shore asset, these consortia are attempting to attract long-term investors as part of the development team, and that requires financial experts to have a better understanding of the technical issues concerned. Hardisty (2009) coined the term device and site developers for these types of consortia.

A major part of the deployment of off-shore wind farms depends upon the marine installation assets, which include port and docking facilities, installation vessels, maintenance vessel fleets, and the manpower and infrastructure to manage and operate these assets. The latter ones are usually provided by civil and marine engineering businesses, who are starting to become important members of off-shore wind farm developer consortia.

Developers of marine energy extraction farms are emerging as consortia of investors, energy companies, off-shore wind manufacturers and operators. Their objective is to gain a financial return when the assets are subsequently sold on to long-term operators, such as the main electricity generating companies.

### *1.8.9 Original equipment manufacturers*

The principal OEMs involved in off-shore wind farms in the past were the wind turbine manufacturers. But the wind farm is a complex generation, collection and transmission asset with a substantial balance of plant, which is drawing in cable as well as transmission equipment OEMs.

The principal OEMs involved may be owned by larger energy groups, such as Siemens-Gamesa or Mitsubishi–Vestas, with large balance sheets and the potential to invest in promising new technologies for installation, deployment and operation.

The actions of the regulator are encouraging transmission OEMs to participate in the OFTO activity, but they still have a significant financial, management and technical role in the collection and off-shore substations of the off-shore wind farm.

### *1.8.10 Operators and asset managers*

The operators of off-shore wind farms are large energy companies providing electricity into the transmission grid. Most of these operators are broad-technology generators with fossil- and nuclear-fired and renewable generation assets, although recently there has been a rationalisation of such operators into those who specifically specialise in renewables and in wind power.

In view of the technical complexity of off-shore wind assets, a few specialised off-shore operators are developing, particularly in the Scandinavian market. A good example was DONG Energy, which in 2005 absorbed the Danish nationalised electricity producers Elsam and Energi E2 and in 2017 renaming itself Oersted, concentrating on the development of wind power in Europe and internationally.

These operators are building their expertise to match existing assets in on-shore wind, hydro and gas-fired generation.

It seems likely that, in the future, additional specialised operators will come into the market, but the size and complexity of marine energy extraction assets means that these will be large companies with diverse international portfolios of assets, developed to balance their exposure and risk in this sector. As the industry matures, the current certification- and safety-oriented approach is likely to change: more stringent demands for return on the larger capital outlays will encourage a more vigorous production-oriented approach.

In this stage of industry development, the interaction between operators, asset managers, certifiers, insurers and investors will be strengthened.

### *1.8.11 Maintainers*

Off-shore wind turbine OEMs have developed service departments and have access to the SCADA data streaming from their wind turbines during commissioning and warranty periods. They also have detailed knowledge of the development of their own equipment through prototype tests, supply chain development and production tests. Their staff are trained on their machines and have built up a detailed personal knowledge of the idiosyncrasies of individual wind turbine types. This expertise is deployed during the warranty period, regulated by the project contract. For some off-shore wind OEMs, this may change with time as they recognise the benefit to their business of the O&M market, and the importance to the developers and operators of through-life performance.

Operators gradually gain experience of off-shore wind farm operation, different in nature to the wind turbine OEM, being more focused on production needs and through-life performance of the asset. They will have their own management and some of their own O&M staff but may rely upon sub-contractors and the off-shore wind turbine OEM for some of that support. However, they frequently lack detailed knowledge of individual off-shore wind farm equipment, especially in the cabling, sub-station and converter station technologies. They therefore rely, in large part, upon the warranty period to gain that knowledge and experience.

Operators may opt to continue with a maintenance contract with the wind turbine OEM after the completion of the warranty period, but off-shore wind farm operators are likely to be large, with experience on other wind farms; many will opt to undertake their own O&M. This will allow them to impose their own asset management objectives on the off-shore wind farm to ensure long life.

Farm maintenance will rely heavily upon the expertise of the management and staff carrying out this highly skilled activity and upon the data collected both historically and currently. Off-shore wind farm design, choice of equipment, availability of appropriate access assets, spares and tools can facilitate the activity, but success is impossible without staff who are well-trained in H&S and the technology of the asset. This is an important issue that will be addressed later.

## 1.9 Summary

The development of large on-shore wind farms has been accelerating around the world over the last 30 years so that wind farms up to 1,000 MW in rating are now becoming commonplace and the world's installed capacity is 600 GW with an AEP of more than 1,200 TWh.

Increasing confidence in the effectiveness of large on-shore wind operations has encouraged governments and developers, over the last 10 years, to start developing larger off-shore wind farms. World-installed off-shore capacity is now 30 GW and that installed capacity may rise by 2030 to over 200 GW.

The lead is currently being taken by Europe, in the North, Baltic and Irish Seas, with the UK currently having the largest installed off-shore capacity of 8.5 GW, potential AEP of 25 TWh, and the largest off-shore wind farm rated at 1,218 MW from 174 wind turbines.

China has also made a large commitment to off-shore wind having installed 1.9 GW of off-shore wind turbines and it seems likely that, with its large South Eastern coastal electrical load, a well-developed grid in those areas and good off-shore wind resource, we are likely to see a large expansion soon.

The USA is now seriously considering exploiting the off-shore wind opportunities on their Eastern and Western seaboards, but the latter will almost certainly need floating off-shore wind. It is likely that this will be a region of high growth.

Economic analyses of European off-shore wind sites to date have shown that the wind turbine installation cost is approximately 100% more for off-shore than on-shore, and the CoE is about 33% more than on-shore, whilst the O&M cost are 18%–23% more than on-shore, all depending upon the off-shore wind location, changing as lessons are learnt in the field.

However, several commentators are reporting that large off-shore and very large floating off-shore wind farms can reduce overall costs below those of smaller, less energetic on-shore wind farms.

To understand off-shore wind reliability and availability issues we need to understand the following:

- Size and scope of the current off-shore wind industry.
- Basic reliability mathematics and how it applies to generating electricity from wind.
- Importance of weather on off-shore wind power reliability and availability.
- Practical off-shore wind farm reliability, considering first wind turbine designs and operation, for which there are much reliability data, second off-shore wind farm layouts and grid connections, for which there are very little reliability data.
- How to build off-shore wind metadata sets from historic reliability data that can allow us to predict future off-shore wind turbine and farm reliability. Then how to create, from that metadata, prospective reliability information for large future off-shore wind farm layouts and grid connections so that the reliability and availability for both fixed and floating wind farms can be predicted.
- Advice on how the reliability and availability of future off-shore wind farm layouts and grid connections could be improved by asset on-line monitoring. Then using that reliability data, combined with this on-line monitoring, to improve performance.
- Finally, how lessons can be learnt from the previously given information and used to improve the success, not only of off-shore wind, but also of wave and tidal energy, the three emerging off-shore renewables industries.

*Chapter 2*

# Reliability theory relevant to off-shore wind

## 2.1 Introduction

A modern, multi-MW off-shore wind turbine is a large steel and supporting structural arrangement on which is mounted a complex robotic electro-mechanical generating machine.

Off-shore wind turbines are then connected by a network of HV collection cables to one or more off-shore AC sub-stations or HVDC converter stations, which in turn are connected via EHV AC or DC export cables to an on-shore AC sub-station or HVDC converter station, where power is transformed or converted to EHV AC and injected into the National Grid.

The reliability of a whole off-shore power station is dependent on epistemic uncertainty affected by the following:

- Structural reliability of the off-shore wind farm components for which predicted failure rates are $<10^{-4}$ failures/year, but the probabilistic time distribution of those low failure rate events needs to be considered.
- Electro-mechanical reliability in the wind turbines and sub-stations, which is subject to the normal vagaries of rotating machinery and electronics, can be predicted using measured constant failure rates for individual sub-assemblies ranging from $10^{0}$ to $10^{-3}$ failures/year.
- Control system reliability in the wind turbines and converters depends upon the environment, electro-mechanical issues and the reliability of the software contained within the control system.

The reliability analysis of such off-shore installations is made more complex because the turbines and off-shore sub-stations are all subject to aleatory uncertainty, due to the stochastic effects of the weather. This relates to the wind itself from which an off-shore wind farm extracts energy, the combined effects of wind, waves and weather on their structures and the corrosion of their submerged or exposed parts.

To understand and predict these effects, there must be an understanding of reliability theory; the relevant textbooks on the subject with electrical engineering content are Goldberg (1981), Billinton *et al.* (1992), Rigdon *et al.* (2000) and Birolini (2007).

To track reliability changes with time of complex off-shore power stations, during different operational phases, reliability growth models should be developed, for example using the Crow-AMSAA model in MIL-HDBK-189 (1981). Such a model can be applied to failure data collected from the field to investigate whether product reliability stays constant or shows improvement or deterioration with time.

The purpose of reliability data and analysis is not only to define the way in which wind power plants should operate but also to provide the numerical data to identify how their operational performance could be improved.

## 2.2 Basic definitions

The reliability of a sub-assembly is defined as the probability that it will meet its required function under stated conditions for a specified period.

This definition of reliability breaks down into four essential elements:

- probability;
- time, and variation of that probability with time;
- required function;
- operational conditions for adequate performance.

Unreliability, the complement of reliability, is related to failure intensity function, $\lambda(t)$, to be defined later. This reliability definition experiences difficulties as a measure for continuously operating systems, such as wind turbines, that tolerate failures which can be repaired.

Then a more appropriate measure is availability, defined as the probability of finding the system in the operating state at some time into the future. This definition reduces to only two elements:

- time;
- operability.

Failure is the inability of a sub-assembly to perform its required function under defined conditions, an item is then in a failed state, in contrast to an operationally working state.

A non-repairable system is one which is discarded after a failure. Examples of non-repairable systems are small batteries or light bulbs.

A repairable system is one that, when a failure occurs, can be restored into operational condition after any action of repair, other than a replacement of the entire system. Examples of repairable systems are wind turbines, sub-stations, cable systems, electrical generators and computers.

Repair actions can involve the exchange of parts, the removal of a damaged part, the addition of a new part, changes or adjustment to settings, software update, lubrication or cleaning.

## 2.3 Random and continuous variables

The random variable, in the context of off-shore wind farm reliability, is failures $X$ recorded discretely against a continuous variable, such as time.

Is it always appropriate to use calendar time as the continuous variable? Calendar time may be convenient but is not necessarily the best for reliability analysis, for example:

- 'Time on test' seems more appropriate.
- Turbine rotations may also be more appropriate, especially for the aerodynamic and transmission sub-assembly reliability.
- Energy generated by the wind turbine, GWh, may also be more appropriate, especially for electrical sub-assemblies reliability.

Operators usually cannot measure 'time on test' because they cannot easily keep track of the date of origin of the wind turbine, but they can easily measure the number of failures in an interval of time, which is called censored data.

An example of such differences in the random variable is shown in Figure 2.1 where identical failure data from the large German WSD survey, referred to in Chapter 3, are in this case plotted against calendar time either in terms of failure/turbine/year or failure/turbine/GWh. The former, in Figure 2.1(a), shows an improving failure rate with time, whereas the latter in Figure 2.1(b) shows a wider variance but an increasing number of failures per GWh generated. The latter probably shows the extent of small but significant failures occurring in the growing number of larger, more technically complex wind turbines.

What this shows us is that the method of collecting and presenting data is important. The choice of continuous variable against which the random variable, *X*, is to be collected should be considered as both continuous and random variables can be presented in different ways:

- Whether continuous variable is to be calendar time, 'time on test', GWh produced or rotations should be selected based on the interpretation to be made.

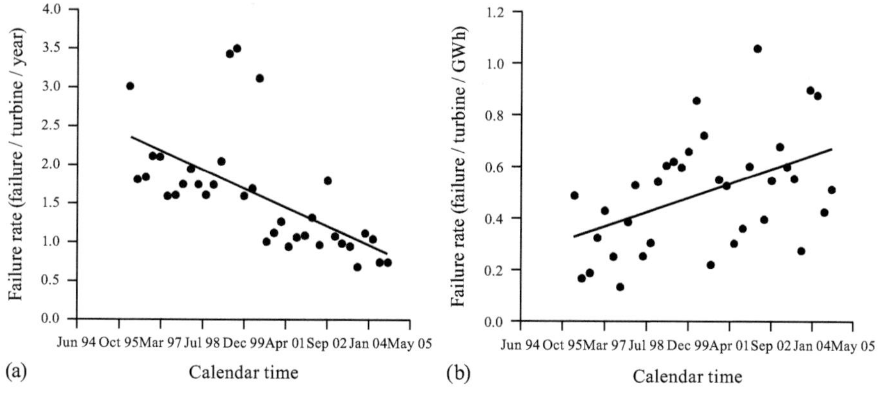


*Figure 2.1 Comparison of plots of identical failure data from the WSD survey plotted as failure/turbine/year or as failure/turbine/GWh, Windstats (WSD and WSDK) (2010): (a) failure/turbine/year versus time; (b) failure/turbine/GWh versus time*

- Plotting the random variable, $X$, in different ways against different discrete or continuous variables reveals different information, depending upon whether the component on which data are being collected is repairable or non-repairable.
- If the data collection method is good and the variable is chosen appropriately, then the statistical data of the random variable, $X$, collected should yield robust reliability information.
- If not, reliability information could be faulty.

Now we can consider probability distributions of a random variable.

## 2.4 Reliability theory

### *2.4.1 Reliability functions*

The following equations and mathematical relationships between the various reliability functions do not assume any specific failure distribution and are equally applicable to all probability distributions used in reliability evaluation. Consider that $N_0$ identical components are tested:

$$N_s(t) = \text{number surviving at time } t \tag{2.1}$$

$$N_f(t) = \text{number failed at time } t \tag{2.2}$$

Therefore,

$$N_s(t) + N_f(t) = N_0 \tag{2.3}$$

At any time, $t$ the survivor or reliability function, $R(t)$, is given by

$$R(t) = N_s(t)/N_0 \tag{2.4}$$

Similarly, the probability of failure or cumulative distribution function or unreliability function, $Q(t)$, is given by

$$Q(t) = N_f(t)/N_0 \tag{2.5}$$

where

$$R(t) = 1 - Q(t) \tag{2.6}$$

The failure density function $f(t)$ is given by

$$f(t) = 1/N_0\left(dN_f(t)/dt\right) \tag{2.7}$$

Failure intensity, hazard rate function or failure density function $f(t)$ (see Figure 2.2) are normalised to the number of survivors, $R(t)$

$$\lambda(t) = 1/N_s(t)\left(dN_f(t)/dt\right) \tag{2.8}$$

$$\lambda(t) = 1/R(t)(dR(t)/dt) \tag{2.9}$$

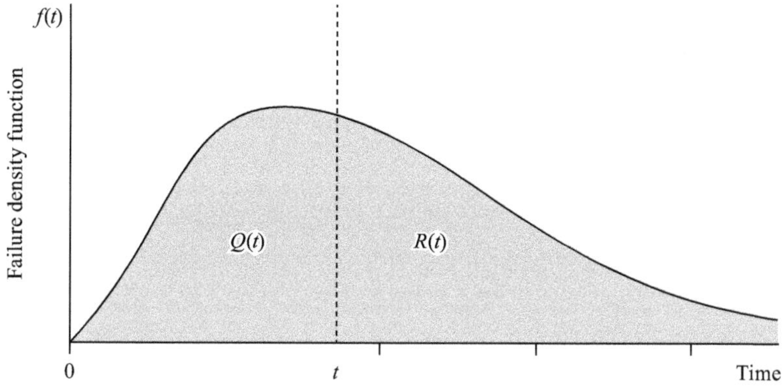


*Figure 2.2 Failure density function against time showing reliability R(t) and Q(t)*

The special case in which $\lambda$ is constant and independent of time is an exponential distribution, and the hazard rate becomes the failure rate.

Here the hazard rate/failure rate $\lambda(t)$ = number of failure per unit time/number of components exposed to failure:

$$R(t) = 1 - Q(t) \tag{2.10}$$

$$f(t) = dQ(t)/dt = -dR(t)/dt \tag{2.11}$$

or

$$Q(t) = \int_0^t f(t)dt \tag{2.12}$$

and

$$R(t) = 1 - \int_0^t f(t)dt \tag{2.13}$$

The total area under the failure density function must be unity. Therefore,

$$R(t) = \int_0^\infty f(t)dt = 1 \tag{2.14}$$

## 2.4.2 *Reliability functions example*

The following is an example of these methods using contrived data from an imaginary large off-shore wind farm, the example of which is based upon data given in Billinton *et al.* (1992) modified to suit the off-shore wind farm situation.

This example considers a large off-shore wind farm of 1,000 wind turbines and for the sake of this example, they are each non-repairable. There is a steady failure of these wind turbines. Table 2.1 records the cumulative failures and survivors over

*Table 2.1 Record of failures of 1,000 non-repairable wind turbines in an off-shore wind farm*

| **Time interval (years)** | **Number of failures in each interval** | **Cumulative failures, $N_f$** | **Number of survivors, $N_s$** | **Failure density function, $f(t)$** | **Unreliability function or cumulative failure distribution, $Q(t)$** | **Reliability or survivor function, $R(t)$** | **Failure intensity or hazard rate, $l(t)$** |
|---|---|---|---|---|---|---|---|
| 0 | 140 | 0 | 1,000 | 0.140 | 0.000 | 1.000 | 0.151 |
| 1 | 85 | 140 | 860 | 0.085 | 0.140 | 0.860 | 0.104 |
| 2 | 75 | 225 | 775 | 0.075 | 0.225 | 0.775 | 0.102 |
| 3 | 68 | 300 | 700 | 0.068 | 0.300 | 0.700 | 0.102 |
| 4 | 60 | 368 | 632 | 0.060 | 0.368 | 0.632 | 0.100 |
| 5 | 53 | 428 | 572 | 0.053 | 0.428 | 0.572 | 0.097 |
| 6 | 48 | 481 | 519 | 0.048 | 0.481 | 0.519 | 0.097 |
| 7 | 43 | 529 | 471 | 0.043 | 0.529 | 0.471 | 0.096 |
| 8 | 38 | 572 | 428 | 0.038 | 0.572 | 0.428 | 0.093 |
| 9 | 34 | 610 | 390 | 0.034 | 0.610 | 0.390 | 0.091 |
| 10 | 31 | 644 | 356 | 0.031 | 0.644 | 0.356 | 0.091 |
| 11 | 28 | 675 | 325 | 0.028 | 0.675 | 0.325 | 0.090 |
| 12 | 40 | 703 | 297 | 0.040 | 0.703 | 0.297 | 0.144 |
| 13 | 60 | 743 | 257 | 0.060 | 0.743 | 0.257 | 0.264 |
| 14 | 75 | 803 | 197 | 0.075 | 0.803 | 0.197 | 0.470 |
| 15 | 60 | 878 | 122 | 0.060 | 0.878 | 0.122 | 0.652 |
| 16 | 42 | 938 | 62 | 0.042 | 0.938 | 0.062 | 1.024 |
| 17 | 15 | 980 | 20 | 0.015 | 0.980 | 0.020 | 1.200 |
| 18 | 5 | 995 | 5 | 0.005 | 0.995 | 0.005 | 2.000 |
| 19 | 0 | 1,000 | 0 | 0.000 | | | |
| **Total** | **1,000** | | | **1** | | | |

Source: Billinton *et al.* (1992).

a period of 19 years upon calculating the failure density function, which sums to 1, and the hazard rate. So Table 2.1 records the reliability of this wind farm while Figure 2.3 plots all these functions so that their nature can clearly be seen.

Parts (c) and (d) of Figure 2.3 are the most interesting as they show, respectively, the failure density function the area under which accumulates to 1, compared with Figure 2.2, and the hazard rate. This clearly shows the bathtub form given in Figure 2.4, with the early failures phase I, steady failure rate phase II and wear-out phase III. Particularly interesting is phase II where Figure 2.3(c) shows the failure density function which decreases exponentially, representing the random nature of failures in that phase. When the failure density function is normalised into the hazard rate in Figure 2.3(d) during phase II, those random failures become a constant hazard or failure rate.

### 2.4.3 *Reliability analysis assuming constant failure rate*

The unreliability of repairable systems can be modelled in terms of failure intensity by the bathtub curve ISO 14224 (2006), which represents the three different phases of a population life, as shown in Figure 2.4.

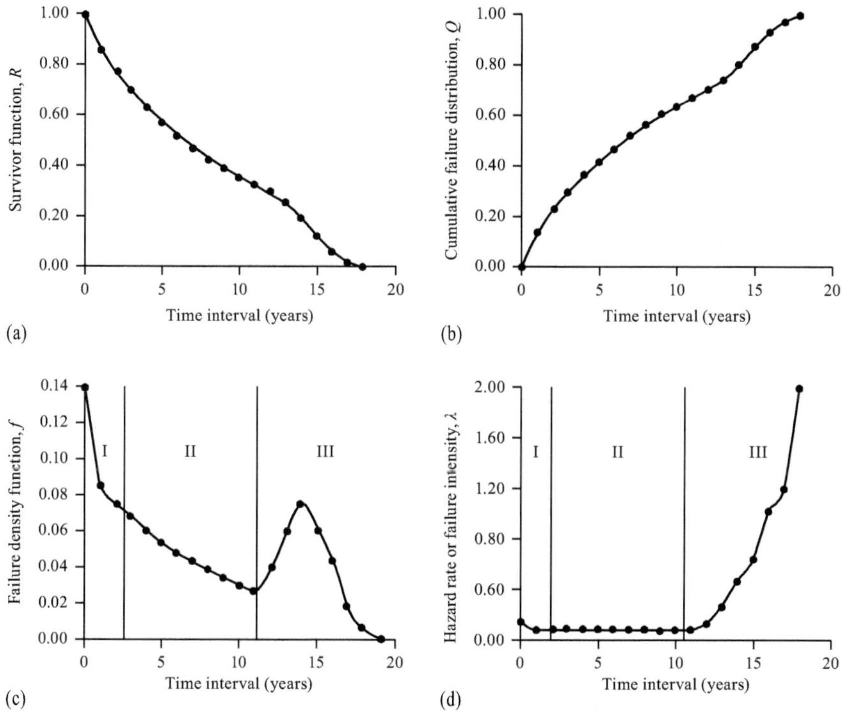


*Figure 2.3 Reliability functions from a wind farm of 1,000 non-repairable wind turbines: (a) reliability or survivors function, R(t), (b) unreliability or cumulative distribution function, Q(t), (c) failure density function, f(t), (d) hazard rate or failure intensity function, λ(t)*

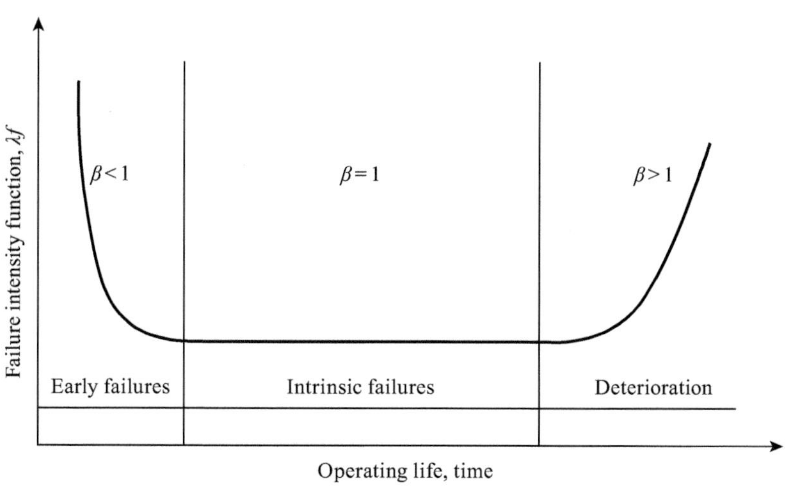


*Figure 2.4 The 'bathtub curve' for the intensity function showing how the reliability varies throughout the life of repairable machinery*

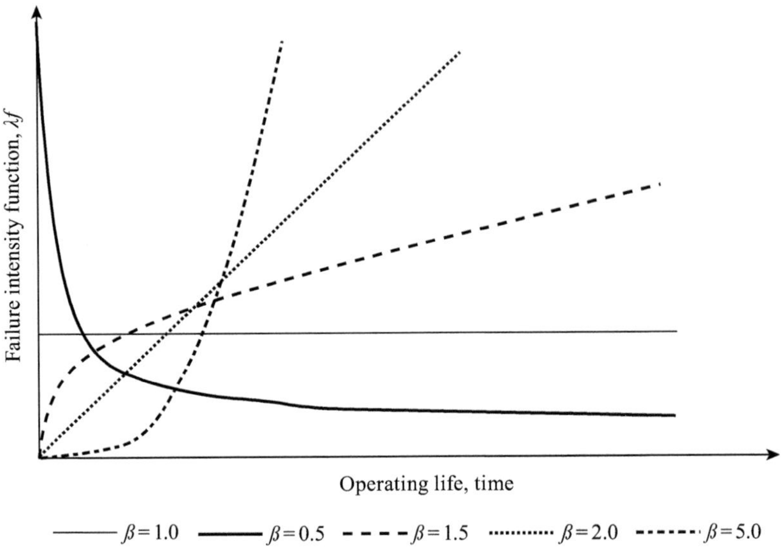


*Figure 2.5 The power law function showing how the failure intensity varies with the shape parameter β*

In turn, each phase of the bathtub curve can be modelled by a failure intensity function as shown in Figure 2.5.

This chapter is based on the bathtub curve concept, Figure 2.4, for a repairable system and its mathematical formulation, the power law process (PLP), which is a special case of a Poisson process with a failure intensity function:

$$\lambda(\tau) = \rho\beta t^{\beta-1} \tag{2.15}$$

$\beta$, determining the trend of the curve, is dimensionless and called the shape parameter, failure intensity changes with the shape parameter, as shown in Figure 2.5.

$\beta$ is a scale parameter which has the units per year $\lambda(t)$ and has units in this chapter of failures per item per year or year$^{-1}$, where an item could be an off-shore sub-station, a wind turbine or a sub-assembly.

For $\beta < 1$ or $\beta > 1$, the curve shows, respectively, a downward or upward trend. When $\beta = 1$, the intensity function of the PLP is equal to $\lambda$, the process represents the bottom of the bathtub curve, called the intrinsic failures phase, and $\lambda$ is described as the average failure rate.

Elements of the reliability theory used to analyse the failure data are given in Birolini (2007) and summarised next.

### 2.4.4 Bathtub curve

One can compare the failure density function, $f(t)$, derived from example data, Figure 2.3, with a theoretical form, Figure 2.2, of different shapes, but the $R(t)$ and $Q(t)$ values, Figures 2.3(a) and (b), can be extracted for any time by integrating under the curve to the left or right of that time, respectively.

*Table 2.2 Values of $\beta$ for different failure intensities*

| Values of $\beta$ | Failure intensity | Reason | Model types |
|---|---|---|---|
| $\beta < 1$ | Decreasing with time | Design weaknesses, improvements and alterations in the field | NHPP |
| $\beta = 1$ | Constant with time $\lambda(t) = \lambda$ | No major design modifications, random failures, wear and tear not apparent yet | HPP |
| $\beta > 1$ | Increasing with time | Normal wear and tear, deterioration of materials and accumulated stresses | NHPP |

The curve, Figure 2.3(d), derived from Figure 2.3(c) as the hazard rate, $\lambda(t)$, for the example data shows how the unreliability of non-repairable and repairable systems can be modelled. This so-called bathtub curve as shown in Rigdon *et al.* (2000) represents the three different phases of a population life in the complete bathtub, shown in Figure 2.4.

One of the advantages of using the PLP model for repairable systems is that its intensity function (2.15) is flexible enough to represent separately the three different phases of the bathtub curve, Figure 2.5, based on shape parameter, $\beta$, values as described in Table 2.2.

The idealised off-shore farm described in Table 2.1 and Figure 2.3 shows that at the bottom of that hypothetical bathtub, $\lambda$ was about 0.09 failure/turbine/year, far better than achieved on any current off-shore wind farms, despite the facts in the following:

- Large numbers of failures are accumulating with time, see second column of Table 2.1.
- Current off-shore wind farms incorporate repairable devices, so when a fault occurs, a wind turbine can swiftly be returned to the operational state by effective maintenance.

## 2.5 Reliability modelling concepts for off-shore wind farms

### *2.5.1 General*

The basis for modelling the reliability of complex off-shore wind farms lies in applying the failure distributions, described in Section 2.4, to the wind farm taxonomy, but several considerations are especially applicable to off-shore wind farms because of the following:

- Lack of operational data.
- Fact that off-shore wind farms incorporate both machinery and structural elements.
- Fact that different parts of an off-shore wind farm will be exposed to different environmental conditions.

All these aspects will be considered in this section to produce a concept for reliability modelling off-shore wind farms.

### 2.5.2 Reliability modelling concepts

When the reliability functions defined earlier are assigned to a total device, its systems, sub-systems, assemblies, sub-assemblies or components could result in different reliability functions. For the purposes of studying off-shore wind farm technology at this stage of development, the author chooses to operate with two reliability characteristics for mathematical modelling:

- Reliability survival function $R(t)$.
- Constant failure rate $\lambda$, describing probabilities of failure of non-repairable parts for comparison.

Mathematically, the reliability survivor function $R(t)$ or probability that a system or part will survive after a specified time $t$ can be expressed as

$$R(t) = 1 - f(t) = 1 - \int_0^1 f(t)dt \equiv \int_1^\infty f(t)dt \tag{2.16}$$

The hazard rate $\lambda(t)$ indicates the instantaneous probability that a given system or part will fail, assuming that the system/part is operational, defined as

$$\lambda(t) = \frac{f(t)}{R(t)} = \frac{f(t)}{(1 - f(t) \cdot R(t))} \tag{2.17}$$

The failure probability is usually high at the start of operational life because of manufacturing defects or mishandling. Towards the end of the system's or part's life, the probability of failure increases due to general wear and tear and the system/part cannot be repaired or replaced. However, there is usually an intermediate period where the probability is constant. As a function of time, this probability has the 'bathtub' shape shown in Figure 2.4. For reliability analysis, the intermediate period where the hazard rate is constant will be used for prediction. If the hazard rate $\lambda(t)$ becomes constant, $\lambda$, as it does during the intrinsic failures period, then

$$f(t) = \lambda e^{\lambda t} \tag{2.18}$$

and

$$R(t) = e^{\lambda} \tag{2.19}$$

The reliability survivor function $R(t)$ shown in (2.19), for a period $t = 1$ year between maintenance services, will be used in this book as the reliability characteristic for off-shore wind farm analysis.

For a device with all parts in series, i.e. where the failure of any one part will cause system failure if there is no redundancy, the total predicted system reliability $R(t)$ over time $t$ is the product of the reliabilities of the component systems:

$$R(t) = \Pi_i R_i(t) \tag{2.20}$$

*Table 2.3 Reliability characteristics for different parallel non-repairable systems*

| Number of parallel systems | Active redundancy | *R(t)* | $\lambda(t)$ |
|---|---|---|---|
| 1 | 1 of 1 must be working | $e^{-\lambda t}$ | $\lambda$ |
| 2 | 1 of 2 must be working | $2e^{-\lambda t}-e^{-2\lambda t}$ | $2\lambda/3$ |
| 3 | 1 of 3 must be working | $3e^{-\lambda t}-3e^{-2\lambda t}+e^{-3\lambda t}$ | $5\lambda/6$ |

and the total predicted system failure rate $\lambda_{tot}$, where part $i$ has a constant failure rate $\lambda_i$, yields

$$\lambda_{tot} = \Sigma\lambda_i \tag{2.21}$$

Different parallel system configurations can be described as Table 2.3, which shows the reliability characteristics of three different system arrangements, assuming each system is non-repairable.

A summary of effective redundancy equations for calculating reliability survivor functions and failure rate estimation for different system configurations is presented in RIAC&DACS (2005).

For non-repairable systems with independent sub-assemblies, when the hazard function or failure intensity $\lambda(t)$ is constant $\lambda$ with mission time $t$, an exponential probability distribution is the most appropriate to describe time to failure. This distribution has been chosen by the author due to its simplicity, also because part field failure rate data are not available and surrogate constant failure rates are used, which represent random variables.

Surrogate sub-assembly constant failure rates are related to the useful lifetime of the hazard failure rate curve, Figure 2.3, and these data can be applied to wind turbines and are analysed in Chapter 4.

Modarres *et al.* (2010) argue that the exponential distribution is a good model for representing systems and complex, non-redundant components consisting of many interacting parts.

Wind turbine and off-shore sub-station sub-assemblies should have long random failure regions, by design. Their architecture will consist of complex, non-redundant electronic, electrical and mechanical components; therefore, it is reasonable to use the exponential time-to-failure distribution rather than other distributions. For example, a Weibull time-to-failure distribution model would require more data than are currently available for many wind turbine sub-assemblies.

### 2.5.2.1 Point processes

A point process is a stochastic model describing the occurrence of discrete events in time or space. In reliability analysis, failures of repairable systems can be described with point processes in the calendar time domain, for example hourly, quarterly or annually, or using an operational variable, like run-time or number of turbine rotations.

A random variable $N(t)$ that represents, for example, the number of failure events in the interval $[0,t]$ is called the counting random variable. Subsequently, the number of events in the interval $(a,b]$ will be

$$N(a,b) = N(b) - N(a) \tag{2.22}$$

The point process mean function $\Lambda(t)$ is the expected number of failures, $E$, in the interval throughout time $t$:

$$\Lambda(t) = E[N(t)] \tag{2.23}$$

The rate of occurrences of failure $\mu(t)$ is the rate of change of the expected number of failures:

$$\mu(t) = \frac{d\Lambda(t)}{dt} \tag{2.24}$$

The intensity function $\lambda(t)$ is the limit of probability, $P$, of having one or more failures in a small interval divided by the length of the interval:

$$\lambda(t) = \lim/\Delta t \rightarrow 0P(N(t, t+\Delta t]) \geq 1/\Delta t \tag{2.25}$$

If the probability of simultaneous failures is zero, which occur only where the mean function $\Lambda(t)$ is not discontinuous, then

$$\lambda(t) = \mu(t) \tag{2.26}$$

#### 2.5.2.2 Non-homogeneous Poisson process

Assuming minimal repair that is failed, sub-assemblies are brought back to the same condition as just before the failure, the non-homogeneous Poisson process (NHPP) can be used to describe changes in reliability of repairable systems (Rigdon *et al.*, 2000). A counting process $N(t)$, which is the cumulative number of failures after operational or calendar time $t$, is a Poisson process if

$$N(0) = 0 \tag{2.27}$$

For any $a<b\leq c<d$, the random variables $N(a,b]$ and $N(c,d]$ are independent. This is known as the independent increment property. There is an intensity function $\lambda$ such that

$$\lambda(t) = \lim/\Delta t \rightarrow 0(P(N(t, t+\Delta t]) = 1)/\Delta t \tag{2.28}$$

Note that if $\lambda$ is constant then the process is homogeneous Poisson process (HPP).

Simultaneous failures are not possible:

$$\lim/\Delta t \rightarrow 0(P(N(t, t+\Delta t]) \geq 2)/\Delta t = 0 \tag{2.29}$$

The main property of NHPP is that the number of failures $N(a,b]$ in the interval $(a,b]$ is a random variable having a Poisson distribution with mean:

$$\Lambda(a,b] = E[N(a,b]] = a\lambda(t)dt \tag{2.30}$$

#### 2.5.2.3 Power law process

An NHPP is called a PLP if the cumulative number of failures through time *t*, $N(t)$ is given by

$$N(t) = \rho t^{\beta} \tag{2.31}$$

Therefore, the expected number of failures for a specific time interval $[t_1,t_2]$ will be

$$N[t_1, t_2] = N(t_2) - N(t_1) = r(t_2^{\beta} - t_1^{\beta}) \tag{2.32}$$

The intensity function is then

$$\lambda(t) = dN(t)/dt = \rho(t_2^{\beta} - t_1^{\beta}) \tag{2.33}$$

One of the advantages of using the PLP model for repairable systems is that its intensity function (2.33) is flexible enough to represent separately the three different phases of the bathtub curve, see Figure 2.4, based on the value of the shape parameter $\beta$, as described in Table 2.2.

### *2.5.3 Total time on test*

The variable *t* that appears in the various equations of the Crow-AMSAA model in MIL-HDBK-189 (1981) represents the time to a point process, but it differs from calendar time, as reported in the failures tables of WSDK, WSD and LWK (2010).

Reliability growth, as well as other reliability analysis, is normally carried out based on specific tests made on sub-assemblies under investigation. For a repairable system, the test is stopped after a failure or a planned inspection and the number of running hours elapsed since the previous failures are recorded. After several failures have been accumulated, failure data are interpolated with a mathematical model, like the Crow-AMSAA model, to verify the achieved reliability or, using the terminology of the military standard, the 'demonstrated reliability'. The independent variable *t* of the plot is the cumulative quantity called the total time on test (*TTT*) that is the integral of the number of running hours of the entire population for the observed period. In this way the hours of inactivity are not included in the evaluation of the *TTT*.

Using *TTT* rather than calendar time presents advantages and disadvantages and the meaning of *TTT*, for wind turbine failure data, must be clarified, as described in Spinato (2008).

First, it is usual in reliability engineering to deal with running hours rather than calendar time. This distinguishes a reliability analysis from an availability analysis. In this case the age of many electromechanical systems can be measured with the number of cycles completed or the total running hours and often this differs substantially from the calendar age.

Nevertheless, the calendar time plays an important role in reliability studies where chemical–physical properties deteriorate with time, for example the insulation of electrical machines or devices.

For data sets like LWK, WSD or WSDK (2010), the *TTT* in a certain interval *i*, $\Delta TTTi$ is calculated by multiplying the number of wind turbines, *Ni*, by the number of hours in the interval, *hi*. The recorded total hours lost from wind turbine production, *li*, in that interval are then subtracted, when this information is available. In these surveys these data included only out of service time, rather than time when the wind turbine was unable to operate for lack of wind. The aggregated *TTT* up to an arbitrary time cell *k, tk*, is then

$$tk = \Sigma_{i=1}\Delta TTTi = \Sigma_{i=1}Ni(hi - li) \tag{2.34}$$

To calculate the *TTT* for the LWK, WSD or WSDK data, three considerations are necessary.

For each time interval, the wind turbines in the survey are considered the representative of the entire of population. Therefore, the sample reliability for each time interval is assumed to represent the reliability of the entire population. This hypothesis is necessary to overcome one of the major deficiencies of the data, the variable number of wind turbines in each time interval. Any reliability improvement or deterioration spreads throughout the population with a certain rate, indicated by the shape parameter *b*, as long as sample wind turbines are assumed randomly chosen from the entire population, and the usage of each wind turbine in the population is similar.

Using *TTT* has the effect of stretching the curve on the abscissa. Since *TTT* depends on the number of turbines considered, it has no absolute meaning, as calendar time would have. The abscissa *t* has significance only for the wind turbine population being examined, however, by showing the cursor at the right of Figure 2.6, see above, calendar time can be inferred.

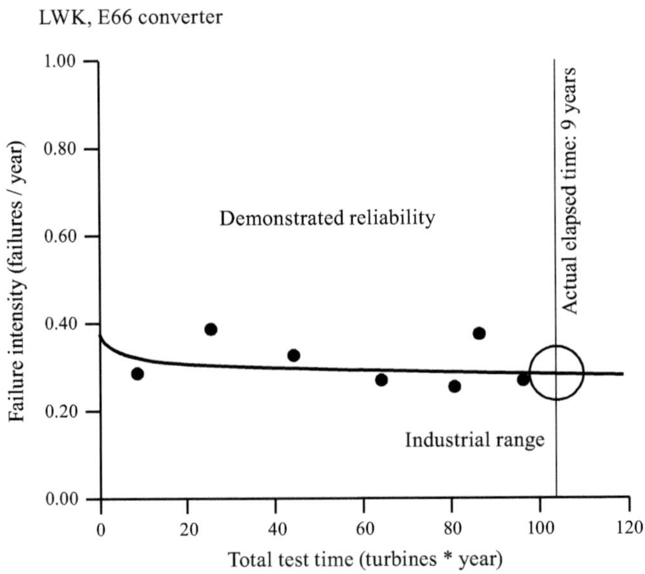


*Figure 2.6 Presentation of failure intensity using total time on test, TTT, showing demonstrated reliability for a sub-assembly with early failures, Spinato (2008)*

As the intensity function interpolates data on *TTT* rather than calendar time, the fit produced is intrinsically weighted by the number of turbines in each period. A larger number of wind turbines result in a larger *TTT* interval and the fit constraint is stronger. When *TTT* is used rather than calendar time, the abscissa stretches to a longer interval for more wind turbines surveyed and the scale parameter increases. In cases of early or constant failures, the most important result is the demonstrated reliability, as shown in Figure 2.6.

## 2.6 Reliability block diagrams

### *2.6.1 General*

Individual sub-assemblies can be represented in the process of reliability modelling and prediction, using the previous methods, by reliability block diagrams in a set and then connected in series or parallel to represent their functionality. Figure 2.7 shows possible arrangements for two reliability blocks.

### *2.6.2 Series systems*

Sub-assemblies in a set are said to be in series, from a reliability point of view, if they all must work for system success and only one needs to fail for system failure. Consider a system consisting of two independent series connected components A and B, for example a gear train.

$$R_p = \Pi R_i \tag{2.35}$$

This equation is referred to as the product rule of reliability.

Let $R_a$ and $R_b$ be the probability of successful operation of the individual sub-assemblies A and B, respectively, in Figure 2.7(a) and $R_s$ be the probability of successful operation of the series set, then

$$R_s = R_a \times R_b \tag{2.36}$$

---

**Example:** A gearbox consists of six successive identical gear wheels, all of which must work for system success. What is the system reliability of the series set if each gearwheel has a reliability of 0.95? From the product rule,

$$R_s = 0.95^6 = 0.7350$$

---

### *2.6.3 Parallel systems*

Sub-assemblies in a set are said to be in parallel, from reliability point of view, if only one needs to be working for system success or all must fail for system failure.

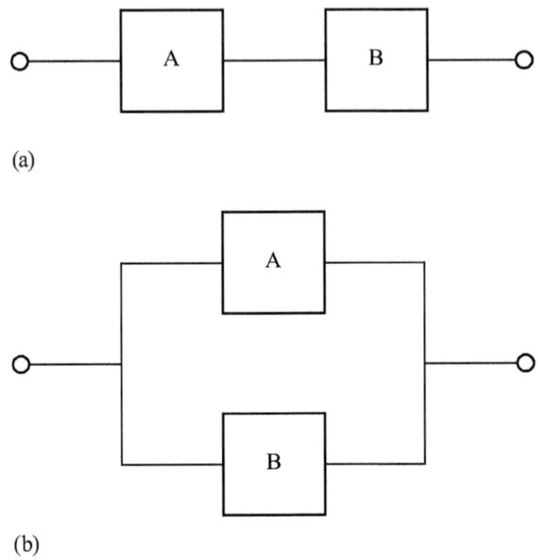


*Figure 2.7 Representation of sub-assemblies in a reliability block diagram: (a) series components, (b) parallel components*

Consider a system consisting of two independent sub-assemblies A and B, connected in parallel, Figure 2.7(b), for example two lubrication oil pumps for a gearbox connected in parallel. From a reliability point of view, the requirement is that only one sub-assembly must be working for system success. Again, let $R_a$ and $R_b$ be the probability of successful operation of individual sub-assemblies and $R_p$ be the probability of successful operation of the parallel set. Let $Q_a$ and $Q_b$ be the probability of failure of sub-assemblies A and B, respectively,

$$Q_p = \Pi Q_i \tag{2.37}$$

$$R_p = 1 - \Pi Q_i \tag{2.38}$$

---

**Example:** A system consists of four pumps in parallel each having reliabilities of 0.99. What is the reliability and unreliability of the parallel set?

$$Q_p = (1 - 0.99)^4 = 0.01^4 = 0.00000001$$

$$R_p = 1 - Q_p = 0.99999999$$

---

## 2.7 Summary

This chapter has presented the essential reliability mathematics necessary to understand the data collected from wind turbines and wind farms and presented in this book. It shows that simple methods can be used to extract essential information and the overall results which can be obtained.

However, care must be taken in manipulating the data to ensure that interpretations are sound.

# Chapter 3

# Weather, its influence on off-shore reliability

## 3.1 Wind, weather and large off-shore wind farms

### 3.1.1 Introduction

Having considered in Chapter 2 the mathematics needed to understand the reliability of wind turbines, we now need to consider the random effects of the weather.

Weather conditions are difficult to describe succinctly for engineering purposes and it is still not clear which aspects are important for reliable wind turbine operation. But the weather has been measured at sea since 1805, using the Beaufort scale summarised in Table 3.1, and this is a helpful starting point for understanding the impact of weather on off-shore wind farms.

It is important to appreciate from Table 3.1 the very large range of weather conditions to which large wind turbines, remote unmanned robotic power units operating 24/7, 365 days a year, are exposed and operate successfully.

Compare this to the relatively benign environmental conditions prevailing in enclosed conventional fossil- and nuclear-fired or hydro-power stations.

The impact of the sea and wind conditions, over the ranges shown in Table 3.1, on the foundations, base, tower, nacelle and operational components of operating off-shore wind turbines and their associated sub-stations and cable need to be borne in mind by all parts of the wind industry, especially with regard to O&M, Figure 3.1.

It should be particularly noted that the wind speed and wave heights or sea condition, shown on the Beaufort scale, are not necessarily contemporaneous because wind speeds may be rising before a storm when wave heights are not fully established, while after the storm a large swell may persist when wind speeds have moderated.

Weather conditions are now studied further under the headings as follows, considered important for off-shore wind farms at the current time.

The range of wind turbine operational wind speeds, with cut-in at 2–3 m/s and cut-out at 26 m/s, is highlighted in Table 3.1 in light grey ranging from Beaufort force 2–10, which is from light breeze to storm, whole gale.

Furthermore some wind turbines, notably the Enercon large wind turbine range, utilise storm control and do not cut-out sharply at 26 m/s but rather adopt a reducing power control from full power at 28 m/s to zero power at 34 m/s, extending their operation into Force 11, violent storm.

*Table 3.1 The Beaufort scale, Wikipedia Beaufort scale (2012) wind turbine operating range, highlighted in lighter grey small off-shore maintenance vessel accessibility, highlighted in darker grey*

| Beaufort number | Description | Wind speed (m/s) | Wave height (m) | Sea conditions |
|---|---|---|---|---|
| 0 | Calm | < 0.3 | 0 | Flat |
| 1 | Light air | 0.3–1.5 | 0–0.2 | Ripples without crests |
| 2 | Light breeze | 1.6–3.4 | 0.2–0.5 | Small wavelets. Crests of glassy appearance, not breaking |
| 3 | Gentle breeze | 3.4–5.4 | 0.5–1 | Large wavelets. Crests begin to break; scattered whitecaps |
| 4 | Moderate breeze | 5.5–7.9 | 1–2 | Small waves with breaking crests. Fairly frequent whitecaps |
| 5 | Fresh breeze | 8.0–10.7 | 2–3 | Moderate waves of some length. Many whitecaps. Small amounts of spray |
| 6 | Strong breeze | 10.8–13.8 | 3–4 | Long waves begin to form. White foam crests are very frequent. Some airborne spray is present |
| 7 | High wind, Moderate gale, Near gale | 13.9–17.1 | 4–5.5 | Sea heaps up. Some foam from breaking waves is blown into streaks along wind direction. Moderate amounts of airborne spray |
| 8 | Gale, Fresh gale | 17.2–20.7 | 5.5–7.5 | Moderately high waves with breaking crests forming spindrift. Well-marked streaks of foam are blown along wind direction. Considerable airborne spray |
| 9 | Strong gale | 20.8–24.4 | 7–10 | High waves whose crests sometimes roll over. Dense foam is blown along wind direction. Large amounts of airborne spray may begin to reduce visibility |
| 10 | Storm, Whole gale | 24.5–28.4 | 9–12.5 | Very high waves with overhanging crests. Large patches of foam from wave crests give the sea a white appearance. Considerable tumbling of waves with heavy impact. Large amounts of airborne spray reduce visibility |
| 11 | Violent storm | 28.5–32.6 | 11.5–16 | Exceptionally high waves. Very large patches of foam, driven before the wind, cover much of the sea surface. Very large amounts of airborne spray severely reduce visibility |
| 12 | Hurricane Force | ≥ 32.8 | ≥ 14 | Huge waves. Sea is completely white with foam and spray. Air is filled with driving spray, greatly reducing visibility |

*Figure 3.1 Jacket mounted off-shore 6-MW wind turbines encounter Force 7 wind conditions, see Table 3.1, © US Department of Energy*

### *3.1.2 Wind speed*

From the work in Figure 1.2, see Chapter 4, the author first noticed that high wind turbine failure rates were related to high wind speed, particularly in the stormy winters of 1998 and 1999 in Denmark and Germany. This was studied by Tavner, Edwards *et al.* (2006), particularly across Denmark and in a more precise study later by Tavner *et al.* (2013), of three specific wind farms in Germany, the reliability results from which are presented in Section 3.3.1. These three wind farms were located at the following points:

- Fehmarn, on a small island in the Baltic Sea, off the coast of Schleswig Holstein, Germany.
- Krummhörn, on the North Sea Coast in Lower Saxony, Germany.
- Ormont, inland in the highlands in Rhineland Palatinate, Germany.

An indication of the annual variation of wind speed at these three disparate sites is shown in Figure 3.2.

### *3.1.3 Wind turbulence*

Wind speed of itself is a significant influence on reliability but probably more important to wind turbine reliability is the wind turbulence.

Wind turbulence refers to wind speed fluctuations on a short time scale. However, there is no established time period over which such wind speed variations are officially classed as turbulent. Indeed as McIlveen (1992) explains '[although] turbulence ... has been studied for over a century ... it is surprisingly difficult to define precisely what we mean by turbulence'.

Turbulent eddies are formed in the atmosphere due to thermal gradients or when the wind flow passes over a rough surface or is disrupted by obstacles such as trees, hills and buildings. Wake effects from neighbouring wind turbines can also significantly contribute to the turbulence experienced by a wind turbine. Due to a lack of obstacles and a relatively smooth surface for the wind to pass over,

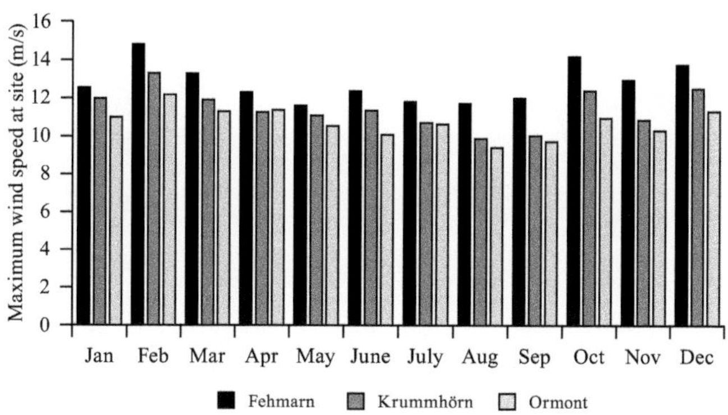


*Figure 3.2 Annual variations in horizontal wind speeds at three disparate German on-shore wind farms, Tavner et al. (2013)*

off-shore sites typically experience less turbulence than on-shore installations, although this does depend on the above-sea temperature gradient and sea state.

Although McIlveen (1992) acknowledged the difficulty in defining turbulence, it does attempt to give an idea of the size of the turbulent eddies formed in the wind. It states that the largest eddies have scales of approximately 100 m and that the smallest are approximately 1 mm. This translates to most turbulent variations lasting less than 100 s at a fixed position.

Power spectral analysis shows the wind speed variation timescales which contain the most energy. An early work in this area, van der Hoeven (1957), identified a 'turbulent peak', as shown on the right in Figure 3.3, at a period of about 1 min for horizontal wind at a height of 100 m. It is probable that the eddies, which have the most effect upon the wind turbine and its drive train, known to be vulnerable to fatigue caused by turbulence, are at turbulent dimensions of significance compared to blade lengths or disc diameter, i.e. 100 m for large off-shore wind turbines.

Nandi *et al.* (2017) set out, in a study of GE 1.5-MW variable speed wind turbines, three important timescales or frequency bands in the interaction between a wind turbine and atmospheric turbulence, as follows:

- Advective energy-dominant eddy passage timescales
  ≈25–50 s, 0.02–0.04 Hz;
- Blade rotation once per revolution timescales
  ≈4–6 s, 0.17–0.25 Hz for modern large off-shore wind turbines;
- Sub-blade pitch rotation timescales.

<1 s, 1 Hz arising from the interaction between a blade and the turbulence eddy's internal structure, as the blade rotates in the wind.

The measured frequencies in Figure 3.4, correlating anemometer and generator power readings from operating wind turbines, reflect the graph in Figure 3.3 and the box indicated.

This approach is expanded in Lavely (2017), Doctoral Thesis.

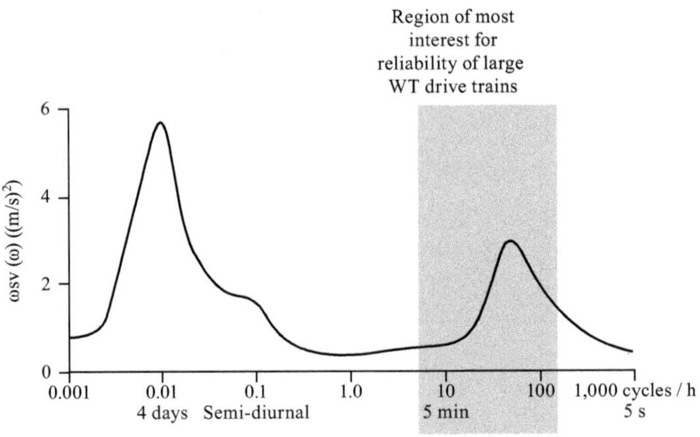


*Figure 3.3 Power spectrum of horizontal wind speeds, van der Hoeven (1957)*

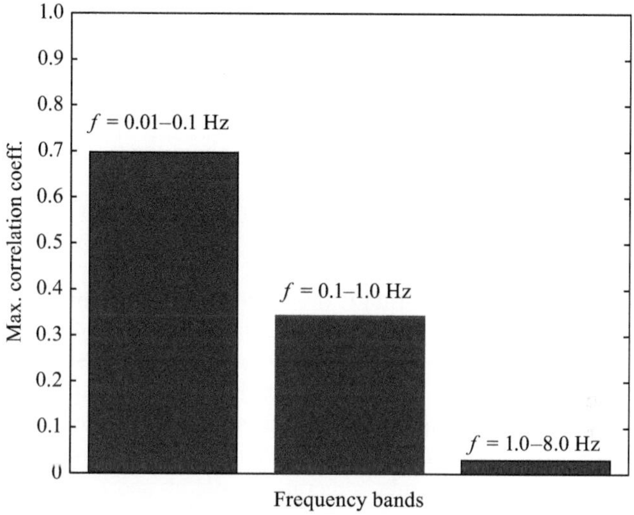


*Figure 3.4 Frequency bands correlating anemometer and generator power readings from a 1.5-MW GE wind turbine; the vertical axis scale is suppressed for proprietary reasons, Nandi et al. (2017)*

The standard IEC 61400-1 uses a measure called the turbulence intensity, $I$, which is the ratio of the wind speed standard deviation, $\sigma$, to the mean wind speed, $u$, for each 10-min reporting period. This is the measure of turbulence used throughout the wind energy industry; both $I$ and $u$ are readily available from most wind turbine supervisory control and data acquisition (SCADA) systems and met masts.

$$I = \frac{s}{u} \tag{3.1}$$

Due to the definition of this measure, when $u$ is small, $I$ becomes large but is physically insignificant. Therefore, some advocate that $I$ for wind speeds below 8–10 m/s are not load relevant.

When describing $I$ over a period greater than 10 min, it is necessary to perform an average using the SCADA 10-min values.

Several subtly different terms are used in the industry for turbulence. Characteristic turbulence intensity, $I_{char}$, was used in an earlier version of standard IEC 61400-1.

Representative turbulence intensity, $I_{rep}$, is used in a later version. They differ in that $I_{char}$ is defined as the mean plus the standard deviation, whereas $I_{rep}$ is the 90% percentile. Both $I_{char}$ and $I_{rep}$ are used in the wind industry, but $I_{char}$ is more popular and will be used for the following cross-correlations.

There is also ambiguity over the term 'gust'. Short-term extreme events would fit with the use of 'gust' in IEC 61400-1, which provides an extreme operating gust model that simulates a rapid wind speed increase, for example 24–36 m/s over 5 s.

For the purposes of this section, gusts will be assumed to be special cases within a wind velocity spectrum, which may be short-term, extreme turbulence events.

### 3.1.4 Wave height and sea condition

The range of wave heights and sea conditions for which smaller access vessels, as shown in Figure 10.5, can transfer personnel safely in an off-shore wind farm is highlighted in darker grey in Table 3.1. It ranges from Beaufort force 0–4 that is up to a moderate breeze. There is as yet no information on the systematic effects of sea state on off-shore wind turbine reliability.

### 3.1.5 Temperature

The variations of maximum, minimum and variance of temperature at the three on-shore German sites described earlier are shown in Figure 3.5(a)–(c), which demonstrates the annual variation in temperature variance. It should be noted that the island location, Fehmarn, which has the least temperature variance, as one would expect surrounded by the sea and this is what we may expect at off-shore locations. Temperature is of significance to wind turbines since it can affect the reliability of electrical components, generators, motors and converters.

### 3.1.6 Humidity

Similarly, the variation of humidity at these three on-shore German sites is shown in Figure 3.6. Again it should be noted that it is the coastal location, Krummhörn, which has the highest mean humidity and least humidity variance, and this is what we may expect at off-shore locations. Humidity is of significance to wind turbine since it can affect the reliability of electronic systems, particularly the power converters.

## 3.2 Mathematics to analyse weather influence

### 3.2.1 General

References Tavner, Xiang *et al.* (2006) and Tavner *et al.* (2013) used failure time periodograms and failure-environmental cross-correlograms to analyse the influence of weather phenomena on failures.

### 3.2.2 Periodograms

This approach transforms the time domain data into the frequency domain using Fourier analysis. If a signal, $f(t)$, is periodic, i.e.

$$f(t) = f(t + T) \tag{3.2}$$

Then it is possible to represent it in the frequency domain by
This may be restated by

$$F(s_k) = \frac{1}{T}\int_{-T/2}^{T/2} f(t)e^{-j2\pi s_k t}dt \tag{3.3}$$

where $k = 0, \pm 1, \pm 2, \ldots$ and denotes the $k$th harmonic of the fundamental frequency $(1/T)$.

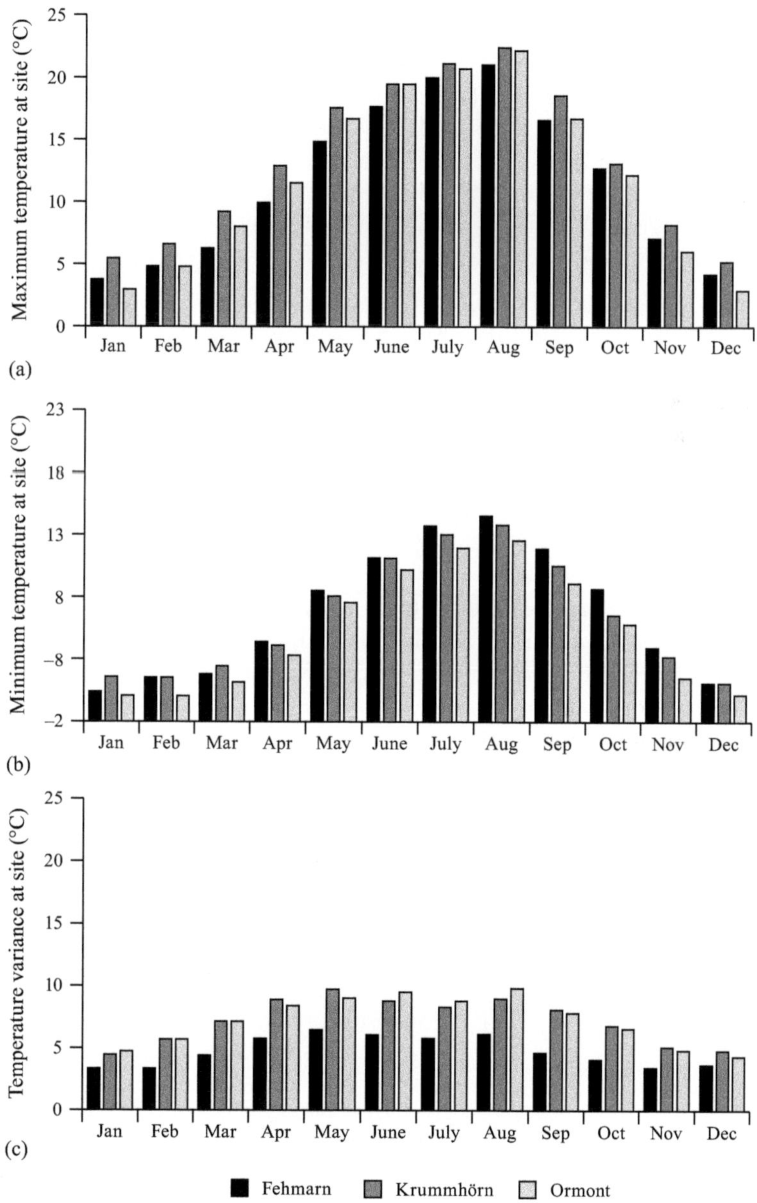


*Figure 3.5 Annual variation of temperatures at three disparate German on-shore wind farms, Tavner et al. (2013). (a) Maximum temperature range, (b) mean temperature range and (c) minimum temperature range*

In this case the time domain data are sampled, so the transformation from the time domain to the frequency domain is expressed by

$$F(s_k) = \frac{1}{N}\sum_{n=0}^{N-1} f(t_n)e^{-j2\pi nk/N} \tag{3.4}$$

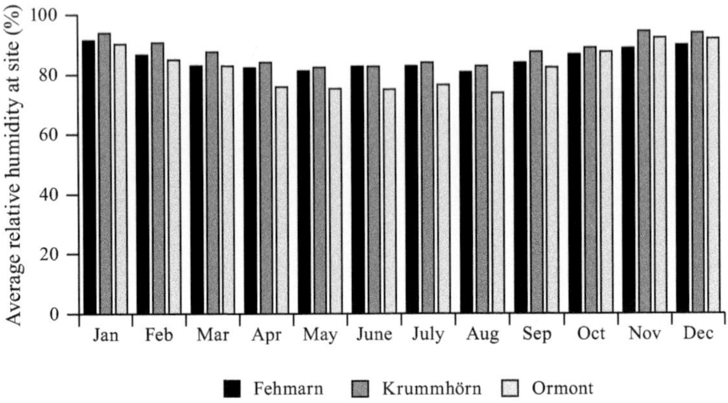


*Figure 3.6 Annual variation of humidity at three disparate German on-shore wind farms, Tavner et al. (2013)*

The transformation was computed using the fast Fourier transform (FT), which is a well-established and computationally efficient way of obtaining a discrete FT. It is only strictly valid to carry out an FT on a periodic signal. When taking the signal FT, it is assumed that the fundamental is the reciprocal of the signal length. If this requirement is not met, there will be a discontinuity in the signal resulting in harmonic leakage in the frequency domain. For the present purposes, this assumption is unlikely to be valid; therefore, it was necessary to apply a Hanning window to minimise the harmonic leakage.

### 3.2.3 Cross-correlograms

Cross-correlation is a time domain technique used to measure the extent to which two signals are linearly related. The cross-correlation function of two stationary time domain signals, $f(t)$ and $g(t)$, is given by

$$R_{fg}(\tau) = \int_{-\infty}^{\infty} f(t)g(t+\tau)dt \tag{3.5}$$

This may be restated as

$$R_{fg}(\tau) = \lim_{T\to\infty} \frac{1}{2T}\int_{-T}^{T} f(t)g(t+\tau)dt \tag{3.6}$$

where $T$ is the period of observation, which is the signal length, $\tau$ is the time lag between the signals. For sampled signals, this is written as

$$R_{fg}[m] = \lim_{N\to\infty} \frac{1}{2N+1}\sum_{-N}^{N} f[n]g[n+m] \tag{3.7}$$

where $N$ is the number of data points, and $m$ is the lag. Note that in order to interpret this lag as a time shift, the time series must be uniformly sampled.

The cross-correlation function can now be estimated where the signals $f(t)$ and $g(t)$ are of finite length. For sampled signals the biased cross-correlation is computed by

$$R_{fg}[m] = \frac{1}{N} \sum_{n=1}^{N-m+1} f[n]g[n+m] \tag{3.8}$$

While the unbiased form is

$$R_{fg}[m] = \frac{1}{N-|m|} \sum_{n=1}^{N-m+1} f[n]g[n+m] \tag{3.9}$$

where $m = 1, \ldots, M + 1$.

### *3.2.4 Concerns*

A serious concern about these analyses has been the relative frequencies of the failure and meteorological data.

Failure data are usually collected daily or weekly and meteorological data may be collected at 1-min intervals. This immediately creates a problem when trying to cross-correlate such widely disparate frequencies.

The physical reality is that the wind turbine failure mechanisms, Figure 3.2, are essentially cumulative or integrative and this needs to be considered in the analysis methods.

The analysis also indicates correlation rather than necessarily causation. However, such analysis should lead to better understanding of causation.

## 3.3 Relationships between weather and failure rate

### *3.3.1 Wind speed*

Based upon the analysis methods described before the effect of wind speed on wind turbine failure rates was investigated in Denmark, where detailed failure and weather data were available, in Tavner (2008) and Tavner, Xiang *et al.* (2006). This showed a significant correlation between wind turbine failure rate and months when the average wind speed was higher, with a cross-correlation factor of 44%. This is shown graphically in Figure 3.7, where Danish wind turbine failure rates during the 1994–2004 period peaked each year in February and October, while wind speeds peaked in February.

Even more illuminating was that the cross-correlation of failures in different sub-assemblies with wind speed varied across those sub-assemblies, as shown in Figure 3.8.

Surprisingly in this survey, the generator proved the most sensitive sub-assembly to higher wind speeds rather than aerodynamic sub-assemblies, for example yaw or pitch control systems. The pragmatic explanation for this behaviour could be that the generators for these wind turbines were commercially procured, standard sub-assemblies not necessarily hardened for the wind industry environment.

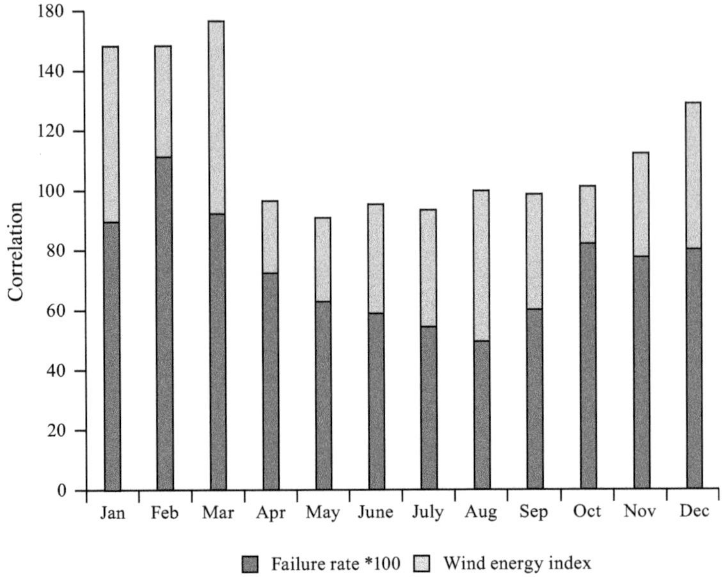


*Figure 3.7 Average on-shore Danish wind turbine monthly failure rates and WEI, related to wind speed, for each of the 12 months, 1994–2004, Tavner et al. (2006)*

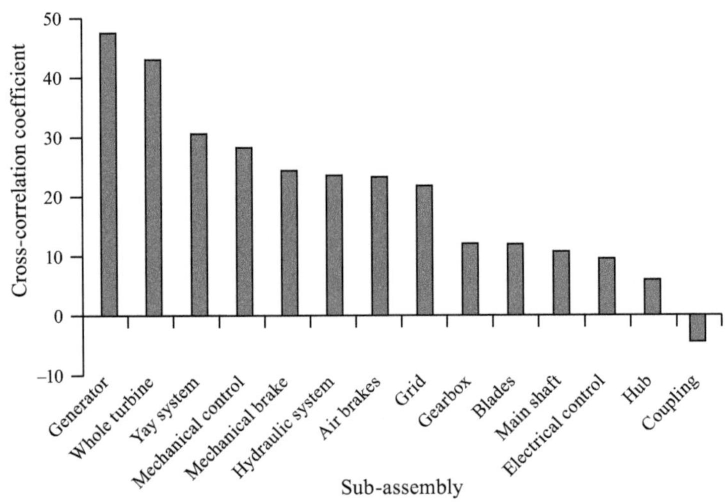


*Figure 3.8 Summary of cross-correlograms of Danish on-shore wind turbine sub-assembly failure rates to wind speed, 1994–2004, Tavner et al. 2006*

The strength of this study was the large number of turbines considered over an extensive period of time and the large number of failures involved. The weakness of the study was that it blurred the reliability of many different turbine designs and considered a monthly average wind speed over the whole of Denmark, thereby concealing, by averaging, more detailed wind speed effects.

An improved study by Tavner *et al.* (2013) of German data has been prepared, which concentrates on the failure rate of one particular wind turbine type located at on-shore wind farms in locations where accurate weather data were available. The authors identified three locations, as described in Section 3.1.2, with different climatic and operating conditions, operating the same type of turbine.

By taking a more focused approach, this paper corrects some of the shortcomings of the previous study of Danish failure data and reveals more significant effects of weather and location on reliability. However, it does not show a cross-correlation of failures with wind speed but does show cross-correlations with other weather factors as described next.

### *3.3.2 Temperature*

The results of Tavner *et al.* (2013) show an interesting consequence of temperature on wind turbine failures, visible in Figure 3.9.

At all three sites the failure rates were affected by temperature, all showing annual variations with the seasons but with a cross-correlation phase variation between them and the general effect of temperature on failure rate being higher in summer than winter.

A more detailed observation was also shown at one site by separating failures between electrical and mechanical sub-assemblies as shown in Figure 3.10.

Figure 3.10 shows that the cross-correlation between temperature and failures is dominated by the electrical rather than mechanical sub-assemblies. The use of sealed, environmentally controlled nacelles off-shore will counter this issue.

### *3.3.3 Humidity*

The results of Tavner *et al.* (2013) showed that there was a significant cross-correlation, 23%–31%, with failures at the island and coastal site with a much

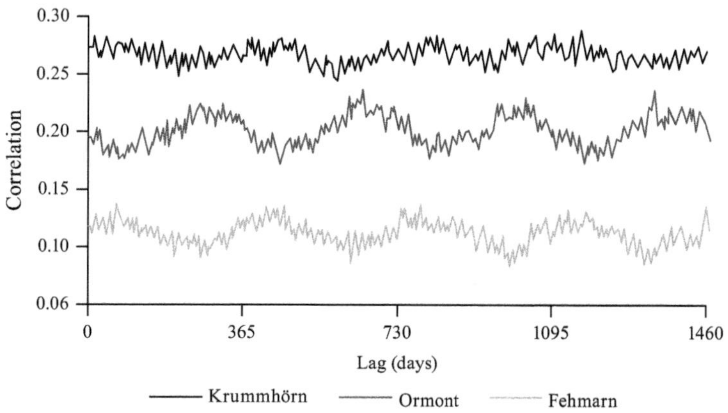


*Figure 3.9 Wrapped correlograms of failure variation at three on-shore wind farms due to daily maximum temperature, Tavner et al. (2013)*

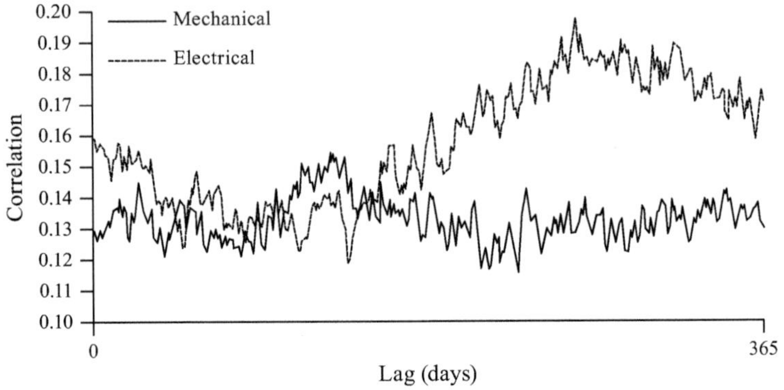


*Figure 3.10 Wrapped correlograms of failure variation at Fehmarn due to daily maximum temperatures, split between electrical and mechanical failures, Tavner et al. (2013)*

lower cross-correlation at the inland site, suggesting that for off-shore the issue of humidity is significant and again this is being handled by the adoption of sealed, environmentally controlled nacelles.

### 3.3.4 Wind turbulence

Analysis performed on SCADA data from 3 off 2-MW wind turbines with hydraulic pitch system faults, Figure 3.11, shows significant correlation between failures and turbulence, represented by $I_{char}$, measured when the daily mean wind speed exceeded the wind speeds shown. This indicates that turbulence is driving pitch failures in this case, although the results are difficult to interpret.

Another analysis on SCADA data from 6 off 1.6-MW wind turbines in three wind farms with electric pitch system faults, considering a different measure of turbulence to $I_{char}$, again shows significant correlation between failures and turbulence, see Table 3.2, using a turbulence measure related to wind speed, $ku_2$, $ku_5$, $ku_8$, $ku_{10}$, compared to the correlation with wind speed itself. First, the analysis shows clearly that failures are more sensitive to turbulence than wind speed; second, it shows a difference between different wind farms, mentioned by the operator, Wind Farm 2 being known to experience more turbulent conditions and being prone to pitch failures.

A further analysis on these wind turbines considered the effect of gusts upon the pitch failures and does not show any particular sensitivity for Wind Farm 2 but rather confirms the correlations of Figure 3.3.

These three sets of results demonstrate that it is possible to detect a sensitivity to wind speed, turbulence and gusts of wind turbine pitch mechanisms failures, known to be significant, see Table 3.1. More work needs to be done to establish the link but operators need to realise that measurements made from SCADA could unlock these root causes.

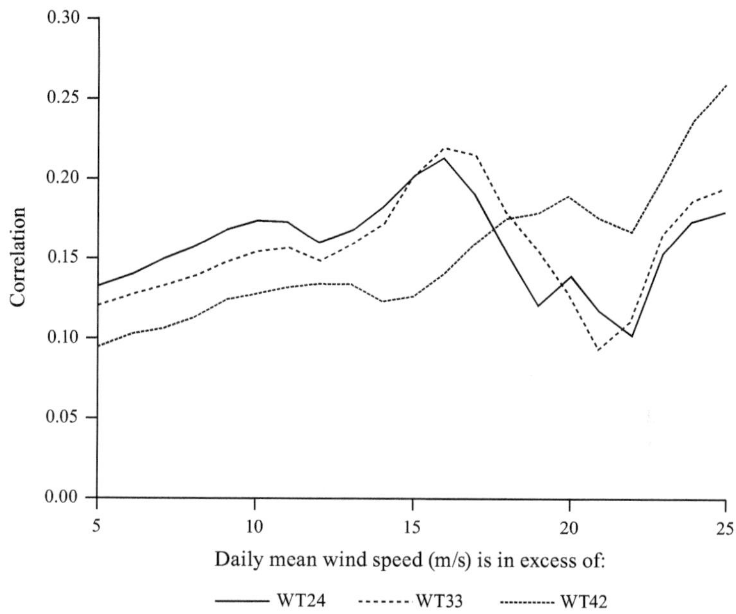


*Figure 3.11 For three wind turbines correlograms of pitch failure with $I_{char}$ on days when the daily mean wind speed exceeded the figures shown*

*Table 3.2 Cross-correlation of pitch failures with turbulence on 1.6-MW wind turbines*

| **Variables** | | **Wind Farm 1** | | | **Wind Farm 2** | | **Wind Farm 3** |
|---|---|---|---|---|---|---|---|
| | | **Wind Turbine 1** | **Wind Turbine 2** | **Wind Turbine 3** | **Wind Turbine 4** | **Wind Turbine 5** | **Wind Turbine 6** |
| Mean wind speed | $u$ | 0.15 | 0.24 | 0.23 | 0.15 | 0.11 | 0.18 |
| Wind speed turbulence coefficients | $ku_2$ | 0.20 | 0.41 | 0.33 | 0.17 | 0.17 | 0.36 |
| | $ku_5$ | 0.23 | 0.41 | 0.28 | 0.21 | 0.28 | 0.33 |
| | $ku_8$ | 0.32 | 0.35 | 0.27 | 0.40 | 0.34 | 0.30 |
| | $ku_{10}$ | 0.34 | 0.31 | 0.28 | 0.45 | 0.50 | 0.24 |

Much more detailed turbulence analysis has been done on data from DE 1.5-MW wind turbines by Nandi *et al.* (2017).

### 3.3.5 Sea surface effects

Sea measurements were made by the FINO project (2014) at the German prototype off-shore wind site, Alpha Ventus, in the Heligoland Bight, near the island of Sylt, in the North Sea, referred to by Krogstad *et al.* (1999), which could illuminate these effects.

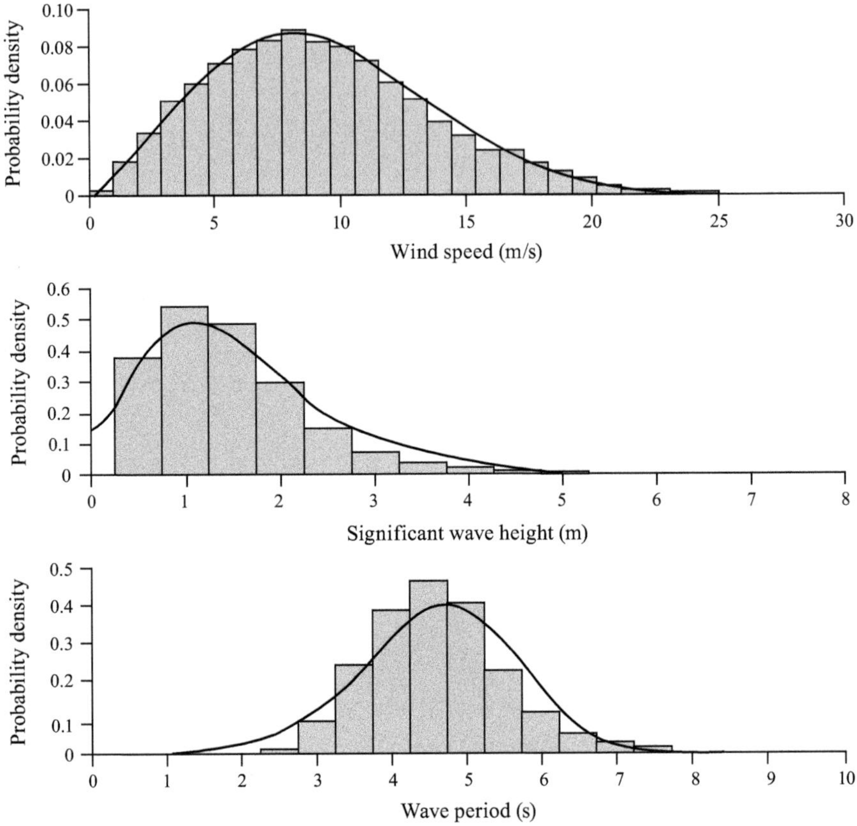


*Figure 3.12 Probability distribution of wind speed, wave height and wave period over 8 years close to Alpha Ventus site, FINO (2014)*

*Table 3.3 Cross-correlation of pitch failures with gusts on 1.6-MW wind turbines*

| **Variables** | **Wind Farm 1** | | | **Wind Farm 2** | | **Wind Farm 3** |
|---|---|---|---|---|---|---|
| | **Wind Turbine 1** | **Wind Turbine 2** | **Wind Turbine 3** | **Wind Turbine 4** | **Wind Turbine 5** | **Wind Turbine 6** |
| Wind gusts over 2 m/s | 0.28 | 0.24 | 0.23 | 0.23 | 0.18 | 0.19 |
| Wind gusts over 5 m/s | 0.33 | 0.47 | 0.37 | 0.31 | 0.31 | 0.34 |
| Wind gusts over 10 m/s | 0.30 | 0.33 | 0.33 | 0.25 | 0.37 | 0.71 |

This was a valid off-shore wind site and enables a closer view of the interaction of wind and wave conditions, Figure 3.12, specific to that site. The clear cross-correlation of pitch failures to wind gusts is also shown in Table 3.3.

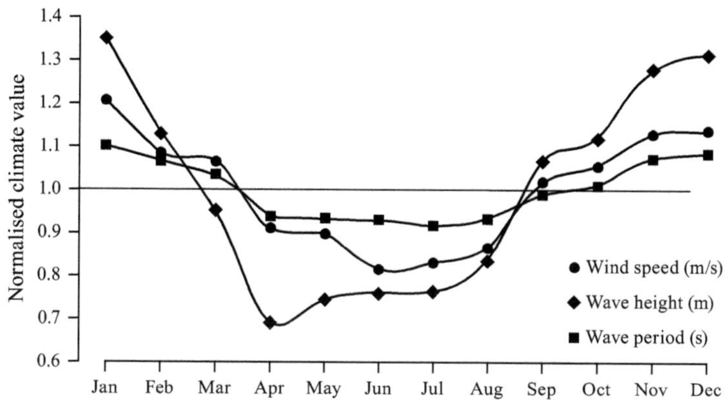


*Figure 3.13 Seasonal variation of wind speed, wave height and wave period close to Alpha Ventus site, FINO (2014)*

An interesting observation is that the wind resource variation distribution is much wider than that of the wave resource, as one would intuitively expect of a less dense and less constrained medium but the denser liquid medium imposes larger turbulent excursions, Figure 3.13.

Seasonal variations in averaged wind speed, wave height and period probabilities will be of interest to MEC farm operators. The normalised, relative variations at the Alpha Ventus site of these three parameters, averaged over an 8-year period, are shown in Figure 3.11. These indicate typical North Sea seasonal variations, where 1.0 is the norm.

## 3.4 Resource, location, reliability and capacity factor

This chapter is showing that wind power device reliability depends upon resource influences, for the following reasons:

1. Complex but predictable aerodynamic and hydrodynamic resources are a fluid continuum. Wind resource shows meteorological-scale or synoptic-scale dynamics and fast-fluctuation small-scale turbulence, Milan *et al.* (2013), affecting device forces, torques and power flows, exemplified for wind by a conventional, simplified image of that process, van der Hoeven (1957), showing a meso-scale, 2D inverse vortex cascade, and a short time-scale, 3D, downward vortex cascade, with hypothetical spectral separation, Figure 3.2. But recent wind studies, Fitton (2013), propose a universal vortex cascade, containing intermittent fluctuations down to small timescales; a typical multi-fractal process. More recent wind studies, Nandi *et al.* (2017), on data from a group of GE 1.5-MW wind turbines, provide more detailed practical information.

   Tidal resource is also predictable for a given site, but affected by the interaction of wind, wave and water, each having a probabilistic distribution, all three depending upon the specific site. Reports from the tidal industry, Delorm (2013), Doctoral Thesis, Okorie (2011) and Wood *et al.* (2010), show an increasing understanding of the importance of small timescale turbulent aspects, as in the wind industry. Figure 3.14 demonstrates resource turbulence effects on a tidal device, off-shore wind turbine and wave device and their effects on the drive train.

2. Each wind turbine, or marine renewable device, delivering power is subject to these turbulent resources and experiences a series of irreversible deteriorating processes or failure mechanisms due to rapidly changing forces, torques and

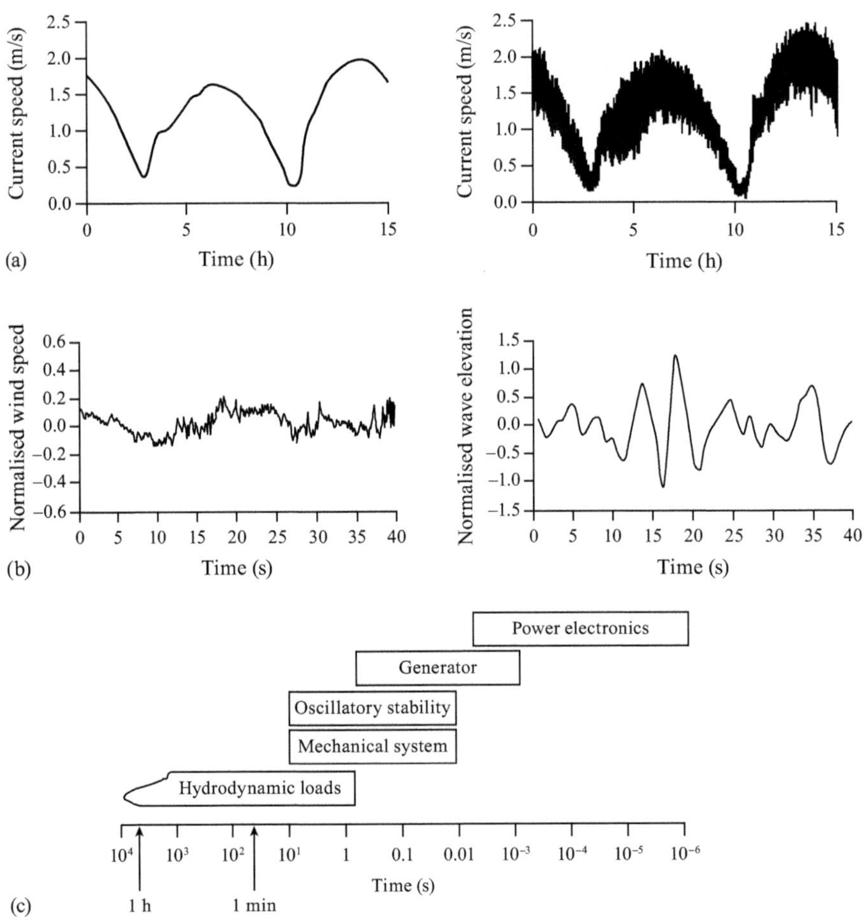


*Figure 3.14 Effects of turbulence on off-shore wind turbines and other marine energy converters: (a) tidal current at a tidal turbine site, smoothed left and unsmoothed with turbulence right; (b) comparison of effect of resource on an off-shore wind turbine (left), and wave device (right); (c) contribution of turbulent effects to a wave or tidal device drive train, Alcorn et al. (2014)*

power flows. These create stress, thermal and mechanical wear, yield, fatigue and corrosion, which may be controlled by device design but interfere with the energy conversion process.

It is clear from Figure 3.14 that an off-shore wind turbine may experience weaker turbulent effects than submerged devices, although that will be aggravated by pitch, roll and yaw motions for floating off-shore wind turbines.

3. In time, the failure mechanisms of (1) and (2) accumulate to cause component failure, a stochastic integration process, worsening device reliability and ultimately causing failure. Wind turbine reliability is fundamentally dependent on the following:
   (i) Design: by virtue of the turbine's materials, load and rating, which stress aerodynamic, mechanical and electrical components, when subject to excessive wind speeds, torques, temperatures, voltages or currents. These may be due to variability or discontinuity in turbine load, or changes in the ambient environment, including generator, gearbox or converter temperature, affected by lubricants, coolants or maintenance state.
   (ii) Application: for an off-shore wind turbine due to natural environmental changes in its particular location and the wind and wave profiles there.
   (iii) Physics: torque being developed by thermodynamically reversible aerodynamic, electro-magnetic and mechanical processes accompanied, due to changing torque, force, current, voltage, speed and temperature, by irreversible loss processes in blades, bearings ($\alpha$ to $\omega r_{brg} F_{brg}$), generator windings ($\alpha$ to $I^2 R_{wind}$) and insulation ($\alpha$ to $V^2/R_{ins}$), core ($\alpha$ to $V^2/R_{core}$) and coolant ($\alpha$ to $\Delta T_{fc}$).

       These irreversible processes constitute mechanical, electrical, thermal and chemical degradation, wear, corrosion and fatigue root causes leading to yield, which may be mitigated by design but ultimately interfere with energy conversion.
   (iv) In time, these root causes accumulate to trigger failure, a stochastic integration process, worsening turbine reliability and leading to failure.

4. The impact of the effects of (1), (2) and (3) on the reliability of off-shore assets will be the subject of the following chapter, see Tavner and Zappalá (2021).

Figure 3.15 shows the gross consequences of these turbulence effects, when applied to large off-shore wind farms over a long period of time, demonstrating how wind farm capacity factors in the United Kingdom have developed with the introduction of both on-shore and off-shore wind farms. This figure used data taken from Crabtree *et al.* (2015) but brought up-to-date by Dr D. Zappalá.

Figure 3.15 demonstrates how UK wind power output, on-shore and off-shore, has developed and changed over the last 14 years.

Note that capacity factor increases from on-shore to off-shore, then off-shore from Rounds 1 to 2. Off-shore capacity factor increases because of the greater wind resource, but also note that this increases with time as wind farm availability improves.

These changes not only depend upon resource itself but also upon wind turbine and farm reliability, incorporating the random effects of (1), (2) and (3) described previously, but also upon the maintenance that is applied to them.

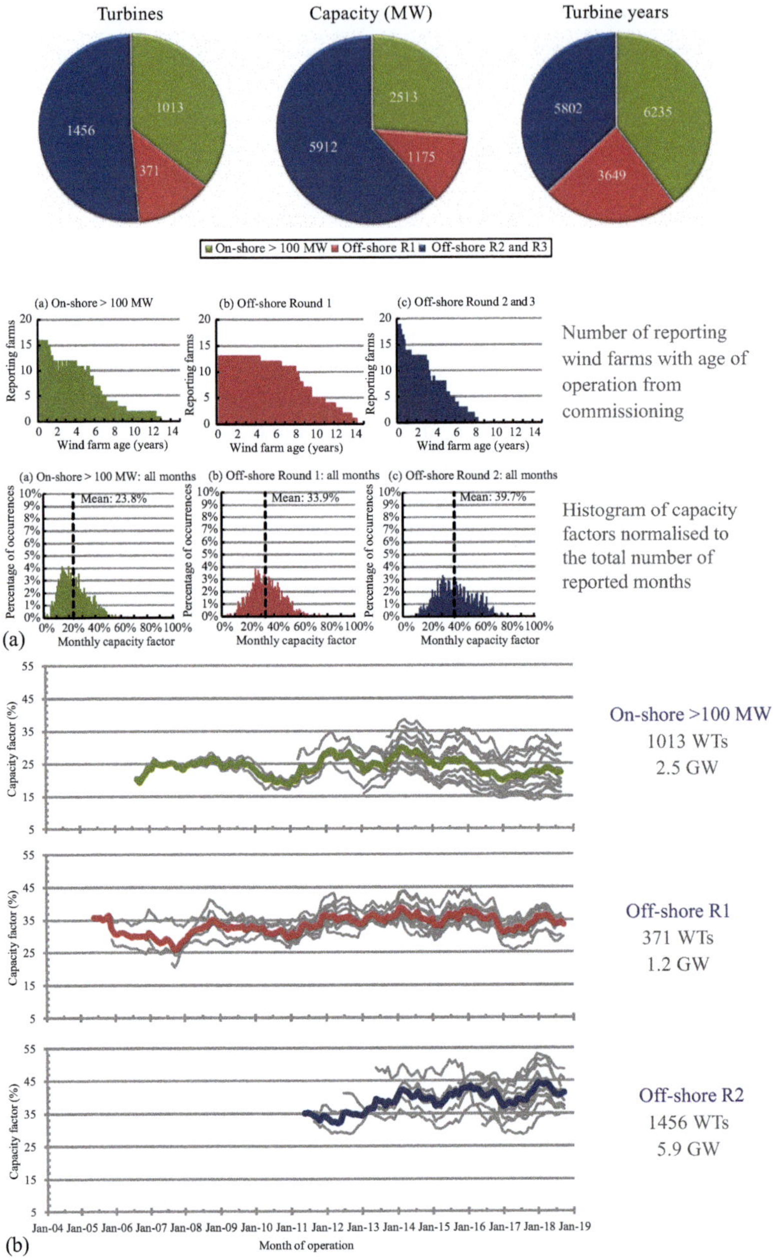


*Figure 3.15 Development of off-shore wind in United Kingdom and developing capacity factors, on-shore and off-shore, figures courtesy of Dr D. Zappalá: (a) number, capacity and turbine-years operation for on-shore, off-shore Round 1 and Round 2 wind farms in United Kingdom; (b) capacity factor for on-shore, off-shore Round 1 and Round 2 wind farms in United Kingdom from 2004 to 2019*

## 3.5 Summary

### *3.5.1 Wind turbine design*

Establishing the influence of weather conditions on wind turbine reliability is at an early stage of development, both in terms of data and analysis methods.

However, the previous work has clearly shown that high wind speeds, wind turbulence, gusts, temperature variance and humidity all affect reliability and can be expected to affect different sub-assemblies in different ways. The following can be concluded:

- High wind speeds, turbulence and gusts deteriorate wind turbine blade, pitch and mechanical drive train reliability.
- Temperature and humidity variance deteriorates electrical rather than mechanical sub-assembly reliability, for example generators and converters.
- Sealed, environmentally controlled nacelles are essential for off-shore wind turbines if electrical equipment, particularly generators and converters, are to have a high reliability, especially following periods of shut-down when there is a risk of condensation.
- There is little information to date on the systematic effects of sea state or icing on fixed or floating off-shore wind turbine reliability.

These issues must form a part of future wind farm design processes, in years to come around the world, as we flex these new renewable technologies to deal with different ambient environments.

### *3.5.2 Wind farm operation*

High wind speeds and turbulence will affect wind farm reliability and consequently availability. Wind turbines further back in an array from the wind front may be less reliable than those near the wind front's leading edge, due to effects of turbulence. But then that depends upon the wind farm layout and the weather pattern a wind farm is exposed to, as the wind veers and backs with respect to the wind farm axis. However, wind farm operational control strategies can be adjusted to accommodate wind direction changes to optimise operational performance and maintenance schedule execution.

Attempts are being made to predict off-shore wind farm operational performance by modelling, particularly by using reduced order models to improve modelling speed, Lin *et al.* (2020), but this is as yet still very early work.

It is unclear what the effects sea state may have upon wind farm performance. For fixed arrays, its influence is primarily upon the access and maintenance regime that a wind farm can adopt before or after foul weather. However, sea-state effects will become very important for the deployment of larger floating wind farm arrays, see Leimeister and Kolios (2018), Musial *et al.* (2018, 2019, 2020). These issues will affect both access and the reliability of structures and moorings and it may have some effect on floating device integrity but our ship-borne design experience will surely minimise such effects. Particular attention is being paid to this issue in the USA and Japan where off-shore wind may be dominated by large floating arrays.

The reliability of floating off-shore wind turbine and wind farm systems should be a future area for research, understanding and implementation. Whilst there may be some lessons from on-shore and off-shore wind, there will be important inputs from the maritime and off-shore oil and gas industries.

The issues of icing will be significant for wind turbines and wind farm operation in Northern areas, and some experience has been built up in the Scandinavian countries, particularly on the Baltic sea, Hochart *et al.* (2007), where the effects of performance changes due to icing, deterioration in blade efficiency and the maintenance access difficulties in icing conditions have been encountered. It can be expected that more experience on this subject will be gained in Russia and China.

# Chapter 4
# Practical off-shore wind farm reliability

## 4.1 Introduction

This chapter describes the reliability of current wind turbines and farms, using research from Tavner *et al.* (2006), Ribrant *et al.* (2007) and Spinato (2008), based upon on-shore and some off-shore data but completes with the most up-to-date data from Cervasco *et al.* (2021).

Figure 1.2 showed the gross failure rate trend for on-shore turbines and at this stage, it is important to define what a failure can be.

Wind turbines are effectively unmanned robotic devices and it is relatively rare that a stoppage can be readily classified as a failure, with the possible exception of a major gearbox, generator or blade event, where the cause of failure is obvious.

More normally the wind turbine is stopped because its controller has detected an operational condition outside the wind turbine's safe envelope. This is usually the result of an unacceptable operational condition, such as overspeed, pitch problems or over-temperature. The control system disconnects the wind turbine from the grid, puts the turbine into the emergency feather condition and the turbine comes to a stop. The fault can either be resolved by the following:

- An automatic restart.
- A manually initiated remote restart.
- A site visit by a wind turbine technician, who may merely initiate a local restart.
- Trigger a repair operation, which then allows the wind turbine to be restarted in a fully repaired condition.

In each case, the stoppage cause may not be known and the figures in Figure 1.2 can really be regarded as stoppage rates rather than failure rates.

The surveys referred to in Tavner *et al.* (2006), Ribrant *et al.* (2007) and Spinato (2008) were concerned with stoppages lasting more than 24 h. Therefore, they constituted serious events, which could not be resolved by an automatic, remote or local restart. They usually involved some form of damage, the exact nature of which could not be identified by the wind turbine original equipment manufacturer (OEM) or operator until a faulty sub-assembly had been replaced or repair work done.

Therefore, to determine a wind turbine's reliability, we must have a working knowledge of the measured stoppage or failure rate, $\lambda$, which allows us to determine an $MTBF = 1/\lambda$. To understand availability, we need to know the stoppage or downtime, which makes up the logistic delay time, *LDT*, and $MTTR = 1/\mu$, from the repair rate, $\mu$, allowing us to determine the availability, $A = MTBF/(MTBF + MTTR + LDT)$, see (1.1)–(1.8).

Knowledge of wind turbine failure rates allows us to compare wind turbine reliability performance and calibrate the contribution made to their unreliability of particular sub-assemblies. In this way, the future wind turbine performance could be improved by maintenance or design.

Interestingly, if a survey shows a low failure rate with high mean time to repair (*MTTR*) or stoppage time, this may result in the same wind turbine availability as much higher failure rates with low *MTTR*. For example, a survey showing 97% availability of wind turbines with a failure rate of 1 failure/turbine/year for $\geq$24-h stoppages will show the same availability as a survey of wind turbines with a failure rate of 24 failure/turbine/year for $\geq$1-h stoppages.

## 4.2 Taxonomies and data from wind turbines and off-shore farms

The basic structure of a modern three-blade, upwind HAWT is exemplified by Figure 4.1.

The main assemblies are shown and it is important to capture failures, their precise locations in the structure and record their effect on a wind turbine's reliability and availability. For operational purposes, the wind turbines are fitted with SCADA and CMS systems that automatically collect data from transducers and alarm circuits distributed around the wind turbine structure and enable the wind turbine to operate automatically within its operating envelope.

The main assemblies are shown but there is a large variety of modern designs.

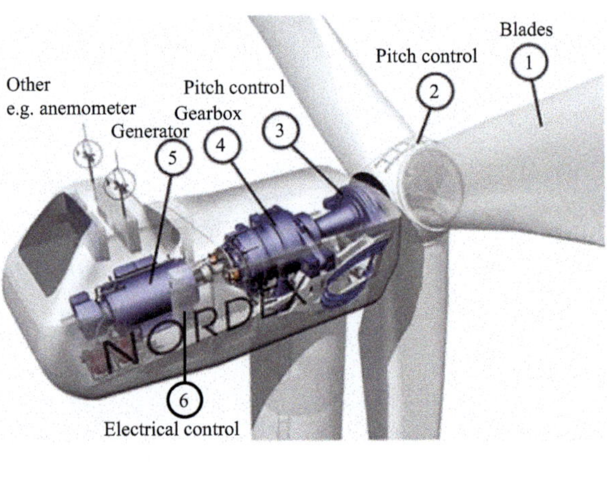


*Figure 4.1 Structure of a modern three-bladed, upwind HAWT. Source: Nordex*

The wind turbine taxonomy is the standardised structure needed so that we can define accurately failure locations and identify where we are to concentrate maintenance and repair activity to maximise availability.

VGB PowerTech, Guideline (2014) applied a power industry standard to wind turbines to derive a taxonomy and naming of parts for the wind industry.

ReliaWind also developed a standardised taxonomy that reflects standards and caters specifically for large wind farms. That taxonomy, see Appendix B, was based upon a five-level system as follows:

- System, which could be the wind farm, including wind turbines, substation and cables.
- Sub-system, which could be an individual wind turbine in that wind farm.
- Assembly, which could be, for example, the gearbox in that wind turbine.
- Sub-assembly, which could, for example, be the high-speed shaft in that gearbox.
- Component, which could be the high-speed bearing on that shaft.

That document also prescribed the way in which reliability data should be collected from wind farms, based on the approach of Thorsen *et al.* (1995). This taxonomy will be used throughout the rest of this book.

The taxonomy of a wind farm is also important in recording the structure and collecting reliability information. The following briefly describes how that taxonomy changes from fixed to floating off-shore wind farms, Figure 4.2.

### *4.2.1 Fixed*

Figure 4.2(a) shows that fixed off-shore wind turbines can be positioned closer together and MV collector cables run accurately from the J-tube of one turbine embedded in the sea bottom and are connected to other turbines and then to the off-shore sub-stations, which, in turn, connect to shore sub-stations by extra high voltage (EHV) export cables.

Therefore, the overall footprint for a given number of fixed off-shore wind turbines will be smaller than for a moored off-shore wind farm.

### *4.2.2 Floating*

Figure 4.2(b) shows that moored off-shore wind turbines must be positioned further apart to accommodate the footprint of their mooring chains, anchors and collector cables.

The MV collector cables must also be flexible as they leave the turbine and are suspended with flotation devices. Eventually, they are connected to the less-flexible MV cables embedded in the sea bottom. Their connection to floating off-shore sub-stations also extends their footprint, see Figure 4.2(c) and the sub-station is connected to shore sub-stations by EHV export cables.

Therefore, the overall footprint for a given number of floating off-shore wind turbines and sub-stations must be much larger than for a fixed off-shore wind farm; however, the power output of each wind turbine is likely to be much larger.

*Figure 4.2 Typical off-shore wind farm taxonomies, Musial et al. (2019): (a) fixed off-shore turbine arrangements, from left to right, monopile, jacket and screwed jacket; (b) floating off-shore turbine mooring arrangements, from left to right, spar buoy, semi-submersible and tension leg platform; (c) floating off-shore turbine and sub-station cable arrangements*

### *4.2.3 Reliability data*

There are very little reliability data in the public domain about off-shore wind farm layouts; however, Richardson (2010) relates on-shore wind turbine reliability to off-shore application.

However, there is a great deal of data available from similar structures in the off-shore oil and gas and electrical power industries.

Readers are recommended to investigate those sources and use that information to create surrogate reliability metadata for the expanding off-shore wind industry.

An example of such data, collected for investigations into the reliability of off-shore wave and tidal devices, is given in this book in Appendix E, taken from Delorm (2013).

## 4.3 Failure location, failure mode, root cause and failure mechanism

The wind turbine taxonomy will allow us to identify accurately in a reliability survey a failure location but from a reliability point of view, we need also to understand the root cause of failure and the failure mechanism which links the two. Figure 4.3 shows the relationship between root cause and failure mode, while Figure 4.4 helps to make this clear showing an example of the linkage between root cause and failure mode of a wind turbine main shaft failure.

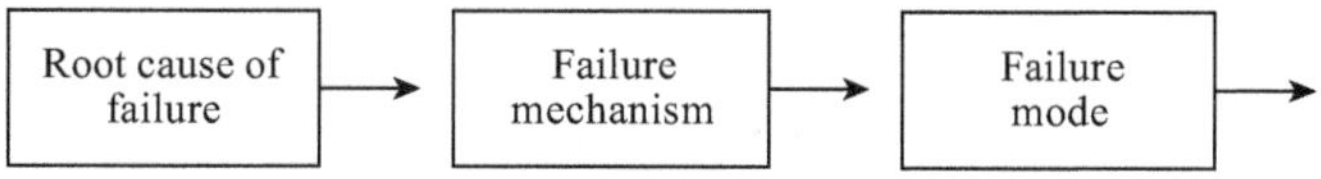


*Figure 4.3 Relationship between root cause and failure mechanism*

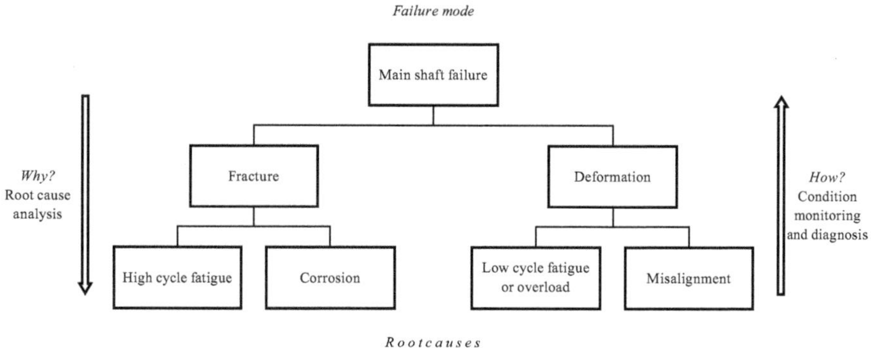


*Figure 4.4 Relationship between failure mode and root cause for a wind turbine main shaft failure*

The importance of this relationship is that we can generally obtain good evidence of failure location, from which we can infer the failure mode, but for operations and maintenance (O&M) purposes, it is much more valuable to identify the root cause that can be tracked by operator or maintainer to predict the progress of incipient failure. This knowledge becomes invaluable to plan maintenance and reduce downtime. Figures 4.3 and 4.4 suggest that monitoring data is a key ingredient to that tracking process.

## 4.4 Reliability field data and collection

Once a wind turbine taxonomy has been defined and the parts of the wind turbine are named in a standardised way, data can be collected on wind turbine reliability. A number of surveys of wind turbine reliability exist in the public domain, including the following:

1. Windstats surveys in Denmark and Germany, termed WSDK and WSD (2010), respectively, containing data on failure rates fixed and variable speed wind turbines with geared or direct drives over 25 years of operation.
2. Various Swedish and Finnish surveys mentioned by Ribrant *et al.* (2007).
3. LWK (2010) survey in Germany containing data on failure rates from fixed and variable speed wind turbines with geared or direct drives over 15 years of operation from 5,800 wind turbine years.
4. WMEP survey (1996) in Germany, WMEP (1996), containing data on failure rates from fixed and variable speed wind turbines with geared or direct drives over 15 years of operation from 15,400 wind turbine years.
5. ReliaWind survey in Europe of 450 wind farm months of data, comprising around 350 on-shore variable speed wind turbines with geared drives operating for varying lengths of time $\leq 4$ years, in the form of 35,000 downtime events, each tagged within the standard taxonomy described previously.

In general, the data from WSD and WSDK (2010) do not segregate failures between different types of wind turbine or into different wind turbine assemblies, whereas data from LWK and WMEP do.

In addition, data from ReliaWind subdivide data from non-specific types of wind turbines into assemblies, sub-assemblies and some components, as prescribed by the taxonomy described in Section 4.2. Therefore, the data sources can be viewed in more detail as one progresses down the list. The only exception is that ReliaWind data do not identify individual wind turbine types, to preserve confidentiality, whereas WMEP and LWK data do.

At the date of writing, very little field data exist in the public domain for off-shore wind farms, although there are a number of reports published from the early publicly funded projects in Europe, see Feng *et al.* (2010).

The wind industry has not yet standardised its methods of reliability data collection whereas the oil and gas industry has done so, OREDA (2015).

However, an early wind industry reliability study from WMEP, Faulstich, Durstewitz *et al.* (2008), developed a prototype data collection system described in the reference.

Each incident occurring on a wind turbine was described by a standardised operators' report form, which is given in Appendix A, the author considers this to be valuable.

ReliaWind developed a standard approach to data collection, based upon work by WMEP, Garrad Hassan and others, described in Appendix B, catering specifically for larger wind farms and making use of both automatic but filtered SCADA data and maintainers' logs, rather than operators' reports. Within a data collection system, it is necessary to define the structure or taxonomy of the plant from which data are to be collected, for the wind farm and the individual wind turbines. The taxonomy will define the detail of the data to be collected, the more detailed that taxonomy the more detailed will be the data collected. Wind turbines are fitted with SCADA systems that collect data from around the wind turbine structure, as described in the previous section. This structure should coincide with the planned taxonomy for collecting reliability data as the SCADA system is already collecting such data automatically, albeit in greater volume per unit time than that collected by operator reports. The wind turbine taxonomy has been described in an earlier section.

Crabtree *et al.* (2015) have described the early experience of off-shore wind farm performance in the United Kingdom compared with on-shore performance, with a detailed analysis of capacity factors and comparison of the balances between capital expenditure (CAPEX) and operational expense (OPEX).

Delorm (2013) is a useful resource when compiling a reliability prediction system for off-shore wave and tidal devices. These data are given in Appendix E and include many valuable results of use in predicting the reliability of off-shore wind farms.

## 4.5 Mathematical concerns about field data

Mathematical concerns have been raised about wind turbine data collection, associated with these methods, and the precise nature of a failure and whether turbine stoppages can be defined as a failure.

Coolen *et al.* (2010) were concerned that wind turbine data had not considered total time on test, as would be necessary for an individual component, such as an electrical transistor or mechanical valve, but rather on censored data measured every month or year for a complete machine.

Whereas Dent *et al.* (2011) were concerned about the validity of partial data from wind farms.

Furthermore, Guo *et al.* (2009) analysed data, also presented in this book, to confirm whether wind turbine reliability could be inferred from incomplete data collected after installation.

It is clear that these detailed mathematical questions are important, as wind farm and wind turbine failure data have been based almost entirely on turbine stoppages. A stoppage cannot signify the failure of a specific turbine component but rather deterioration of the turbine's performance until its control system shuts the turbine down.

The purpose of collecting reliability data for wind farms is not to identify the exact time or cause of a failure but rather to obtain information to allow operators and maintainers to intervene before failure or manage maintenance resources. Precise failure rate is not the maintainers' objective but information with which to control a reliable maintenance regime.

The author reminds readers of his own experience, monitoring large turbo-generators in the UK electrical system in the 1980s, when the resolution of the problems depended on accurate analysis of the available data. In other words, analysis of practical engineering data may contain mathematical or statistical inconsistencies but that is the only data available and analysis is an essential route for the resolution of reliability problems.

But we must also remember Sam Clemens (Mark Twain) who said, 'there are lies, damned lies and statistics', incorrectly attributing this quotation to Benjamin Disraeli. Modern epidemiologists and engineers would advise, however, that we cannot hope to resolve complex failure phenomena without statistical analysis.

## 4.6 Comparative data analysis

The simplest comparison of on-shore wind turbine reliability results was done by the author in Tavner, Xiang *et al.* (2006) based on WMEP and LWK mixed wind turbine data, and an extracted example is shown from LWK data in Figure 4.5.

Comparisons have also been published in Ribrant *et al.* (2007), Wilkinson *et al.* (2011), Faulstch *et al.* (2011) and Crabtree *et al.* (2015).

Figure 4.5 shows how in general failure rates for stoppages >24 h seem to be increasing with increasing wind turbine rating.

The results of Figure 4.6 showed that wind turbine electrical sub-assemblies appear to have the higher failure rates but the highest downtimes are in the drive train due to the blades, gearbox and generator sub-assemblies. From the failure rates, this is clearly not due to their intrinsic design weakness but rather the complexity of changing them in the field, entailing the use of cranes and the need for prior planning.

It is also interesting to note from Figure 4.7 the differences between downtime recorded by the two surveys. LWK represented the total downtime, whereas WMEP tried to record MTTR itself. MTTR is shorter than the downtime, as shown in (1.1), generally confirmed by Figure 4.7.

Figure 4.6 shows the results from the ReliaWind survey that had a much more detailed breakdown of wind turbine sub-assemblies and data collected for stoppages >1 h. The failure rate lessons from ReliaWind, Figure 4.6(a), are rather

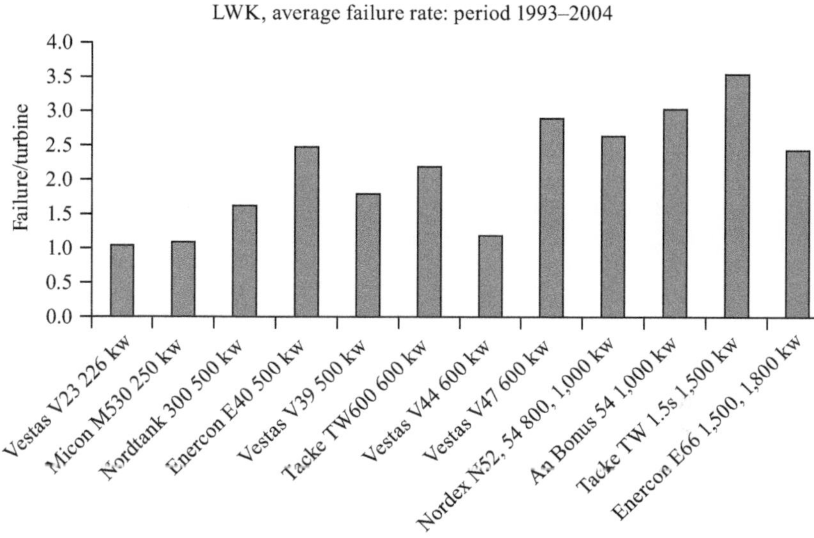


*Figure 4.5 Variation of wind turbine failure rates with rating from the 5,800 turbine year LWK survey between 1993 and 2004. Source: LWK (2010)*

similar to the public domain surveys but the downtime structure, Figure 4.6(b), is different showing a much greater emphasis on the rotor and power modules, because it is believed these newer wind turbines have not experienced any major gearbox, generator or blade failure to date in service.

Figure 4.7 from extensive surveys in three countries shows the least reliable sub-assemblies in modern wind turbines, which sub-assemblies cause unreliably long downtimes indicating a linkage between those failures and their root causes.

The most up-to-date wind turbine reliability data come from Cevasco *et al.* (2021), concentrating on off-shore wind, but take account of a wider range of 24 surveys, both off-shore and on-shore, than previously set out in this chapter. The results of this work will be summarised together with the information here at the end of the chapter.

## 4.7 Current reliability and failure mode knowledge

On the basis of the above-mentioned results, our current knowledge of on-shore wind turbine reliability is as follows:

- Wind turbine failure rates are generally falling with time, so the industry is producing more reliable wind turbines as time progresses. This process is described elsewhere as the learning process, see Mukora (2013), Doctoral Thesis and Junjinger *et al.* (2010).

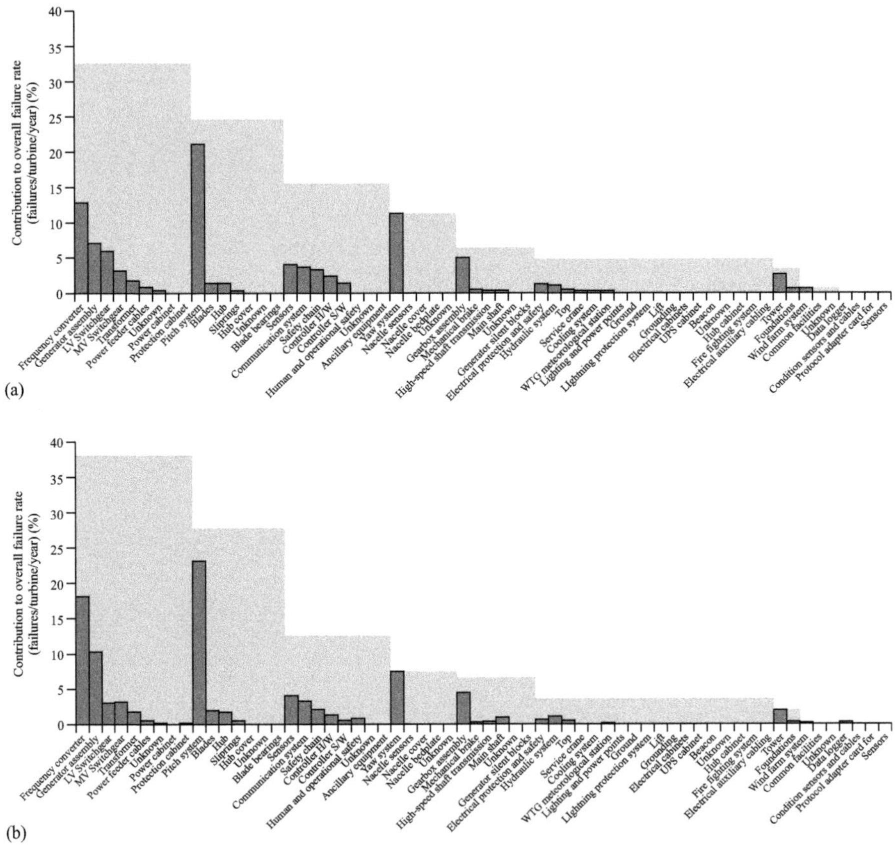


*Figure 4.6 Wind turbine sub-assembly reliability information from the 1,400 turbine years ReliaWind survey, 2004–10, Wilkinson et al. (2011): (a) sub-assembly failure rate distribution; (b) sub-assembly downtime distribution*

- Failure rates of 1–3 failure/turbine/year for stoppages $\geq$24 h are common on-shore, depending on the definition of a failure.
- Off-shore, a failure rate of 0.5 failure/turbine/year is likely to be necessary, where planned maintenance visits need to be kept at or below 1 per year, if possible.
- Failure rates vary with wind turbine configuration but there is, as yet, no clear advantage in any one technology. The impression is given that any technology can achieve a reasonable reliability provided it was well designed, has had sufficient operational experience and competent maintenance.
- Failure rates of wind turbines generally rise with wind turbine size. This can be ascribed to rapid increases in wind turbine design sizes over the last 20 years and their increasing complexity.

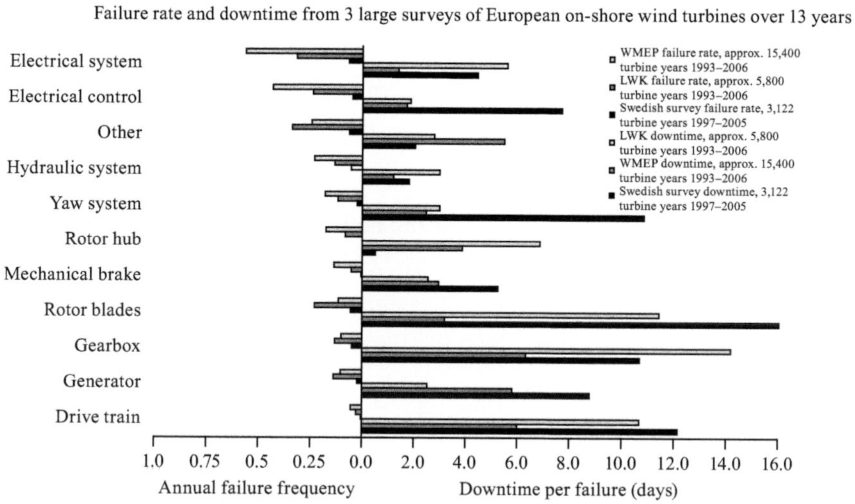


*Figure 4.7 Wind turbine sub-assembly failure rate and downtime per failure for 20,000 turbine-years of data from Swedish, LWK and WMEP surveys. Source: Tavner (2017)*

- Wind turbine sub-assemblies with the highest failure rates have been shown from public domain surveys to be in descending order of significance:
  - rotor, pitch system;
  - converter, i.e. electrical control, electronics and inverter;
  - electrical system;
  - rotor blades;
  - generator;
  - hydraulics;
  - gearbox.
- Sub-assemblies with the highest downtimes have been shown to be in descending order of significance:
  - gearbox;
  - generator;
  - rotor blades;
  - pitch system;
  - converter, i.e. electrical control, electronics, inverter;
  - electrical system;
  - hydraulics.

The relative standing of these lists will vary with wind turbine type and configuration and may be altered in the future, as a result of wind turbine O&M and asset management strategies.

The study by Faulstich, Lyding *et al.* (2011) has shown the following:

- On-shore 75% of all faults cause 5% of the downtime, whereas major failures cause 25% of the faults but 95% of the downtime. Downtime on-shore is dominated by a few large faults, many associated with gearboxes, generators and blades, requiring complex and costly replacement procedures.
- But the 75% of faults causing 5% of the downtime are mostly associated with the electrical plant, the converter, electric pitch systems, control equipment and switchgear, the defects of which are relatively easy to fix in an on-shore environment. It is known that a large proportion of off-shore wind turbine alarms originate in the electrical systems.

The cost of off-shore wind turbine operations will be profoundly affected by these figures. It is likely that the off-shore failure rates will be largely similar to on-shore but that downtimes will be affected by off-shore wind farm location and lower accessibility, increasing the 5% of on-shore downtime arising from 75% of faults.

ReliaWind presented data in Figure 4.6, determining the six least reliable sub-assemblies in a 1,400-turbine-year survey, summarised as follows in descending unreliability:

- pitch mechanism, electric or hydraulic;
- power electronic converter;
- yaw system;
- control system;
- generator;
- gearbox.

ReliaWind also conducted a failure modes and effects analysis of the wind turbine-type covered by the survey and this revealed the most important failure modes identified in those six sub-assemblies, as set out in Table 4.1. Their FMEA methodology was set out in Arabian-Hoseynabadi *et al.* (2010) *Int J Electrical Power & Energy Systems*.

The unreliable sub-assemblies were identified objectively from measured data whereas the failure modes were identified subjectively by ReliaWind partners.

Figure 4.7, from a three-source survey, showed the least reliable sub-assemblies in modern wind turbines at that time, which sub-assemblies cause long downtimes and the linkage between those failures and their root causes. But Figure 4.8 now summarises, from an up-to-date nine-source survey, Cervasco *et al.* (2021), how this understanding has developed over larger wind farms, including some off-shore. Note that these data show failure rates ranging from 4 to 20 failure/turbine/year, higher than the author's desire for off-shore wind of 0.5 failure/turbine/year.

However, Table 4.2 demonstrates the similarities between these three- and nine-source surveys.

*Table 4.1 Unreliable sub-assemblies and major failure modes identified in ReliaWind project FMEA*

| **Sub-system/ assembly** | **Failure mode 1** | **Failure mode 2** | **Failure mode 3** | **Failure mode 4** | **Failure mode 5** |
|---|---|---|---|---|---|
| Pitch system | Electrical (5 out of 13) | Battery failure | Pitch motor failure | Pitch motor converter failure | Pitch bearing failure |
| | Hydraulic (5 out of 5) | Internal leakage of proportional valve | Internal leakage of solenoid valve | Hydraulic cylinder leakage | Position sensor degraded or no signal |
| Frequency converter | Generator- or grid-side inverter failure | Loss of generator speed signal | Crowbar failure | Converter cooling failure | Control board failure |
| Yaw system | Yaw gearbox and pinion lubrication out of specification | Degraded wind direction signal | Degraded guiding element function | Degraded hydraulic cylinder function | Brake operation valve does not operate |
| Control system | Temperature sensor module malfunction | PLC analogue input malfunction | PLC analogue output malfunction | PLC digital input malfunction | PLC in-line controller malfunction |
| Generator assembly | Worn slip ring brushes | Stator winding temperature sensor failure | Encoder failure | Bearing failure | External fan failure |
| Gearbox assembly | Planetary gear failure | High-speed shaft bearing failure | Intermediate shaft bearing failure | Planetary bearing failure | Lubrication system malfunction |

## 4.8 Linkage between failure mode and root cause

The failure information in Figures 4.5–4.8 shows the failure locations whilst Table 4.1 identifies failure modes. But to raise reliability, it is necessary to identify and if possible eliminate root cause. The linkage between those two, described in Figures 4.3 and 4.4, depends upon the sequences shown in Figures 4.8(b) and 4.9.

Because of the distributed nature of wind power and the relatively low rating of individual wind turbines it is rare for the wind turbine OEM or operator to perform failure root cause analyses. Therefore, the knowledge of root cause must be built up in the industry by relying upon the monitoring available from the wind

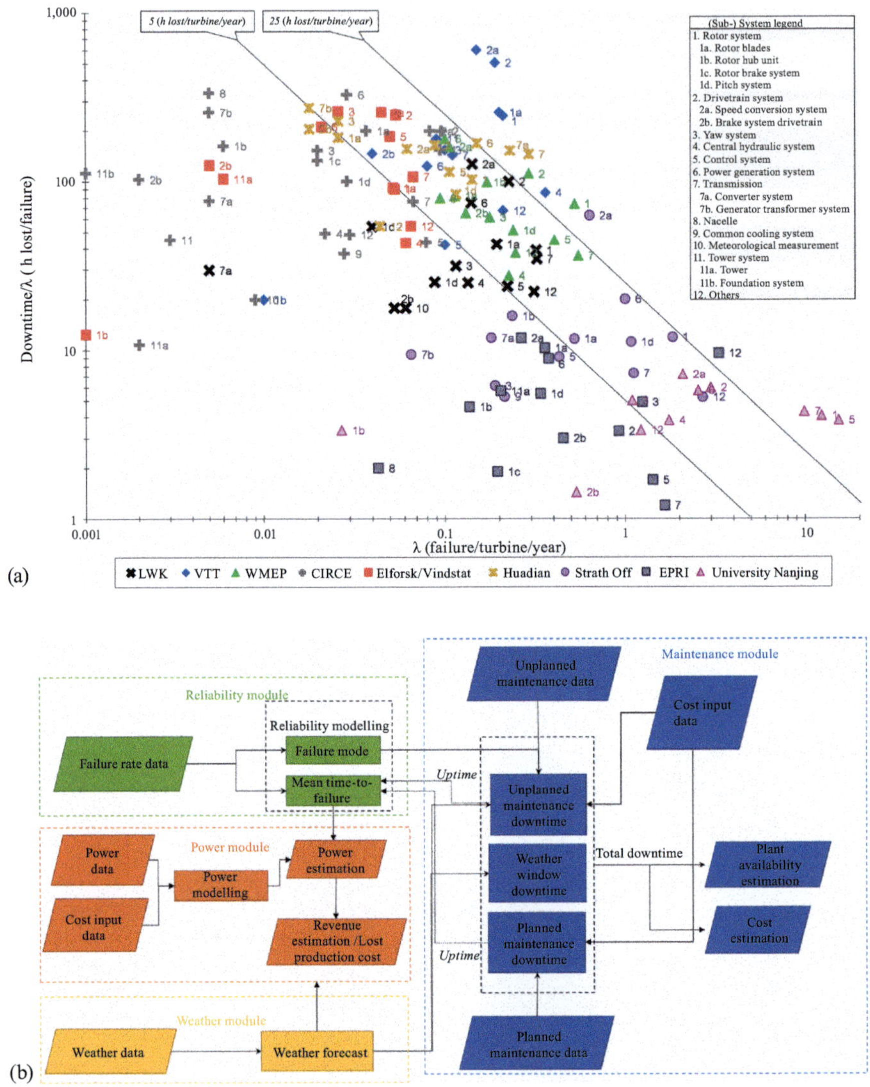


*Figure 4.8 Data from the most up-to-date and complete reliability statistics from 24 international sources, Cervasco et al. (2021): (a) current reliability statistics from nine international sources, covering on-shore and off-shore wind turbines, identifying the most significant contributor sub-assemblies with downtime >25 h lost/turbine/year, rotor (1), drive train (2), hydraulics (4), control (5) and transmission (7); (b) summary of a proposed open operation and maintenance management workflow tool, based on the above-mentioned data, cf. Figure 4.9*

*Table 4.2 Comparison of reliability statistics, wind turbine sub-assemblies with longest downtimes in descending order*

| **Three-source survey, Tavner (2012)** | **Nine-source survey, Cervasco *et al.* (2021)** |
|---|---|
| Gearbox | Drive train |
| Generator | Rotor |
| Rotor blades | Hydraulics |
| Pitch system | Control |
| Hydraulics; converter, i.e. electrical control, electronics and inverters; electrical system | Transmission, i.e. converter and transformer |

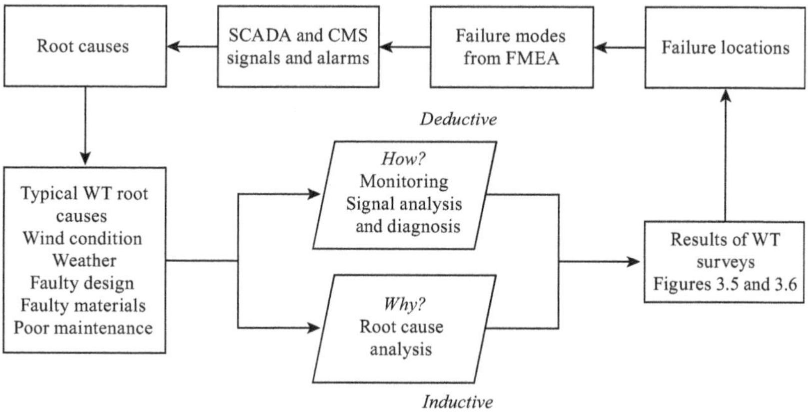


*Figure 4.9 Linkage between wind turbine failure location, failure mode and root cause, cf. Figure 4.8(b)*

far, a topic which will be developed in Chapter 9. It is important to note from Figure 4.9 that the weather plays a significant role in wind power reliability, not only as the resource for energy conversion but also as a root cause for failure. This was developed and discussed in Chapter 3.

## 4.9 Reliability analysis, machinery versus structure versus taxonomy

The reliability of off-shore wind farms does not depend only upon the wind turbine machinery but also upon the reliability of their major structural components, for example:

- off-shore turbine towers, piles and foundations;
- off-shore sub-station and HVDC station structural supports, piles and foundations;
- floating off-shore turbine hull structures;

- undersea cable systems;
- moorings and anchors.

The probability density curve for an off-shore wind turbine machinery component is likely to be a normal distribution with a life expectancy less than the expected life of the whole wind farm.

However, the probability density curve for structural and flotation components, piles, foundations and moorings will be a normal distribution that will be designed to safety cases with a life extending beyond the off-shore wind farm expected life.

An example comparison between the probability of failure of machinery and structural components is shown in Figure 4.10.

The machinery component, for example a generator, has a failure probability curve with an $MTBF_{machine}$ up to perhaps half the expected wind turbine life.

Whereas the structural component, for example a monopole, has an $MTBF_{structure}$, much greater than the expected wind turbine or even wind farm life.

The issue, for example for a wind turbine generator, is to ensure that it can be replaced before $MTBF_{machine}$, whereas for the monopile the issue is to ensure that the area to expected failure, under the failure probability density curve, for wind farm life is very low, that is an analysis based upon extreme risk, described by Sorensen *et al.* (2014).

These issues determine the design of the wind turbine and the probable through-life reliability of a complete wind farm.

Concentration on device design in machinery components with $MTBF_{machine} <$ expected wind turbine life may lead to high failure rates and low availability but

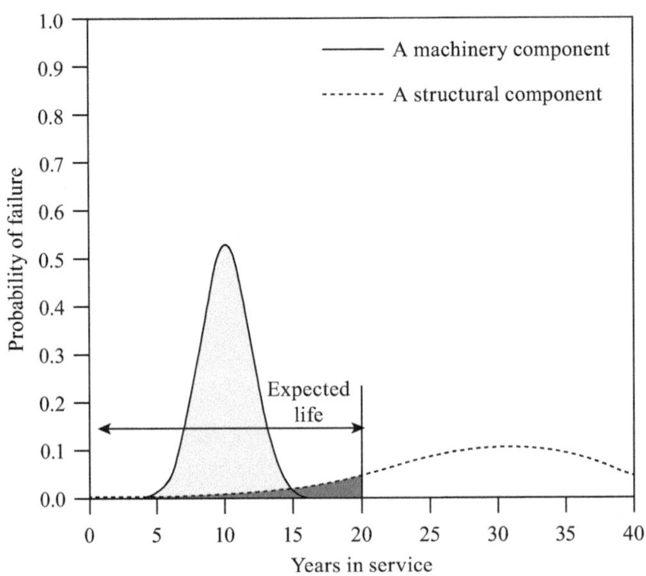


*Figure 4.10 Comparing machinery and structural component continuous failure probability density functions*

low capital cost, whereas concentration on device and structural component design will lead to a low failure rate and high availability but a higher capital cost.

## 4.10 Premature series failures

Premature serial failure (PSF) was a phrase used by industrial sources to rationalise early generator, gearbox and converter failures in earlier European on-shore and off-shore wind farms. To take account of PSFs, a modification was proposed to the bathtub curve, see Figure 4.11, where that distribution was added to a regular wind turbine failure rate bathtub curve.

However, Figure 4.11 could not possibly be correct for the following reasons:

- The bottom of this WT bathtub indicates a failure rate of 0.01 failure/turbine/year, 100 times better than any European wind farm reported, off-shore or on-shore, see, for example, Figure 2.1(a) in 2005 showing 1.0 failure/WT/year, for less arduous on-shore WT conditions. Such a result would even be nine times better than our hypothetical example, Figure 2.3, 0.09 failure/turbine/year.
- Published measurements, Faulstich, Lyding *et al.* (2011), suggested that off-shore wind turbine failure rates could be 4.0–5.0 failure/turbine/year, although this did depend on failure definition and outage length. Of course, such high implicit failure rates could not be economic long-term off-shore, and the author suggests elsewhere that the industry should realistically be aiming for 0.5 failure/turbine/year.
- Nor is Figure 4.11 a valid justification for the existence of early off-shore wind farm failures, as practice in other industries, such as off-shore oil and gas, rail and air travel, could not countenance such early failures, from PSF or any other cause. These failures have to be eliminated by thorough pre-testing, commissioning and early operational maintenance management, rather lacking in the off-shore renewable energy industry.

Perhaps PSF should be re-described as potentially self-inflicted failures.

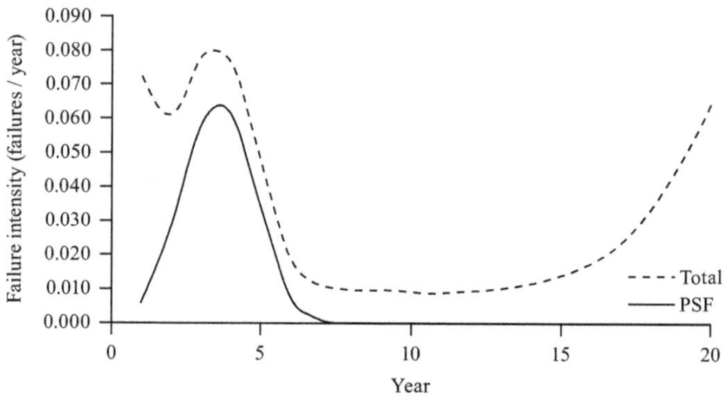


*Figure 4.11 A WT 'bathtub curve' incorporating premature serial failure (PSF) of some sub-assemblies, Stiesdahl et al. (2005)*

## 4.11 Summary

This chapter has demonstrated how off-shore wind farm reliability data can be used to benchmark wind turbine and wind farm performance for organising and planning future O&M, particularly off-shore.

Data need to be collected carefully and the taxonomy of off-shore wind turbines and farms must be defined in a common way, for which standards exist in other industries and are being prepared for the wind industry. It is also clear that the definition of failures, or indeed stoppages, need to be standardised to ensure that data can be compared in a useful engineering and management way.

Some data are available in the public domain and give clear indications of the major reliability problem areas within the wind turbine taxonomy.

Failure rates of 1–3 failure/turbine/year are common on-shore for stoppages of $\geq$24 h.

On-shore 75% of the faults cause 5% of the downtime, whereas 25% of the faults cause 95% of the downtime.

It is likely that off-shore the figure of 5% of downtime due to 75% of faults will rise due to increased off-shore access times and this will be due in large part to relatively minor faults that on-shore were repaired by short and easy to arrange visits to site, whereas access off-shore will extend downtimes.

Failure rates off-shore of 0.5 failure/turbine/year would be desirable but are not yet believed to be anywhere near this level.

The chapter highlights five important reliability issues:

- An early off-shore wind industry approach was to so-called PSF for wind turbine machinery. The author's view, based on his power industry experience, was that PSF arose largely from the use poorly specified, tested, commissioned and maintained generators, gearboxes and converters. This was a situation that no other power industry could entertain, now is the time for the wind industry to eliminate it.
- Concerns about the statistical significance of wind turbine failure rates based on strict mathematical interpretation. This author suggests that while these concerns may be mathematically valid, the wind industry has no option but to analyse their data on the basis in which it is collected. There is a simple nostrum that analysis based upon real field data must be more valuable to improving wind farm reliability than no analysis at all.
- Figure 4.8(b) and particularly Figure 4.9 show how the information in this chapter could be used to plan future off-shore wind farm maintenance.
- The difference between machinery and structural component reliabilities. This is important off-shore, where a substantial proportion of initial CAPEX investment, is in structural rather than machinery components, which strongly affect the OPEX. Structural reliability is better than machinery reliability but both are important for through-life off-shore wind farm viability.

Finally, Cervasco *et al.* (2021), Figure 4.9 and Table 4.2, show the off-shore wind industry the value of reliability data for identifying the most vulnerable wind turbine and farm components, emphasising the importance of reliability data for planning off-shore repair work and indicating clearly the way that it must go with O&M.

# Chapter 5
# Wind turbine configuration and reliability

## 5.1 Modern wind turbine configurations

### *5.1.1 General*

Chapter 1 has shown that modern electric power HAWTs have evolved in rating over the last 80 years but also in a number and variety of configurations as follows:

- upwind or downwind wind turbine rotors;
- two-, three- or even one-blade rotors;
- fixed speed or variable speed rotors;
- stall-regulated or pitch-regulated;
- direct drive or geared drive.

These changes are clearly visible in Appendix F at the end of this book.

More recently, they have standardised towards three-blade, upwind, pitch-regulated HAWT rotors, growing in size exemplified by Figure 5.1, but we are now regularly installing machines of between 8- and 12-MW off-shore.

The variations in wind turbines are now more concentrated on the drive train itself and the electrical arrangements of these configurations, and these features affect the turbine performance and therefore its reliability. So when considering reliability, a clear understanding of the configuration and its strengths and weaknesses is very important. Some parts of the industry have perceived certain configurations to be more reliable than others but as yet no clear measured data seem to point in that direction. In fact, recent experience has emphasised that any configuration can achieve reliability providing that the component sub-assemblies are well manufactured, well installed and well maintained.

The reliability of wind power has been a subject of intense interest to the author since he was first involved in 2002 in the acquisition of the German wind turbine designer and manufacturer, DeWind.

In Figure 5.2, based on the nomenclature used in Hansen *et al.* (2007), the main wind turbine drive train configurations currently in use in the industry can be summarised as follows:

- Type A for fixed- or dual-speed, stall-regulated wind turbines with a geared-drive LV squirrel cage induction generator (SCIG) connected directly to the MV grid through a transformer, with power factor correction and a soft starter to reduce synchronisation inrush current.

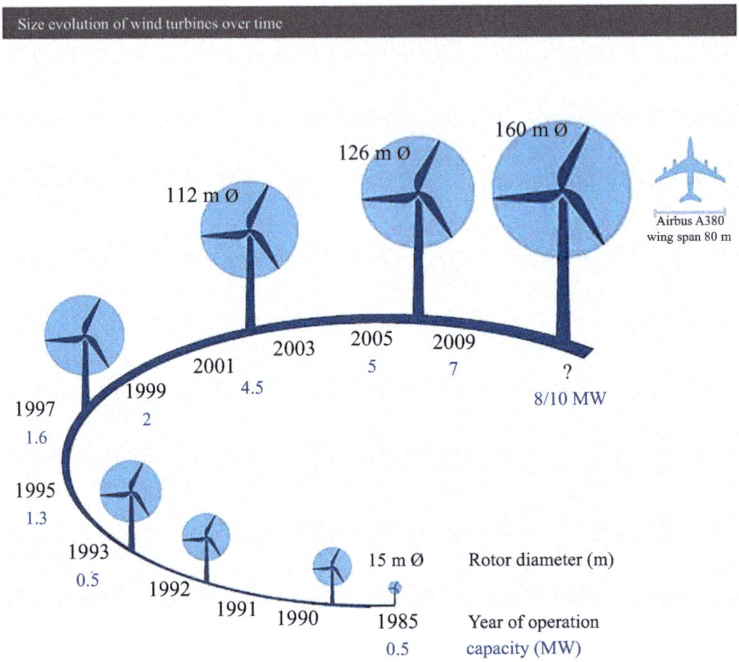


*Figure 5.1 Growth in size of commercial wind designs 1985–2009. Source: EWEA*

- Type B for fixed- or dual-speed, stall-regulated or variable-speed, controlled-stall-regulated wind turbines with a geared-drive LV wound rotor induction generator (WRIG) with variable rotor resistance connected directly to the MV grid through a transformer with power factor correction and a soft starter to reduce synchronisation inrush current.
- Type C for variable-speed, variable-pitch wind turbines with a geared-drive LV WRIG and partially rated, four-quadrant converter connected to the WRIG rotor, the stator of which is connected to the MV grid through a transformer. This is the so-called doubly fed induction generator (DFIG) scheme which is the most widely fitted in the wind industry for wind turbines $\geq$1.5 MW.
- Type D for variable-speed, variable-pitch wind turbines with a direct-drive LV wound rotor synchronous generator with exciter (WRSGE), permanent magnet synchronous or SCIG with a fully rated four-quadrant converter connected to the stator connected to the MV grid through a transformer.

### *5.1.2 Background*

Chapter 4 has shown that the reliability of large modern on-shore wind turbines is improving but the wind industry must have a clear understanding of the factors driving this reliability to face the economic challenges of off-shore installations, where the wind energy harvest is greater but the conditions are more inclement.

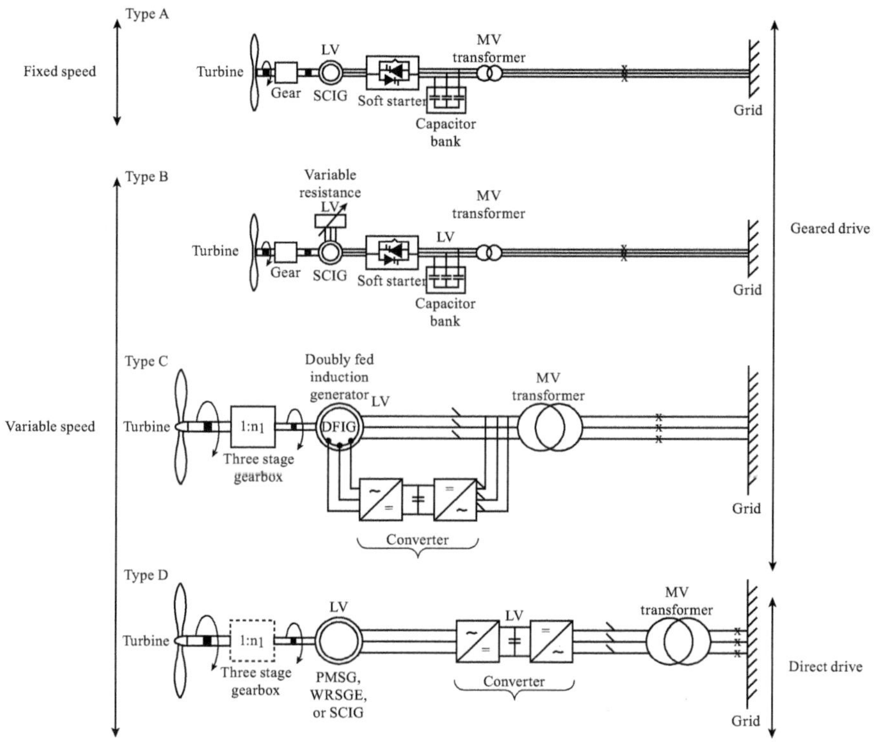


*Figure 5.2 Summary of main electrical configurations for current wind turbine drive trains*

It will be necessary to increase reliability further, because access to those wind turbines may be more limited.

Chapter 4 showed that a failure rate of 1–3 failures/turbine/year is common on-shore and some would argue that real failure rates are much higher than that if all stoppages are considered. Off-shore a failure rate of 0.5 failures/turbine/year is likely to be necessary, where planned maintenance visits need to be kept at or below 1 per year if availability and low cost of operation and maintenance (O&M) and energy are to be achieved.

This section considers the unreliability, or failure intensity function, $\lambda(t)$, of wind turbine sub-assemblies rather than the wider issue of availability and capacity factor (CF), because reliability depends primarily on wind turbine construction and is intrinsically predictable. On the other hand, availability, yearly production and CF depend not only on reliability but also more strongly on wind conditions and the consequences of faults, which in turn depend on turbine location, access logistics and maintenance regime, not primarily to the wind turbine construction. This section carries forward the analysis of Chapter 4 on public domain data, paying particular attention to vital wind turbine sub-assemblies, the gearbox, generator and power electronic converter.

The foundation of these analyses has been the population of wind turbines of known model and design covered by the LWK (2010) survey. They will show striking differences between the reliability characteristics of the selected sub-assemblies over the period. Some of the results can be related to experience with such sub-assemblies outside the wind industry.

Considerable interest has also been shown in the industry about differences in cost and performance achieved by different wind turbine architectures, see for example Polinder *et al.* (2006), but reliability information was lacking. The analysis here sheds light on the effect of wind turbine configuration on reliability and identifies specific reliability behaviours of selected sub-assemblies, where work could be done to improve overall wind turbine.

### *5.1.3 Concepts and configurations*

As the technology of modern wind turbines has matured, the construction has become standardised around the three-bladed, upwind, variable speed concept. But within this concept, there are different architecture and Figure 5.2 (Types C and D) shows two, as follows:

- Geared wind turbines with a gearbox, high-speed asynchronous generator and a partially rated converter (DFIG).
- Direct-drive wind turbines with no gearbox but a specialised direct-drive, low-speed synchronous generator and fully rated converter.

The anticipated benefit of the geared concept is that it uses a more standardised, high-speed generator and a partially rated converter, thereby saving cost as shown in Polinder *et al.* (2006). An anticipated benefit of the direct-drive concept is that by avoiding the use of the gearbox, it should prove to be more reliable but there are other potential benefits, for example lower losses in low wind. There are also a number of control configurations which need consideration and these are listed in Table 5.1. This chapter will investigate the reliability of a number of these turbine concepts where a concept means the sum of the wind turbine architecture and control configuration.

### *5.1.4 Sub-assemblies*

To understand wind turbine drive train configuration reliability, we need to break down the wind turbine into more detail than Figure 2.1 using the nomenclature in Section 4.2:

- System, the whole wind turbine.
- Sub-systems of the wind turbine, such as the drive train, consisting of rotor hub, shaft, bearing, gearbox, couplings and generator.
- Assemblies, such as the gearbox.
- Sub-assemblies, such as the high-speed shaft of the gearbox.
- Components, such as the high-speed coupling of the gearbox.

This chapter focuses on sub-assemblies recorded in the surveys WSDK, WSD, LWK (2010) and the sub-assembly breakdown is shown in Figure 4.7. The terminology used by these surveys was not consistent and it has been necessary to aggregate sub-assemblies as shown in Table 5.2.

*Table 5.1 Wind turbine control concepts considered in this chapter*

| **Speed control** | **Pitch control** | **Power control** | **Wind turbine models considered in this section** |
|---|---|---|---|
| Fixed- or dual-speed | None | Passive stall regulation geared-drive train with SCIG generator | NEG Micon, M530, Tacke TW600 |
| Fixed speed | Yes, pitch to stall | Active stall regulation geared-drive train with SCIG generator | Vestas V27, Nordex N52/54 |
| Limited variable speed | Yes | Geared-drive train with WRIG control using variable rotor resistance | Vestas V39 |
| Variable speed | Yes | Geared-drive train with DFIG control using partially rated converter | Tacke TW1500, Bonus 1 MW, 54 |
| Variable speed | Yes | Direct-drive train with synchronous generator control using fully rated converter | Enercon E40, E66 |

*Table 5.2 Naming of wind turbine sub-assemblies, used in this chapter and elsewhere*

| **This chapter** | **WSD** | **WSDK** | **LWK** |
|---|---|---|---|
| Rotor | Rotor | Blades, hub | Blades |
| Air brake | Air brake | Air brake | Rotor brake |
| Mechanical brake | Mechanical brake | Mechanical brake | Brake |
| Main shaft | Main shaft, bearings | Axle, bearing, coupling | Shaft, bearings |
| Gearbox | Gearbox | Gearbox | Gearbox |
| Generator | Generator | Generator | Generator |
| Yaw system | Yaw system | Yaw system | Yaw system |
| Converter | Electrical control | Electrical control | Electronics, inverter |
| Hydraulics | Hydraulics | Hydraulics | Hydraulics |
| Electrical system | Electrical system | Grid | Electrics |
| Pitch control | Pitch adjustment | Mechanical control | Pitch mechanism |
| Other | Anemometry, sensors, other | Other | Anemometry, sensors, other |

### *5.1.5 Populations and operating experience*

WSD, WSDK and LWK (2010) data were collected by operators on hand-written or computer-written report sheets, rather than generated automatically, and the data have some limitations, as follows:

1. They gather the failures in a given period for each turbine and sub-assembly within the population without giving details of failure modes.

2. The periods of data collection differ for each population as follows: WSDK monthly, WSD quarterly, LWK annually.
3. These periods have affected the results presented.
4. There are other differences between the populations as follows:
   (i) WSDK is a large mixed population decreasing in wind turbine numbers (2,345–851 over the period), with turbines of average age of 14 years, mostly of stall-regulated configuration. Their technology is consolidated as confirmed by their failure intensities, approaching a constant average failure rate. The failures of individual turbine models cannot be distinguished in these data.
   (ii) WSD is a larger mixed population growing in number (1,295–4,285 over the period) and includes larger turbines, with an average age of 3 years, including a variety of turbine models with different control configurations, but their failure intensities are also approaching a constant value, although at a faster rate than WSDK. The failures of individual turbine models again cannot be distinguished in these data.
   (iii) LWK is a smaller, segregated, more static population in number (158–643 over period) and includes larger turbines of average age up to 15 years, with fixed and variable speed configurations, both with geared and a significant number with the direct-drive concept. The failures of individual turbine models can be distinguished in these data.

### *5.1.6 Industrial reliability data for sub-assemblies*

Some wind turbine sub-assemblies, such as the rotor and pitch control, are specialised for the wind power application. But some, such as the gearbox, generator and converter, can be found in similar form, albeit in different sizes and designs, in other power conversion machinery. The usefulness to the industry of reliability figures presented in this chapter is enhanced by comparing them to values from other industries, as tabulated in Table 5.3.

## 5.2 Reliability analysis assuming constant failure rate

The previous work by the author, Tavner, Xiang *et al.* (2006), concentrated on the average wind turbine failure rate, assuming that the systems were at the bottom of the bathtub, Figure 5.4. This shows for WSD, WSDK, Figure 1.2, the overall trend

*Table 5.3 Reliability of generators, gearboxes and converters from industrial experience*

| Sub-assembly | Failure rate (failures/ sub-assembly/year) | MTBF (h) | Source |
|---|---|---|---|
| Generator | 0.0315–0.0707 | 123,900–278,000 | Tavner (2008), IEEE Std 493 (1997) Gold Book |
| Gearbox | 0.1550 | 56,500 | Knowles, private communication |
| Converter | 0.0450–0.2000 | 43,800–195,000 | Spinato *et al.* (2009) |

in wind turbine failure intensities against calendar time since the days of the early expansion of wind power in California in the early 1980s.

The results of the LWK survey have been added to Figure 1.2 with the measured failure rates from other mature power generation sources, largely extracted from IEEE Std 493 (1997) Gold Book, showing that wind turbine reliability is becoming better than some other generation sources, notably diesel generator sets. However, this graph needs to be treated with caution for the following reasons:

- The wind turbine data are taken from mixed and changing wind turbine populations. Because the ratings of newly introduced wind turbines are increasing and their failure rates are generally rising, the averaging implicit in the HPP process tends to underestimate the failure rates of these newer, larger, more complex wind turbines, at least during the early failures period.
- The other, mature power generation source, failure data came from historic surveys of limited size, which do not represent the reliability improvement to be studied in this chapter but which is also inherent in those sources.

The relative unreliability of wind turbine sub-assemblies can also be extracted from the WSD, WSDK and LWK (2010) data as shown in Figure 4.7, where the assumed constant failure rates of 12 major turbine sub-assemblies have been compared. The LWK population has a higher consistency in terms of technology throughout the period, as it is an installed fleet that has remained relatively unchanged. However, the LWK population is much smaller than the WSD and WSDK populations. Figure 4.7 reveals interesting information:

- Overall failure rates in Danish turbines are lower than German turbines, as seen in Figure 4.7. This was attributed to the greater age, smaller size and simpler technology of the Danish turbines resulting in a higher overall reliability.
- The failure rates of sub-assemblies in the two German populations in Figure 4.7, WSD and LWK, are remarkably similar and have more in common with one another than with the WSDK data. This consistency supports the validity of the two German surveys despite their different sizes.
- The results of Figure 4.7 show that the sub-assemblies with the highest failure rates are, in descending order of significance,
  - electrical system;
  - rotor (i.e. blades and hub);
  - converter (i.e. electrical control, electronics, inverter);
  - generator;
  - hydraulics;
  - gearbox.

Similar results have been reported from Sweden, Ribrant *et al.* (2007) and from a different survey in Germany, WMEP (1996), Faulstich, Durstewitz *et al.* (2008).

Table 5.3 showed industrial failure rates for common sub-assemblies found in the wind industry. These can be compared to measured LWK wind turbine sub-assembly failure rates in Figure 5.3. However, this data is also recorded in Figure 4.7 alongside these sub-assembly downtimes or MTTR, which may be more valuable to the wind turbine operator. Here the effects of electrical system,

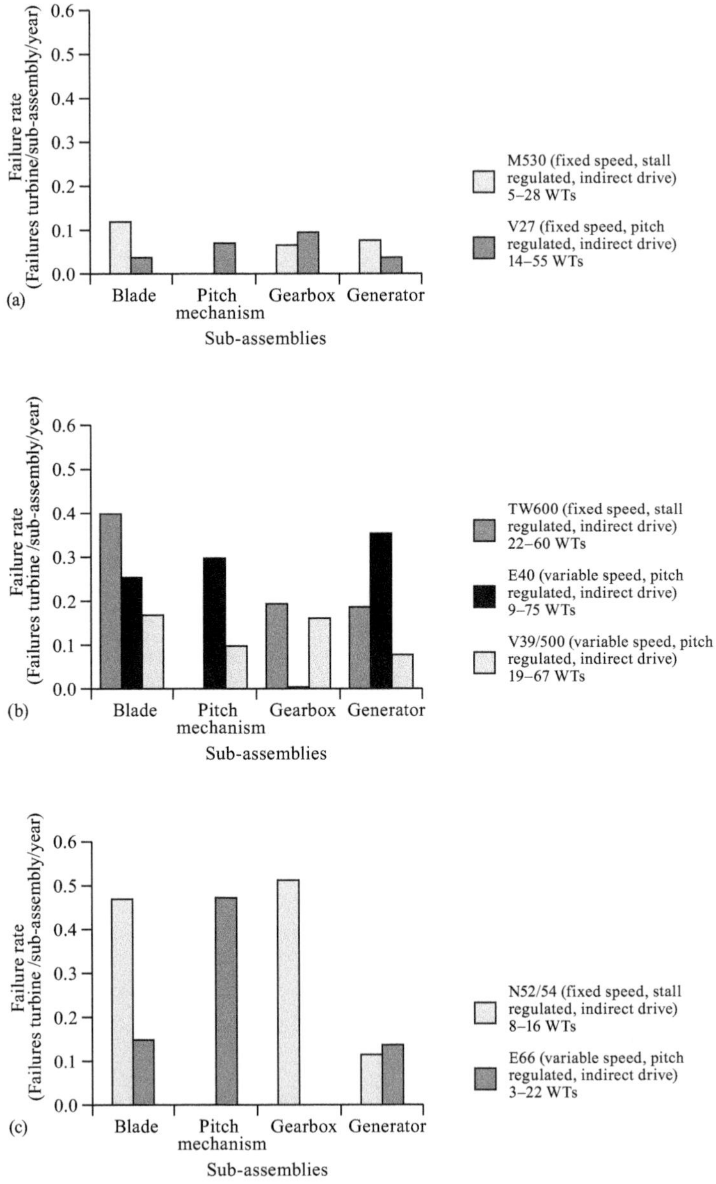


*Figure 5.3 LWK failure intensity distributions focusing on blades, pitch, gearbox and generator: (a) 300-kW fixed speed, geared turbines with pitch- or stall-regulated control; (b) 600-kW fixed speed, geared, stall-regulated, or limited variable speed pitch-regulated turbines or variable speed direct-drive pitch regulated turbines; (c) 1-MW variable speed-geared, pitch-regulated, turbines or variable speed-direct drive, pitch-regulated turbines stall-regulated turbines on left, variable speed, pitch-regulated turbines on right, Spinato (2008)*

generator and gearbox failures are more apparent, in particular the dominance of the gearbox MTTR. It is suggested that this is the main reason for the industry's focus on gearbox failures. Again similar results have also been obtained in Sweden, Ribrant *et al.* (2007).

## 5.3 Analysis of turbine concepts

### *5.3.1 Comparison of concepts*

We now consider the failure rates of individual sub-assemblies most at risk. The LWK data allow turbine models to be grouped according to size and concept. Figure 4.5 summarised the failure rates over 11 years for 12 wind turbine models in the LWK population. This shows the general trend of failure rate rising with turbine rating, reaffirming a conclusion of Tavner *et al.* (2006). The next analysis repeats the approach of Figure 4.6, comparing sub-assembly failure rates for selected LWK turbine models, concentrating on drive train sub-assemblies. This is shown in Figure 5.3, which is segregated by turbine concept and control configuration, see the third column of Table 5.1.

The figure shows the relationship between failure rates of blades, pitch mechanism, gearbox and generator as turbine concepts and control configurations change.

With fixed speed, stall-regulated turbines, a significant number of failures are concentrated in the blades and gearbox. With the introduction of variable speed, pitch-regulated machines, the pitch mechanism now appears as a failure mode, as anticipated.

However, the introduction of the pitch mechanism reduces the blade and generator failure rates, see Figure 5.3(a) for smaller wind turbines. This is confirmed for larger wind turbines in Figure 5.3(b) where blade, generator and gearbox failure rates reduce, with the exception of the E40, direct-drive wind turbine, where the generator failure rate was high. This will be discussed in Section 5.3.2. The reduction in blade failures is even more noticeable with the larger E66 direct-drive wind turbine in Figure 5.3(c).

In other words the technological advance of variable speed and pitch control not only confers energy extraction and noise reduction improvements but also, despite introducing another failure mode, can improve wind turbine reliability with time.

### *5.3.2 Reliability of sub-assemblies*

#### 5.3.2.1 General

The failure data collected exhibit a variation with time and can be represented by a non-homogeneous Poisson process (NHPP), see Section 2.5.2.2. This section will now use reliability growth analysis, based on Figure 4.3, the PLP representation, a specific case of the NHPP, to analyse reliability time trends from the LWK population of wind turbines concentrating on three sub-assemblies identified earlier:

- generator;
- gearbox;
- converter, i.e. electrical control, electronics, inverters.

These sub-assemblies have been chosen because they are crucial to wind turbine operation and are central to the debate about turbine concept, in particular whether to employ direct-drive or geared wind turbines.

The method of presentation is to plot the intensity function obtained from the LWK data against total time on test of the sub-assembly, see Section 2.5.3. Plotted failure intensity points have been aggregated to comply with requirements for valid numbers of failures in an interval and the Crow AMSAA model in MIL-HDBK-189 (1981), as described in Section 2.4. On each graph, the failure rate of that sub-assembly in other industries taken from Table 5.3 is also shown, together with a time cursor to demonstrate the span in years of the data, as described in Section 2.5.3.

For these sub-assemblies, the PLP interpolation of data presented has been tested against two statistical criteria:

- goodness-of-fit;
- null hypothesis of no reliability growth.

Only results complying with those criteria have been presented. Sub-assemblies from specific wind turbine models are selected here but the conclusions drawn later may be generalised to other wind turbines in the LWK population. These results have all been summarised in Spinato *et al.* (2009).

#### 5.3.2.2 Gearboxes

Figure 5.4 shows the results for the reliability of several LWK gearboxes, each of which shows a remarkably similar form with rising failure intensities, which is a PLP with $\beta$ from 1.2 to 1.8, see Figure 2.2. That is the deterioration or wear-out phase of the bathtub curve, Figure 2.4, suggesting steady mechanical wear, one would expect. So wind turbine gearboxes are a mature technology, and machines are operating in the deterioration phase of the bathtub curve. Therefore, substantial improvements in designed reliability for these gearboxes are unlikely.

The reliability data for industrial gearboxes, given in Table 5.2, are an average from a number of sources and have been superimposed on the graphs. It shows that, from this limited extract from LWK data, reliabilities being obtained by these wind industry gearboxes are comparable with those obtained by other industries, apart from the Nordex 52/54 wind turbine data.

#### 5.3.2.3 Generators

Figure 5.5 shows the reliability of a number of LWK generators showing that failure intensities are generally falling, that is a PLP with $\beta < 1$, reflecting that reliability is improving. The early failures of the bathtub curve, Figure 2.4, can clearly be seen in these figures. Industrial generator reliability data, given in Table 5.3, are superimposed on the graphs, and Figure 5.4 shows that both direct- and geared-drive wind turbine generator reliabilities are not as good as these at the start of life. However, the demonstrated reliabilities, as defined in Figure 2.3, achieved by all except the E40 generator show a good result when compared with

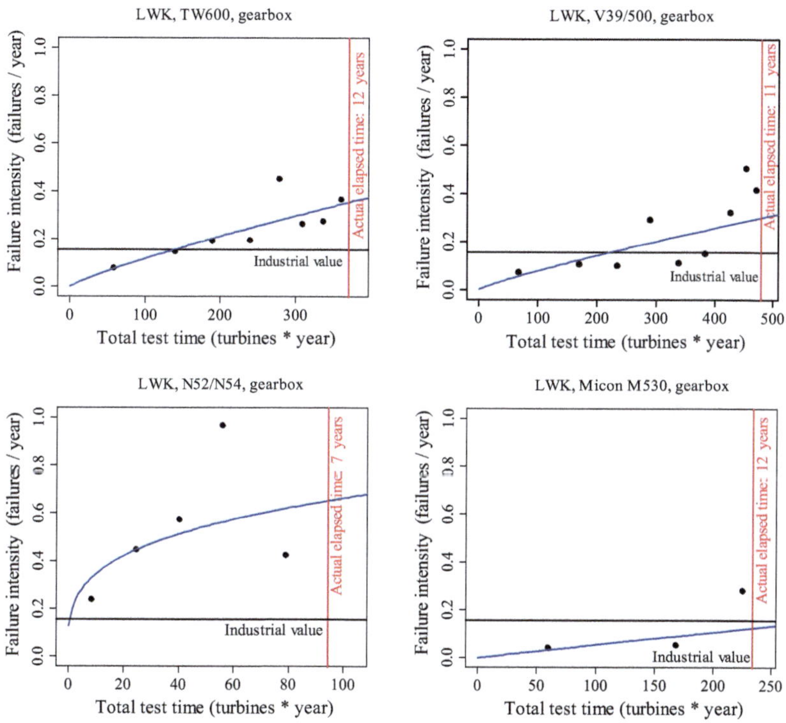


*Figure 5.4 Variation in failure intensities of gearbox sub-assembly, in LWK population, using PLP model, Spinato (2008)*

the industrial failure rate. The failure intensities for both direct-drive generators are higher than the failure intensities of their geared-drive competitors. However, it is clear that the E66 generator is a considerable improvement on the E40 generator.

More recent information has come to light from a wind turbine repair company, Alewine *et al.* (2010), about wind turbine generator failure rates compared to electrical machines in other industries. This confirmed that wind turbine generators are not as reliable as similar-sized electrical machines in other industries, as shown in Figure 5.4, but throws more light on the location of wind turbine generator failures, see Table 5.4, summarised in Figure 5.6. These confirm that the location of wind turbine generator failures are not dissimilar from other electrical machines but are dominated by bearing, slip-ring and brush-gear faults. This is not unexpected as the majority of large wind turbine generators were at the time of this work DFIG.

Important questions are raised by these results as follows:

- Why is there such a large disparity between the reliabilities of direct- and geared-drive generators at the start of operational life?
- Why do the failure intensities of three generators improve with time?

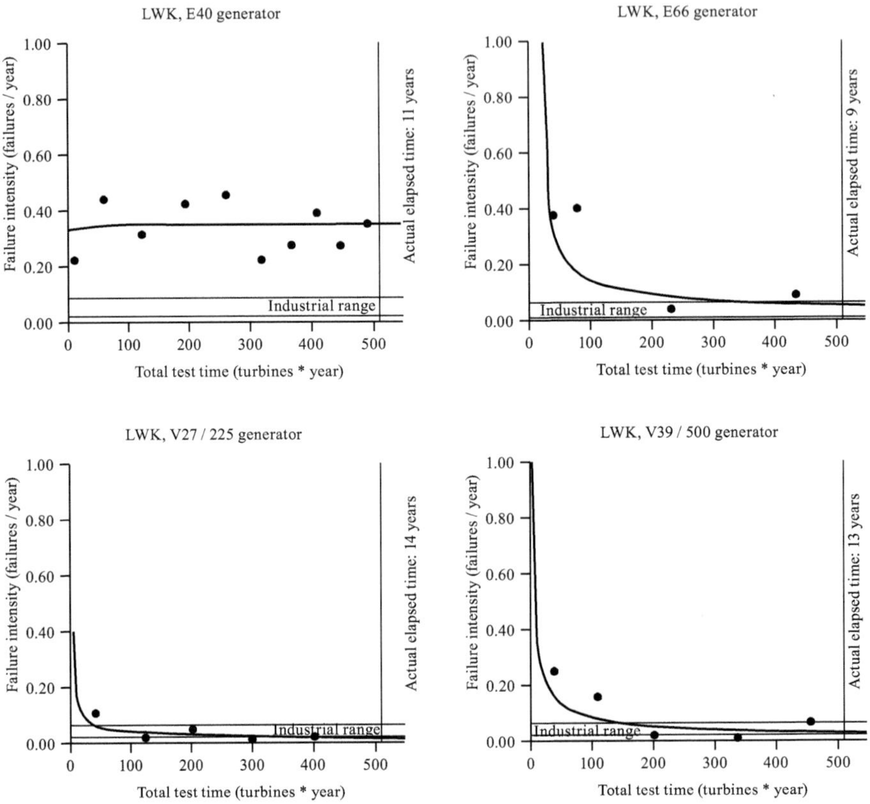


*Figure 5.5 Variation in failure intensities of generator sub-assembly, in LWK population, using PLP model, Spinato (2008): upper graphs show low-speed direct drive, whereas lower graphs show high-speed geared-drive generators*

- Why cannot the wind industry achieve, at the start of operational life, the respectable demonstrated reliabilities ultimately achieved?

These questions suggest, from this limited extract from LWK data, that generators deserve reliability attention from OEMs and operators if we are to achieve higher wind turbine reliability and will be discussed in Chapter 6. Whittle (2013), Doctoral Thesis provides a useful review.

#### 5.3.2.4 Converters

The converter is a complex sub-assembly with a large number of components. There is difficulty in recording failures for converter sub-assemblies as operators may be unable to assign a turbine failure unequivocally to the converter because of sub-assembly complexity. This is in contrast to the generator or gearbox where this is usually straightforward. This means that we must be cautious in considering recorded converter failures.

*Table 5.4 Distribution of electrical machine failed sub-assemblies from different industries, from literature*

| **Surveys** | | **IEEE large motor survey, O'Donnell (1985)** | **Motors in utility applications, Albrecht *et al.* (1986)** | **Motor survey off-shore and petro-chemical, Thorsen *et al.* (1995)** | **Wind turbine generator survey, Alewine *et al.* (2010)** | | |
|---|---|---|---|---|---|---|---|
| **Industry** | | **General** | **Utility applications** | **Off-shore petro-chemical** | **Wind generation** | | |
| Types of machines, rating and voltage | | Motors >150 kW generally MV and MV squirrel cage IMs | Motors >75 kW generally MV and MV squirrel cage IMs | Motors >11 kW generally MV and MV squirrel cage IMs | Wind generators <1 MW, LV, 95%+ wound rotor machines but with electronically controlled rotor voltage rather than collector rings with out-board electronics | Wind generators 1–2 MW, LV, mostly DFIGs | Wind generators >2 MW, LV, mostly DFIGs |
| No. of failed machines in survey | | 360 | 1,474 | 1,637 | 196 | 507 | 297 |
| Sub-assemblies | Bearings | 41% | 41% | 42% | 21% | 70% | 58% |
| | Cooling system | – | – | – | – | 2% | – |
| | Stator wedges | – | – | – | – | – | 14% |
| | Stator related | 37% | 36% | 13% | 24% | 3% | 15% |
| | Rotor related | 10% | 9% | 8% | 50% | 4% | 4% |
| | Collector or slip–rings | – | – | – | 1% | 16% | 4% |
| | Rotor leads | – | – | – | – | 1% | 4% |
| | Other | 12% | 14% | 37% | 4% | 4% | 1% |
| Total | | 100% | 100% | 100% | 100% | 100% | 100% |

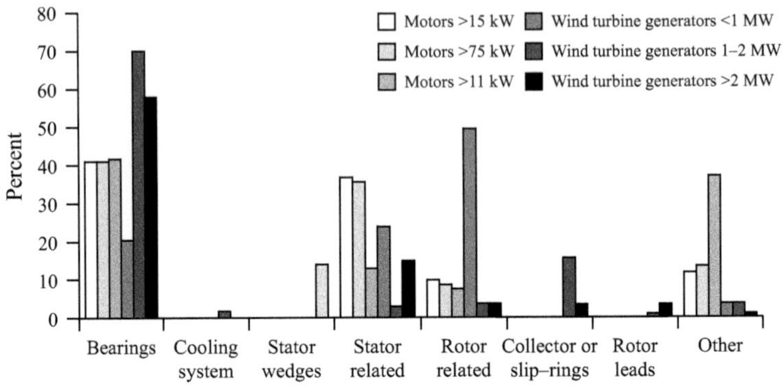


*Figure 5.6 Location of failures in wind turbine generators and other electrical machines, Alewine et al. (2010)*

To overcome this, the failures from inverter and electronics in the LWK survey, see Table 5.4, have been aggregated and the data plotted for specific turbines with the generic sub-assembly name, converter.

Figure 5.7 gives reliability results for three LWK converters. Again, these exhibit the early part of the bath-tub curve, Figure 2.4, but specifically the first curve shows elements of the full bathtub curve with early failures, intrinsic failures and wear-out. For two cases, the Enercon E40 and TW 1,500, the results are similar to the generators, in that failure intensities are falling, that is a PLP with $\beta < 1$, reflecting reliability improvement. However in the case of the Enercon E66 and E40 converters, the failure intensities improve with time but are nearly flat with $\beta = 1$. Industrial converter failure rate data in Table 5.2 range between 0.045 and 0.2 failure/sub-assembly/year. The lower limit arises from a specific analysis of relatively small converters, Spinato (2008), but such a low value of failure rate cannot be applicable to the larger converters in wind turbines; therefore, an upper limit of 0.2 failure/sub-assembly/year is proposed.

More recent work has tracked the distribution of wind turbine failure rates due to the converter, as shown in Table 5.5, where failures due to the converter are compared between different surveys. These show failure rates for converters ranging from 0.22 to 2.63 failures/unit/year, which should be compared to those shown in Figure 5.7. It is important to point out here that the figures in Table 5.5 represent wind turbine stoppages, ascribed by the operator as due to converter faults. These arise from the many alarm signals and trips which the converter produces. That does not locate the faults in the converter, which have been estimated in the lower rows of Table 5.5, based upon knowledge of converter sub-assembly reliabilities. It can be seen that the inverter bridge and DC link failures dominate converter failure rates and downtimes.

Despite its limitations, the data in Table 5.5 give a clear and consistent picture across a variety of surveys of the converter failure rate issue.

Important questions arise from these results as follows:

- Why do the failure intensities of converters improve with time?

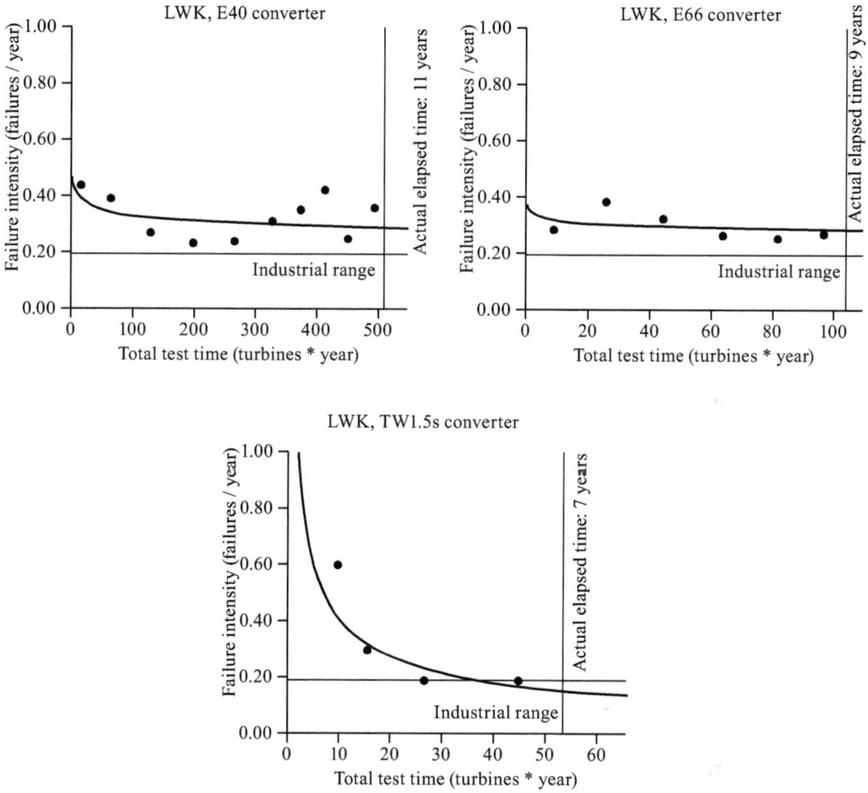


*Figure 5.7 Variation in failure intensities of converter sub-assembly, in LWK population, using PLP model, Spinato (2008)*

- Why are the failure intensities considerably higher than values given for converters in normal industrial use?
- Why is not more attention being placed on reducing the high number of converter failures, perhaps by improving the alarm management and minimising the number of nuisance trips?

These results suggest, from this limited extract from LWK data, that converters deserve reliability attention from OEMs and operators if we are to achieve higher wind turbine reliability and this is investigated in the next section.

#### 5.3.2.5 Measured and predicted reliabilities for generators and converters

Previous sub-sections examined the main wind turbine sub-assemblies and gave clear indications of reliability strengths and weaknesses.

In this sub-section, we compare that wind turbine data for two particularly key sub-assemblies, the generator and converter, with reliability data from other industries.

*Table 5.5 Distribution of WT converter failed sub-assemblies from various surveys*

| **Source** | | **WMEP_D data, Faulstich, Durstewitz *et al.* (2008)** | | **LWK_D, LWK (2010)** | | **Wilkinson *et al.* (2011)** | |
|---|---|---|---|---|---|---|---|
| Turbine-years in the survey | | 209 | 1,028 | 5,719 | 679 | 366 | 1 |
| Additional information | | Large wind turbines | Total wind turbines | Total wind turbines | Specific data from wind turbines, with partially rated or fully rated converter (E40, E66, Tacke 1.5s) | Specific data from wind turbines, about 2 MW with DFIG and partially rated converter | |
| | | 1998–2000 | 1989–2006 | 1993–2006 | 1993–2006 | 2007–11 | |
| | | Measured, failure rate (failures/unit/year) | | | | | FMEA, failure rate (failures/unit/year) |
| Whole wind turbine | | 5.23 | 3.60 | 1.92 | 2.60 | Not disclosed for confidentiality reasons | 23.37 |
| Converter total | | 1.00 | 0.45 | 0.22 | 0.32 | | 2.63 |
| Converter as % of wind turbine | | 19.1% | 12.4% | 11.6% | 12.2% | 11.6% | 11.3% |
| Estimated location of the faults | Converter control unit | 0.070 | 0.031 | 0.016 | 0.022 | – | 0.184 |
| | Series contactor | 0.090 | 0.040 | 0.020 | 0.028 | – | 0.237 |
| | Grid-side filter | 0.030 | 0.013 | 0.007 | 0.009 | – | 0.079 |
| | Grid-side inverter | 0.189 | 0.085 | 0.042 | 0.060 | – | 0.500 |
| | Pre-charge circuit | 0.060 | 0.027 | 0.013 | 0.019 | – | 0.158 |
| | DC link capacitor | 0.110 | 0.049 | 0.024 | 0.035 | – | 0.289 |
| | Chopper circuit | 0.060 | 0.027 | 0.013 | 0.019 | – | 0.158 |
| | Generator-side inverter | 0.189 | 0.085 | 0.042 | 0.060 | – | 0.500 |
| | Crow-bar circuit | 0.060 | 0.027 | 0.013 | 0.019 | – | 0.158 |
| | Generator-side filter | 0.030 | 0.013 | 0.007 | 0.009 | – | 0.079 |
| | Bypass contactor | 0.090 | 0.040 | 0.020 | 0.028 | – | 0.237 |
| | Auxiliaries | 0.025 | 0.011 | 0.006 | 0.008 | – | 0.066 |

Table 5.6 is the most comprehensive collection of past, measured generator and converter data available from IEEE and Wind Industry Surveys, demonstrating what results could currently be expected or improved upon.

The point of Table 5.6, from various references, including Carroll *et al.* (2014), is to show the off-shore wind industry that other industries are able to achieve better generator and converter reliabilities than have currently been available to the wind industry.

Better wind turbine converter performance must be a major target to improve off-shore wind farm reliability. Table 5.7 shows predicted failure rates for LV converters at the present time, based on the measurements in Table 5.5 of 8,000 WT-years from 1997 to 2010, the results of Spinato *et al.* (2009) and ReliaWind.

Finally, Table 5.8 shows results from a recent and substantial industrial survey of in-service reliability measurements of nearly 90,000 converter/inverter bridge modules in WT hosts from 2015 to 2018.

This is for comparison with the earlier results in Table 5.7, 1997–2010.

The purpose of the comparison is to show up-to-date industrial power electronic reliability measurements alongside earlier practical field reliability measurements.

Comparing Tables 5.7 and 5.8 shows that recent inverter bridge failure rates are better than those predicted from past wind turbine converters.

This is because the industry has passed through various bathtub curve starts, for each converter solution, and is reaching a through-life, bottom-of-bathtub performance for many modules.

This cannot easily be explained, because power bridge design has not fundamentally changed; however, what has changed is the maturity of wind industry power-bridge manufacturing, testing, storage and operational processes, but Table 5.8 confirms that higher reliability can be achieved.

There is more information on this subject in Fischer *et al.* (2019), where a consortium of three research institutes and 16 companies, including turbine and component manufacturers, operators and maintainers collectively identify main causes of wind turbine converter failures.

Wyllie (2014), Doctoral Thesis, also considered temperature effects on converter/inverter bridge IGBT module survivability. Chung *et al.* (2015) is another valuable resource on converter reliability. Finally, Mohammed *et al.* (2020) have applied distributed FBG-sensing technology to monitor the temperatures of wind turbine converter/inverter bridge IGBT modules to aid the improvement of device temperature control.

All this new knowledge shows what must be possible to improve future offshore wind turbine converter reliability.

### *5.3.3 Evaluation of current different wind turbine configurations*

Polinder *et al.* (2006) evaluated five current different 3-MW wind turbine drive configurations shown in Figure 5.8:

*Table 5.6 Distribution of generator and converter failure rates in different industries, from literature as quoted*

| | | Measured Industrial Electrical Machine Reliability | | | | | | Measured Wind Turbine Generator Reliability | | | | | | | | | | | | Predicted Wind Turbine Drive Train Reliability | | |
|---|---|---|---|---|---|---|---|---|---|---|---|---|---|---|---|---|---|---|---|---|---|---|
| | Industry | General | | Utility applications | | Offshore petrochemical | | ReliaWind | Wind Turbine Generator Survey 2010 | | | | | | Comparison of DFIG & PMG | | | | | Comparison of Offshore Wind Turbines | Prospective Magnomatic PDD WT | |
| | Survey Unit-Years | 5085 | | 9718 | | 16157 | | 1400 | 73199 | | | | 13200 | | 16955 | | 1200 | | 7400 | | | |
| | Types of machines, ratings, voltages, authors | Motors in Utility Applications | | Motors in Utility Applications | | Motor Survey Offshore and Petrochemical | | 1.5-2.5 MW LV DFIG, ReliaWind [5] | 95% LV WRIG <1MW, electronic-controlled rotor voltage rather than collector rings & converter | | 90% LV DFIG 1-2 MW | | 90% LV DFIG >2MW | | LV DFIG 1.5-2.5 MW | | LV PMG DD 1.5-2.5 MW | | Onshore & Offshore WTs Fischer *et al*. [9] | 2 MW LV Geared PM Generator Delorm *et al*. [8] | 2 MW No Gearbox LV PMG PDD, Generator Tavner [4] | |
| | | Smaller MV IMs O Donnell (1985) [1] | | Large & small MV & HV IMs Albrecht (1986) [2] | | Almost all MV & HV IMs, Thorsen (1995) [3] | | | 1-2 MW DFIG, LV Alewine and Chen [6] | | | | >2 MW DFIG, LV Alewine and Chen [6] | | 1.5-2.5 MW DFIG, LV Carroll *et al*. [7] | | 1.5-2.5 MW DD PM, LV Carroll *et al*. [7] | | | | | |
| Nos | Machines in Survey | 1141 | | 6312 | | 2596 | | 233 | 7173 | | | | 1980 | | 3391 | | 400 | | 2700 | Ersatz data | Ersatz data | |
| | | 10049 | | | | | | | | | | | | | | | | | | | | |
| | Failed Machines in Survey | 360 | | 1474 | | 1637 | | 12 | 196 | | 507 | | 297 | | 417 | | 30 | | - | | | |
| | | 4483 | | | | | | | 703 | | | | | | 447 | | | | | | | |
| | Years in Survey | 14.1 | | 6.6 | | 9.9 | | 6.0 | 10.2 | | | | 6.7 | | 5.0 | | 3.0 | | 2.7 | | | |
| Electrical Machine, Motor/Generator | Cooling system | - | - | - | - | - | - | 0.050 | - | - | 2% | 0.000 | - | - | 4% | 0.005 | 4% | 0.003 | | 0.271 | 6% | 0.005 |
| | Stator winding | 37% | 0.008 | 36% | 0.013 | 13% | 0.008 | | 24% | 0.024 | 3% | 0.003 | 5% | - | 8% | 0.010 | 24% | 0.018 | | | 10% | 0.008 |
| | Stator wedges | - | - | - | - | - | - | | - | - | - | - | 14% | 0.021 | - | - | - | - | | | - | - |
| | Stator magnets | - | - | - | - | - | - | | - | - | - | - | - | - | - | - | - | - | | | 13% | 0.011 |
| | Rotor Winding | 10% | 0.002 | 9% | 0.003 | 8% | 0.005 | | 50% | 0.049 | 4% | 0.004 | 4% | 0.006 | 7% | 0.009 | - | - | | | - | - |
| | Rotor Bearings | 41% | 0.009 | 41% | 0.015 | 42% | 0.027 | | 21% | 0.021 | 70% | 0.069 | 58% | 0.087 | 12% | 0.005 | 58% | 0.003 | | | 33% | 0.027 |
| | Rotor Magnets | - | - | - | - | - | - | | - | - | - | - | - | - | - | - | 14% | 0.011 | | | 13% | 0.011 |
| | High Speed Rotor | - | - | - | - | - | - | | - | - | - | - | - | - | - | - | - | - | | | 7% | 0.005 |
| | High Speed Rotor | - | - | - | - | - | - | | - | - | - | - | - | - | - | - | - | - | | | 17% | 0.013 |
| | Collector or slip rings | - | - | - | - | - | - | | 1% | 0.001 | 16% | 0.016 | 14% | 0.021 | 31% | 0.038 | - | - | | | - | - |
| | Rotor leads | - | - | - | - | - | - | | - | - | 1% | 0.001 | 4% | 0.006 | | | - | - | | | - | - |
| | Other | 12% | 0.003 | 14% | 0.005 | 37% | 0.024 | | 4% | 0.004 | 4% | 0.004 | 1% | 0.002 | 38% | 0.047 | - | - | | | - | - |
| | | 100% | 0.022 | 100% | 0.035 | 100% | 0.064 | | 100% | 0.098 | 100% | 0.098 | 100% | 0.150 | 100% | 0.123 | 100% | 0.076 | | | 100% | 0.080 |
| | Machines Totals and Conclusions | | | | | | | Measured failure rates worse than Industrial Electrical Machines | | | | | | | | | | | | Predicted failure rates worse than measured Industrial Electrical Machines | | |
| | | Measured failure rates better than WT Generators, except Reliawind | | | | | | Reliawind measures better than Carroll measured or Delorm predicted | Mixture of IGs & DFIGs, worse than industrial IMs | | | | Bigger DFIGs worse than smaller DFIGs | | Carroll DFIG figures better than Alewine & Chen larger machines | | Carroll DD PMG figures better than Carroll DFIG | | Fischer *et al*. [9] | Prediction worse than measured DFIGs & Carroll measured DD PMG | Prediction better than measured DFIGs | |
| | | | | | | | | | | | | | | | | | | | | | Prediction similar to measured Carroll DD PMG | |

| | | | | | | | | | | | | | | | | | | | | | | | |
|---|---|---|---|---|---|---|---|---|---|---|---|---|---|---|---|---|---|---|---|---|---|---|---|
| Gearbox failure rates | | - | - | - | - | - | - | 16% | 0.070 | 52% | 0.155 | 31% | 0.155 | 40% | 0.155 | - | | 33% | 0.155 | 16% | 0.134 | - | |
| Motor/Generator failure rates | | - | 0.022 | - | 0.035 | - | 0.064 | 11% | 0.050 | 33% | 0.098 | 30% | 0.150 | 32% | 0.123 | 11% | 0.076 | 32% | 0.150 | 32% | 0.271 | 12% | 0.080 |
| Converter failure rates | Partially-Rated Converter for DFIG | - | - | - | - | - | - | 73% | 0.330 | 15% | 0.045 | 40% | 0.200 | 28% | 0.106 | - | | 34% | 0.160 | - | | - | |
| | Fully-Rated Converter for IG or PMG | - | - | - | - | - | - | - | | - | | - | | - | | 89% | 0.593 | | | 51% | 0.430 | 88% | 0.593 |
| Whole Drive Train failure rate | | - | - | - | - | - | - | 100% | 0.450 | 100% | 0.298 | 100% | 0.505 | 100% | 0.384 | 100% | 0.669 | 100% | 0.465 | 100% | 0.835 | 100% | 0.673 |


[1] O'Donnell P.: IEEE reliability working group, report of large motor reliability survey of industrial and commercial installations. Part I, II and III, *IEEE Trans. Ind. Appl*, 1985, 21, pp. 853–872

[2] Albrecht P.F., McCoy R.M., Owen E.L., Sharma D.K.: Assessment of the reliability of motors in utility applications-updated, *IEEE Trans. Energy Convers*, 1986, 1, pp. 39–46

[3] Thorsen O.V., Dalva M.: A survey of faults on induction motors in offshore oil industry, petrochemical industry, gas terminals and oil refineries, *IEEE Trans. Ind. Appl*, 1995, 31, (5), pp. 1186–1196

[4] Tavner, P.J.: Review of condition monitoring of rotating electrical machines, IET Electr. Power *Appl.*, 2008, Vol. 2, No. 4, pp. 215–247

[5] Wilkinson, M. *et al.*: Methodology and results of ReliaWind reliability field study, EWEC2009, Brussels

[6] Alewine, K., Chen, W., Wind Turbine Generator Failure Modes Analysis and Occurrence, Windpower 2010, Dallas, Texas, May 24-26, 2010

[7] Carroll, J., McDonald, D., McMillan, D.: Reliability Comparison of Wind Turbines With DFIG and PMG Drive Trains, *IEEE Trans. Energy Convers*, 2014

[8] Delorm, T., Lu, Y., Christou, A., McCluskey, P.: Wind Turbine Generator Failure Modes Analysis and Occurrence, *Proc IMechE, Part O: J Risk and Reliability*, pp 1–14, 2016

[9] Fischer, K., Pelka, K., Puls, S., *et al.*: Exploring the Causes of Power-Converter Failure in Wind Turbines based on Comprehensive Field-Data and Damage Analysis, Energies, 2019

[10] Hayward, N., COREL reports, 2020

*Table 5.7 Past wind turbine inverter/converter bridge measured and metadata sub-assembly failure rates*

| **Reliability characteristics for converter and circuit breaker** | | | **Surrogate sub-assembly failure rate estimate, based on an LV converter with 6 IGBT/diodes per inverter, $\lambda i\_FREcon$ (failures/year)** | | | | |
|---|---|---|---|---|---|---|---|
| **Assemblies** | **Sub-assemblies** | **VGB classification code** | **Sub-assembly failure rate estimate $\lambda i\_FREcon$** | **Sub-assembly $Qt$** | **Assembly failure rate source** | **Assembly failure rate estimate $\lambda i\_FREcon$** | |
| Circuit breaker | Circuit breaker | MSC10 | 0.0176 | 1 | IEEE Gold Book (1997) | 0.0176 | |
| Grid-side filter | $dV/dt$ filter | BFA20 | 0.0095 | 1 | LWK_D (2010) | 0.0095 | |
| Grid-side inverter BF | Inverter firing circuits | BF QBA10 | 0.0005 | 1 | LWK_D (2010) | 0.0600 | |
| | Inverter IGBTs | BF QBA11 | 0.0050 | 6 | | | |
| | Inverter switching diodes | BF QBA12 | 0.0050 | 6 | | | |
| DC-link BU | Braking chopper resistor | BUA20 | 0.0085 | 1 | LWK_D (2010) | 0.0190 | |
| | Chopper firing circuit | BU QBA10 | 0.0005 | 1 | | | |
| | DC-link switching IGCT | BU QBA11 | 0.0050 | 1 | | | |
| | DC-link switching diode | BU QBA12 | 0.0050 | 1 | | | |
| | DC-link capacitor | BUA10 | 0.0348 | 1 | LWK_D (2010) | 0.0348 | |
| Generator-side inverter BG | Inverter firing circuits | BG QBA10 | 0.0005 | 1 | LWK_D (2010) | 0.0600 | |
| | Inverter IGCTs | BG QBA11 | 0.0050 | 6 | | | |
| | Inverter IGBTs | BG QBA12 | 0.0050 | 6 | | | |
| Generator-side filter | $dV/dt$ filter | BGA20 | 0.0095 | 1 | LWK_D (2010) | 0.0095 | |
| Cooling unit | Cooling unit | MKY10 EC001 | 0.0058 | 1 | IEEE Gold Book (1997) | 0.0058 | |
| Control unit | Control unit | MKY10 KF001 | 0.0221 | 1 | LWK_D (2010) | 0.0221 | |
| Frame | Frame | UMD10 | 0.0100 | 1 | Ersdal (2005) | 0.0100 | |
| Total converter failure rate consistent with Spinato *et al.* (2009) and ReliaWind experience | | | | | | Converter failure rate (failures/year) 0.2483 | Reliability survivor function, $R$(1 year) (%) 78.0 |

Based on data from Table 5.5 and sources shown, dated 1997–2010.

*Table 5.8 Present measured wind turbine inverter/converter bridge module failure rates, dated 2015–18, compared to failure rate and survivor function in Table 5.7, dated 1997–2010*

| **Industry converter types** | **Inverter configuration** | **Number of WT inverter/converter modules in failure study** | **Inverter/converter bridge module failure rate, λ (failures/year)** | **Reliability survivor function, *R* (1 year) (%)** |
|---|---|---|---|---|
| Highest reliability | Standard industrial AC/DC or DC/AC inverters | 58,104 | 0.0064 | 99.4 |
| Mean reliability | Using half-bridge modules and matched gate-drivers, 2 level | 31,018 | 0.0268 | 97.4 |
| Lowest reliability, due to early deployment | AC/DC/AC single phase units, 3 level, NPC circuit<br>Does not use half-bridge modules but discrete devices in a single package, excepting IGBT and APD. Does not have an integral DC link capacitor or intelligent interface board. Voltage clamp/brake chopper is separate | 803 | 0.0357 | 96.5 |

- The indirect-drive DFIG with three-stage gearbox and partially rated converter (DFIG3G), the turbine speed range being 3:1; therefore the converter rating is usually about one-third that of the generator and gearbox.
- The direct-drive wound synchronous generator with electrical excitation and fully rated converter (DDWRSGE).
- The direct-drive permanent-magnet generator with a fully rated converter (DDPMG).
- The semi-direct-drive permanent-magnet generator with a single-stage gearbox and a fully rated converter (PMG1G).
- The semi-direct-drive DFIG with a single-stage gearbox and a third-rated converter (DFIG1G).

The evaluation, Table 5.9 is based on cost, annual energy yield for a given wind climate and here reliability considerations have been added to the productivity using an approach described in Arabian-Hoseynabadi *et al.* (2010) *Wind Energy*.

The evaluation showed that the indirect-drive DFIG3G was the lightest, lowest cost solution using standard sub-assemblies, explaining why it is most widely used commercially. OEMs use generator and converter sub-assemblies close to industrial standards yielding standardisation, cost and reliability benefits. However, this

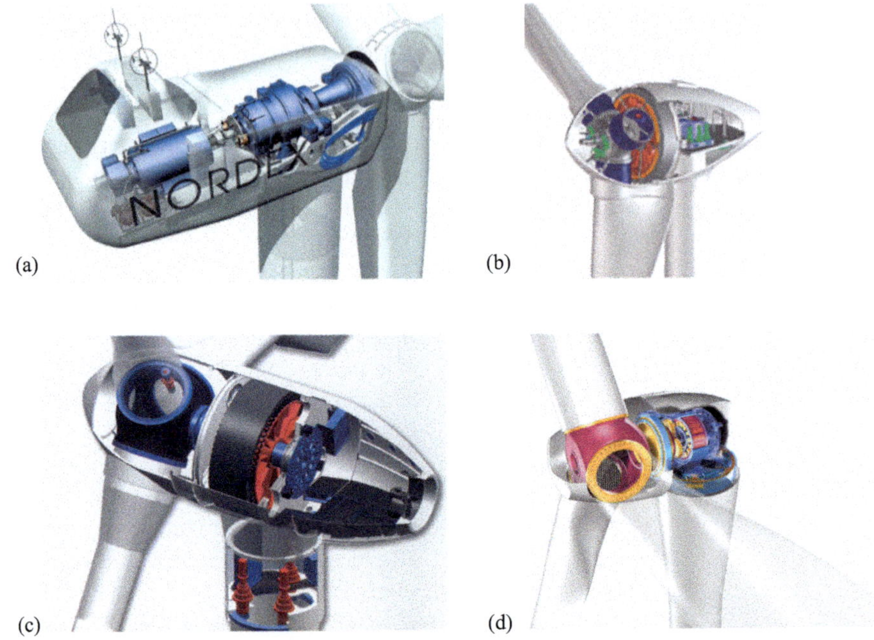


*Figure 5.8* *Pictures of different wind turbine configurations evaluated, Karaki et al. (1999): (a) conventional geared DFIG3G; (b) conventional direct-drive DDWRSGE; (c) permanent-magnet direct-drive DDPMG; (d) integral design PMG1G*

*Table 5.9* *Evaluation of 3-MW drive train configurations with an addition of reliability, Polinder et al. (2006)*

| | **DDWRSGE** | **DDPMG** | **PMG1G** | **DFIG3G** | **DFIG1G** |
|---|---|---|---|---|---|
| Annual energy yield (GWh) | 7.88 | 8.04 | 7.84 | 7.73 | 7.80 |
| Weight ($t$) | 45.1 | 24.1 | 6.1 | 5.3 | 11.4 |
| Relative cost (%) | 115 | 108 | 103 | 102 | 100 |
| Subjective relative reliability (%) | 70 | 80 | 100 | 90 | 80 |

system has wear in the gearbox and generator brush-gear and slip-rings and known unreliability in those areas. It also has a low-energy yield, due to the high gearbox losses. Because it uses a low cost, standard electrical machine and gearbox, future major improvements in performance or cost reduction cannot be expected.

The DDWRSGE appeared to be the heaviest, most expensive alternative and from information later does not necessarily have the best reliability. Prior to 2011, the only commercially successful large direct-drive wind turbine OEM, Enercon, used this configuration, but they claim other benefits from it, including immunity to problems from voltage disturbances due to grid faults, as a result of the use of a fully rated converter. But this sub-assembly is of particular concern having three times the number of parts to the DFIG3G partially rated converter, three times the cost and probably three times the failure rate, Tavner, Higgins, Arabian-Hoseynabadi *et al.* (2010) and IEEE Std 493 (1997) Gold Book.

The wind industry frequently misunderstands that the power converter is one of the highest cost, least reliable drive train items, not the gearbox as frequently quoted.

However, converter faults have low MTTR, unlike the gearbox, and it is also clear that substantial improvements are progressively taking place in power electronics, reducing cost, raising reliability.

In principle, the DDPMG should be the best solution because the generator only has one winding, does not have brushes or a gearbox but has a fully rated converter. An important attraction of this configuration is that the active generator material weight for the same air-gap diameter is nearly halved over the DDWRSGE, while the energy yield is a few per cent higher giving the highest energy yield of the configurations considered. However, compared to indirect-drive systems, it is more expensive. Further improvements of this configuration may be expected because of decreasing power electronics costs and further optimisation and integration of the generator system. However, the rising costs of permanent magnet materials have been a cause of concern.

The PMG1G, with a single-stage gearbox, is an interesting option because the electrical machine size is reduced by the higher speed and it is clear that this generator configuration with single stage rather than three-stage gearboxes will be more reliable. Polinder *et al.* (2006) also made the case that this type of machine is used in other applications, for example ship propulsion, so development costs could be shared.

Surprisingly, the DFIG1G, with a single-stage gearbox, seems to be the most interesting choice in terms of energy yield divided by cost mainly because the lower converter rating results in a reduction of converter cost and losses. However, this system has been too specialised to attract electrical machine and gearbox OEMs, and it is likely that the larger diameter, slower, less standard higher rated DFIG could suffer unacceptable reliability.

Interestingly, the DFIG is being steadily rejected by the wind industry. Its first adoption was because of its cheapness and the fact that wind turbine speed range only required third-rated power converters, which had a dramatic influence on drive train costs. However, although LV DFIGs were cheap, their poor reliability, due to the use low quality, LV wire-wound windings and lack of pretesting, their partially rated converter lead to complex control algorithms to provide the necessary wind farm VAr support, particularly in the event of grid-switching or lightning incidents.

Eventually, increased wind turbine rating, the need for VAr support and the importance of isolating the grid from the turbine have led to the almost universal use off-shore of much more expensive, fully rated converters. These are still not very reliable although their components can be changed-out in-service relatively easily.

There are several new approaches to solve these reliability problems by moving from LV to MV generators and converters, where both MV generator windings and IGBT modules are proving more reliable, and novel converter arrangements could allow a wind turbine to provide more inertial support to the grid, see the next section.

### 5.3.4 *Innovative wind turbine configurations*

Beyond the possibilities of current geared and direct-drive wind turbine configurations, shown in Figure 5.2, a number of innovative new concepts are now under consideration. These are summarised briefly in Figure 5.9, expanding the classification used by the European Wind Energy Association (2009). They are described

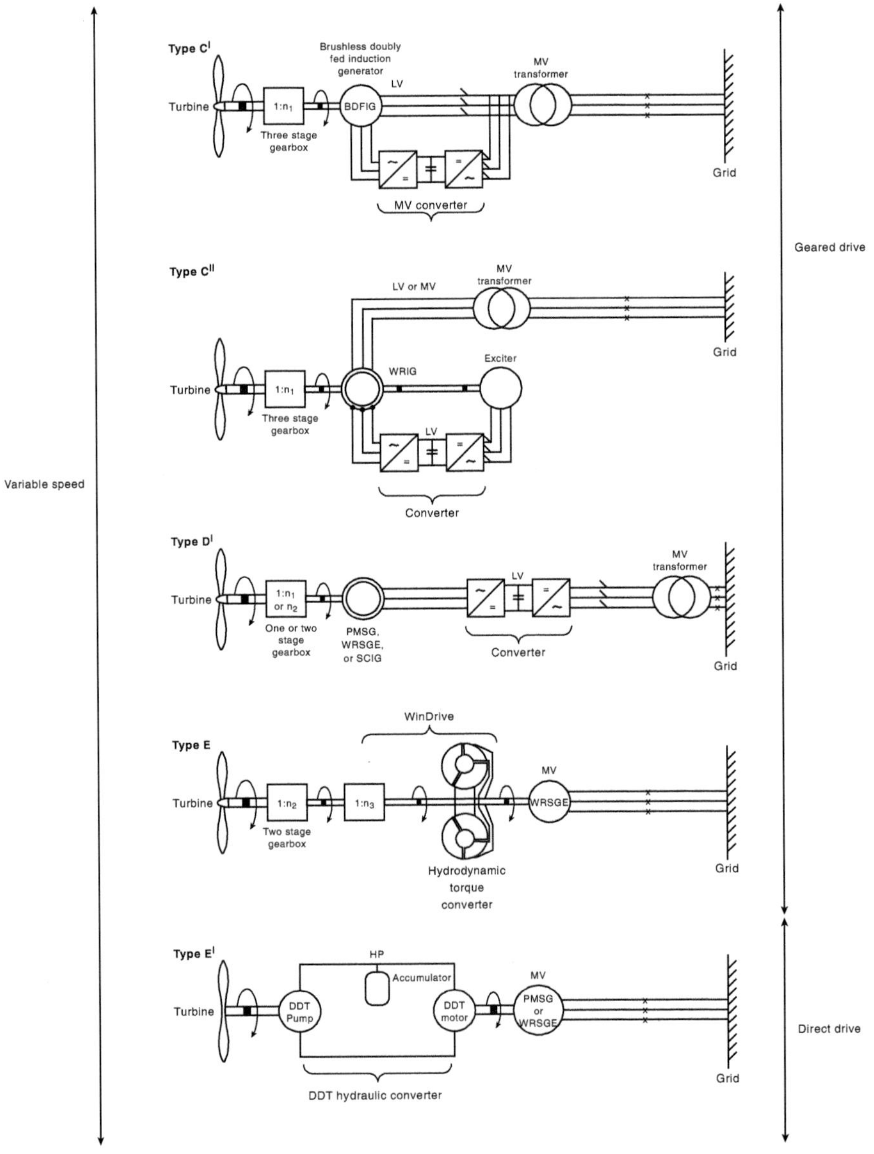


*Figure 5.9 Summary of main electrical configurations for innovative wind turbine drive trains, cf. Figure 5.2*

here being divided into electrical or hydraulic options as follows and all offer potential reliability benefits:

- Type C′, a derivative of Type C using an LV brushless DFIG instead of the DFIG, removing the need for brush-gear and slip-rings, which not only affects the maintenance of DFIGs but also offers a lower speed generator, allowing the use of a two-stage rather than three-stage gearbox.
- Type C″, another derivative of Type C using an LV WRIG with a self-driven three-phase AC brushless exciter, feeding the rotor through a two-quadrant excitation-rated converter, removing the need for brush-gear and slip-rings, so that a synchronous generator is effectively coupled to the grid.
- Type D′, a derivative of Type D, the geared drive wind turbine but using a single- or two-stage gearbox, low-speed generator and fully rated converter, thereby raising the gearbox reliability, gaining the power quality advantages of the power converter and eliminating brush-gear and slip-rings.
- Type E, a hydraulic arrangement based on a conventional geared drive train but using a limited speed range hydraulic torque converter to drive an MV WRSGE. The advantage of this arrangement is that a power electronic converter and transformer are eliminated by synchronous generation at MV, an example drive train is shown in Figure 5.10.
- Type E′, a more innovative hydraulic solution, using digital drive technology (DDT) from Artemis Innovative Power with the turbine driving a slow-speed hydraulic pump feeding a high-speed hydraulic motor with high-pressure hydraulic fluid. That motor drives an MV WRSGE. The advantage of this arrangement with synchronised MV generation is that gearbox, power electronic converter and transformer are eliminated. However, the DDT hydraulic scheme is new and untried and contains a large number of individual hydraulic components, each dependent on separate valves and seals, which would need

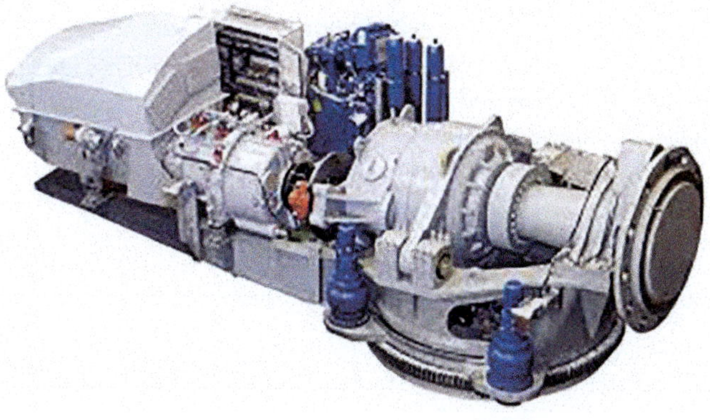

*Figure 5.10 Example drive train Type E, an assembled DeWind D8.2 wind turbine with a Voith WinDrive hydraulic torque converter driving a 2-MW, 11-kV WRSGE. Source: Voith Drives*

to have extremely high reliability performance figures, because they are effectively in parallel, if a hydraulic drive is to have high reliability.

- Types C and C′, E and E′ solutions all offer a grid-connected synchronous generator. All these innovations are important and deserve a second look since their protection and VAr control can all easily accommodate grid-switching or lightning incident issues, no different than conventional grid-connected synchronous generators in diesel, gas or nuclear power plants.

### 5.3.5 *Alternative drive train analysis*

An alternative recent analysis of wind turbine drive trains was performed by Serrano-González *et al.* (2016) using a database containing information about more than 26,000 wind farms worldwide, with a cumulative capacity, to the end of 2014, of 336.4 GW, in Europe, Asia, United States and the rest of the world collected by the European Commission Joint Research Centre.

This gives valuable Table 5.10 to be compared with Table 5.8 from Polinder *et al.* (2006), expanding the classification used by the European Wind Energy Association (2009).

## 5.4 Using data to predict prospective drive train reliability

The work of Polinder *et al.* (2006) and Serrano-González *et al.* (2016) leads the author to consider how we can use past data to make future reliability predictions for upcoming operational off-shore wind farms.

By considering different drive train configurations the O&M engineer could use reliability knowledge from operational experience to predict the reliability that can be expected from improved drive train configurations.

As an example Figure 5.11(a) shows data collected from five separate reliability data surveys from wind turbine drive trains, amounting to 96,349 unit-years of operation, mostly on-shore. These are metadata like that described earlier. The results in Figure 5.11(a) show a drive train failure rate, $\lambda$, with considerable consistency, varying between 0.3 and 0.5 failures/year, that means a wind turbine drive train failure at least every 6 months. Failures could be classified as

- Minor: such as an electronic module or bearing fault requiring a change-out downtime taking no more than 1 day.
- Major: such as gearbox or generator problems requiring components to be removed from the nacelle for replacement, taking 1 week to 1 month.

From these metadata, it is possible to predict the failure rates which could accrue if new wind turbines utilised different drive configurations, such as shown in Figure 5.11(b), or, for example, from the data given by Serrano-González *et al.* (2016).

These graphs used results from two papers, by Delorm *et al.* (2016) for off-shore and Tavner (2017) for on-shore wind turbines, to demonstrate that

*Table 5.10 Advantages and disadvantages of different drive train configurations, Serrano-González et al. (2016)*

| Drive train type using similar nomenclature to Table 5.9 | Advantages | Disadvantages | Subjective relative reliability (%) |
|---|---|---|---|
| Type A SCIG3G | Low up-front cost (compared to Types B, C, D, E and F next), no power converter, squirrel cage induction generator, cheaper than a wound induction generator;<br>Robust electric generator, lower maintenance | No speed control with squirrel cage induction generator;<br>Poor aerodynamic efficiency (as speed control cannot be performed, power coefficient is maximum at only one operating point);<br>Poor grid integration: no fault ride-through capability, no reactive control;<br>Poor mechanical control and demanding mechanical component loads on gearbox and turbine;<br>Three-stage gearbox required | 65 |
| Type B DFIG3G | Low up-front cost (compared to Types C, D, E and F next);<br>Inexpensive power converter and variable rotor resistance;<br>Enhanced speed control capability (compared to Type A) | Higher electrical losses because it is a wound rotor doubly fed induction generator;<br>Variable speed limited to ±10% around synchronous speed;<br>Wound rotor doubly fed induction generator, needs slip-rings, more expensive and higher maintenance than SCIG;<br>Three-stage gearbox required | 70 |
| Type C DFIG3G | Relatively inexpensive power converter;<br>Reactive power control capability, enabling voltage control;<br>Power converter provides smoother grid connection | Variable speed limited to ±30% around synchronous speed;<br>Wider operating speed range but still limited;<br>Limited grid integration (compared to fully rated converter Types D, E and F); | 75 |

*(Continues)*

*Table 5.10 (Continued)*

| **Drive train type using similar nomenclature to Table 5.9** | **Advantages** | **Disadvantages** | **Subjective relative reliability (%)** |
|---|---|---|---|
| | | Wound rotor doubly fed induction generator with expensive slip-rings and higher maintenance than SCIG; Three-stage gearbox required | |
| Type D DDPMG | Full-speed range; No gearbox required; Complete control of reactive and active power; Full control of voltage and frequency | Expensive full-scale power converter; Expensive, heavy low-speed permanent magnet generator required | 80 |
| Type E″ PMG1G | Full-speed range; Full control of reactive and active power; Full control of voltage and frequency; Lighter mid-speed permanent magnet generator using less rare earth permanent magnets (compared with Type D); Reduced gearbox ratio and weight (compared with Types A, B, C and F) | Expensive full-scale power converter; Only one-stage gearbox required | 100 |
| Type F SCIG3G | Full-speed range; Complete control of reactive and active power; Full control of voltage and frequency; Lower up-front cost (compared to Types D and E″ earlier); Robust and cheap squirrel cage induction generator, cheaper and lighter than Type D and E″ permanent magnet generators | Expensive full-scale power converter; Three-stage gearbox required; Heavier, bigger and higher ratio gearbox (compared with Type E) | 105 |

indirect- or direct-drive train arrangements with fully rated converters cannot necessarily deliver higher overall drive reliability, because of converter power electronic unreliability.

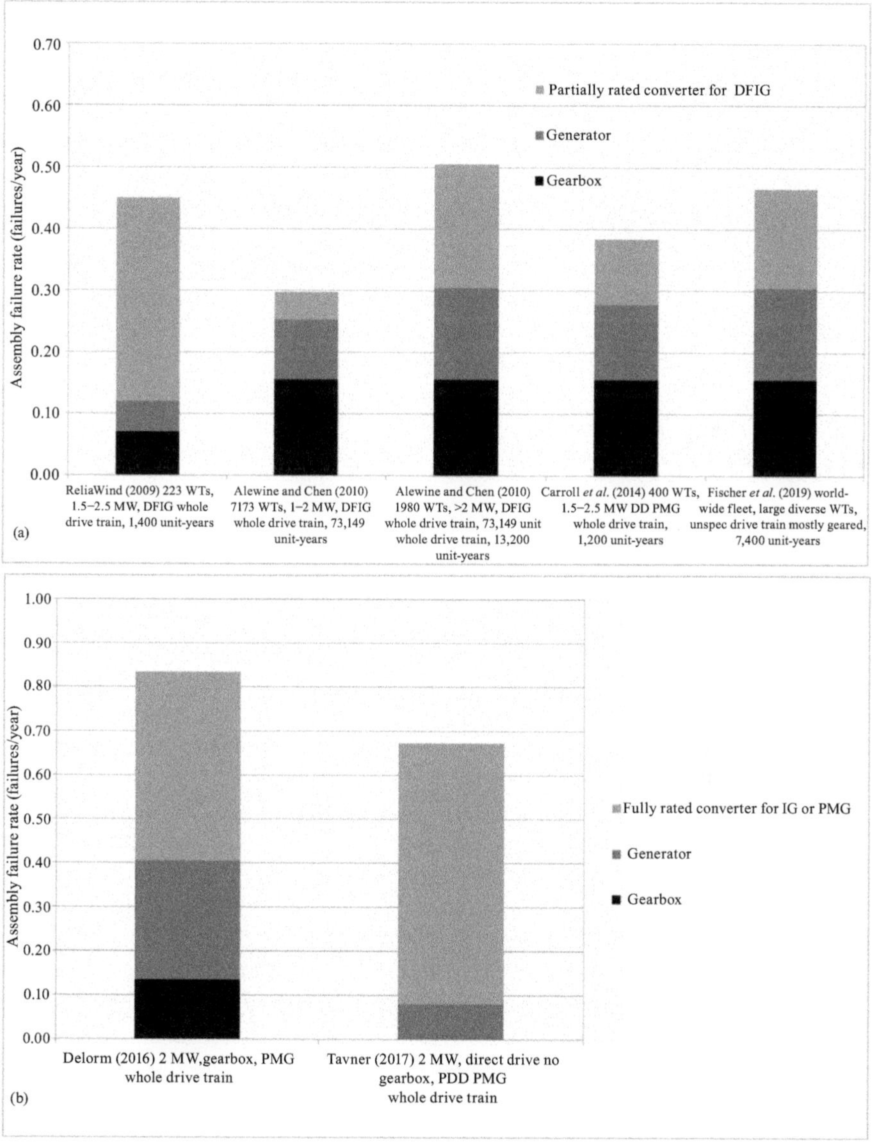


*Figure 5.11 Comparison of wind turbine drive train reliabilities, measured then predicted using metadata: (a) measured reliabilities from 5-WT drive train architecture with or without gearboxes, using partially rated converters, based on 96,349 unit-years operation; (b) predicted reliabilities of WT-geared and direct-drive trains with fully rated converters, using previously measured metadata*

The fact is that direct-drive power electronic systems could lower turbine drive train reliability, unless power electronic reliability is improved. However, fully rated converters have become essential to meet transmission network Grid Code requirements. This a subject to be considered later in the book when dealing with the network, when drive trains like Types C and C′, E and E′, Figure 5.9, could allow wind turbines to make an inertial contribution to the grid, rather than using fully rated converters, which isolate the generator from the grid, but ensure full voltage ride-though capability.

The previous cases are an excellent example of the value of reliability data collection, showing the potential for combining data from different sources to create metadata, then using those metadata to predict future potential operational plant reliability and extract valuable lessons.

## 5.5 Summary

This chapter has shown that wind turbine configuration does have an effect on reliability but there are some industry myths that are not supported by evidence.

For example it is simply not proven that a direct-drive wind turbine is more reliable than a geared-drive machine.

Electrical sub-assemblies, such as the converter, appear to have improving reliability, MTBF, with time and lower MTTRs, whereas heavy mechanical sub-assemblies, such as the gearbox, have high MTTRs and are a mature technology whose reliability is unlikely to improve significantly. This suggests that in the long term, all-electric wind turbines should have a more reliable future.

On the other hand, there are some emerging drive train technologies, such as semi-direct drives with single-stage gearboxes, or hydraulic drive transmissions which allow the use of fixed speed MV generators, showing great promise for reducing weight, reducing balance of plant costs and improving reliability, but their full production capital costs are as yet unknown.

Recent experience has emphasised that any configuration can achieve reliability providing that the component sub-assemblies are well designed, manufactured, installed and maintained.

A conclusion at this stage is that there is no clear ideal off-shore wind turbine configuration for reliability, rather that OEMs should ensure that drive train sub-assemblies are thoroughly tested before installation in the wind turbine and that wind turbine nacelles should be prototype load-tested or even production load-tested if they are to be installed off-shore. Furthermore, the off-shore commissioning of drive train components, turbine and pitch, gearbox, generator, bearings, converter and controls, must be thorough, documented and traceable.

However, a reliable single wind turbine drive train configuration is an important future possibility and this could be an integral turbine, gearbox, generator and converter design, improving stress distribution, bearing performance with integral manufacture, testing, transportation and installation, which may also considerably affect the wind turbine price.

This author believes that the wind industry rejected geared drives because of poor drive train design during the development of 1–2-MW wind turbines, rather than poor-quality gearboxes.

Polinder *et al.* (2006) had it right that for very large off-shore wind turbines with PMG1G, DFIG1G or SCIG1G designs could still deliver light-weight, low-cost, grid-compatible, high-reliability solution. Note that Polinder updated his 2006 work in Polinder *et al.* (2013) and Serrano-González *et al.* (2016) have given us a market-based analysis of wind turbine configurations.

On the other hand, Cevasco *et al.* (2021) say that moving towards larger turbine systems, with direct-drive designs by changing to synchronous permanent magnet generators, will incur higher transmission system failure corrective maintenance costs.

This chapter has shown that metadata, compiled from surveys like those described, can be useful to wind farm designers and their O&M engineers, to predict how reliable particular configurations are likely to be.

Higher reliability would also be conferred by the use of MV generator windings and MV converter IGBT modules with fully loaded, back-to-back generator, gearbox and converter testing at the start of each new turbine class. It is also clear that poor power electronic component reliability is hampering the industry, particularly as we go further off-shore.

Whilst changing power electronic modules is relatively easy on-shore or close inshore, it carries an expensive OPEX overhead in large far off-shore wind farms.

Why cannot wind turbines use more reliable, lower temperature, MV power electronic units?

It is also clear that the influence of low-reliability LV power electronics and generators does not help the reliability case for fully rated converter wind turbines off-shore. A move towards MV generators and converters, which have both been shown to be more reliable in high-performance rail traction and ship propulsion, using technically similar but differently packaged converter technologies, is clearly signalled. There may also be a future trade-off for larger off-shore wind farms where turbine drive considerations may have to provide more grid inertial support, affecting the design and operation of drive-train power converters.

It is also gradually becoming clear that wind turbine technology is less significant for overall off-shore wind farm reliability than wind farm size and layout requiring large sub-station/converter station, collection cabling and export cabling assets.

Therefore, a later chapter will explore the impact of off-shore wind farm layout on reliability.

But before that, we will consider the wind turbine design and testing approach as a way towards higher reliability.

# Chapter 6

# Design and testing for wind farm availability

## 6.1 Introduction

The high penetration of wind power into power systems will have several impacts on their planning and operation. One of these will be the effect on power system reliability, emphasised because wind power is intermittent, so wind turbine reliability delivering this power is becoming an essential consideration. Due to the competitive environment, power generation industry developers and operators usually prefer the most economically productive wind turbine configurations. This must take into consideration work like that shown in Section 6.2.

However, through-life productivity must also be considered, emphasised in off-shore operation, where access is difficult and otherwise productive wind turbines may be unproductive because of small but unresolved faults, see Faulstich, Lyding *et al.* (2011) and the importance of extended downtime off-shore was identified by Faulstich, Hahn *et al.* (2011).

Long-term cost-analysis of wind turbines, including both first investment and operation and maintenance (O&M) costs, will result in better wind turbine configuration choices, but this is only possible if such analysis includes the reliability of the different wind turbine technologies.

Reliability of wind turbines as part of a larger power system has been assessed in several references, Wangdee *et al.* (2007), Karaki *et al.* (1999) and Mohammed *et al.* (2000), considering the wind as a stochastic process, using an appropriate time-series to model the wind resource input, combined with the power–speed curve of an appropriate wind turbine.

There have been few studies of the reliability of wind turbines as isolated systems rather than as part of a large power system, Karaki *et al.* (2004), although Xie *et al.* (2011) do consider the impact of wind turbine reliability on the overall reliability of the power system.

A valuable reference is Hou *et al.* (2019) describing the optimisation of off-shore wind farm electrical system layouts.

This chapter follows the previous one focusing on wind turbine design, consisting of several mechanical, electrical and auxiliary assemblies, as part of a larger wind farm, showing that these methods can be applied to assemble the objective of a complete wind farm's reliability.

Reliability analysis methods in the initial stages of power generation system design are usually qualitative, depending on comparison with data from similar systems, whereas after several years, off-shore wind farm power generation reliability analysis can become more quantitative as large quantities of valid field statistics data are generated.

## 6.2 Methods to improve reliability

### *6.2.1 Reliability results and future turbines*

The results presented in Chapters 4 and 5 were mostly obtained on existing wind turbines of historic designs and sizes ranging from 200 kW to 3.6 MW, mostly from the on-shore environment but with some from off-shore.

To what extent can these data be used to predict the reliability performance of new designs of off-shore wind turbines of much larger sizes, say 3–10 MW, and for placement in the off-shore environment?

Reliability analysis is of necessity backward looking and rarely produces data less than 5 years old; however, its advantage is that data are numerical and comparable.

It is proposed that the wind turbine failure rates shown in Figures 4.5–4.7 could be used as a datum against which future off-shore designs should be measured.

However, while an average failure rate of 1 failure/turbine/year could be acceptable on-shore, it is unlikely to be acceptable off-shore where access may need to be limited to one visit a year.

The wind turbine sub-assembly failure rates can also be used as a datum for comparison between different concepts and designs; however, the mean time to repair (MTTR) must also be considered, as the gearbox data have shown.

Reliability improvement analysis will be useful for wind turbine and sub-assembly original equipment manufacturers (OEMs) to define where design and testing effort should be deployed to improve future reliability.

### *6.2.2 Wind turbine design concepts*

Wind turbine OEMs moving into the off-shore market are concerned to develop designs appropriate to that market. Some OEMs have deployed off-shore wind turbines designed for on-shore and this has given rise to a number of operational problems, for example associated with particular maintenance operations, implicit in their design, which are untenable off-shore.

Therefore, some OEMs have been anxious to develop new or modified wind turbine designs, more appropriate to the onerous off-shore environment. Some wind turbine OEMs and their investors have favoured particular design concepts for this application, for example direct drive as opposed to geared-drive, or hydraulic transmission as opposed to electrical conversion, to avoid perceived on-shore wind turbine reliability problems.

The issue of direct- versus geared-drive was investigated in Polinder *et al.* (2006), their overall reliability was investigated in Spinato *et al.* (2009) and their electrical parts reliability was investigated in Faulstich, Lyding *et al.* (2011).

The issue of hydraulic transmission versus electrical conversion has not been investigated from a reliability point of view, although it could have been using the methods of Spinato *et al.* (2009). One OEM, Mitsubishi, prototyped a hydraulic transmission design after acquiring Artemis but that application does not seem to have been developed further.

As a general principal, it would be unwise to introduce off-shore a new turbine design concept that had not been thoroughly pretested and exposed to on-shore operation. Furthermore, it is a mistake for the wind industry to imagine that any particular wind turbine design concept, whether direct- or geared-drive, all-electric, hydraulic or mechanical transmission, is likely to be a panacea for off-shore operation.

It is clear that there are good examples of all those technologies working reliably on-shore, which can be made to succeed off-shore but only with adequate pretesting and an appropriate O&M regime enforced in the off-shore environment.

As concluded in the last chapter, experience has emphasised that any concept can achieve reliability providing that the component sub-assemblies are well designed, manufactured, installed and maintained. The object here is to highlight precautions that can be taken during design to raise reliability.

One simple, but more costly, approach to improve reliability, taken by Enercon and other wind turbine OEMs, was to remove the gearbox and use a direct-drive configuration.

Enercon also adopted an all-electric approach, avoiding the use of hydraulics for pitch or yaw control. This almost certainly stems from their original owner, Aloys Wobben's experience as a power electronics engineer. Taking up the comparison between direct- and geared-drive wind turbines, raised by Polinder *et al.* (2006), it was shown by Spinato *et al.* (2009) that:

- From Figure 5.5 direct-drive wind turbines do not necessarily have better reliability than geared-drive wind turbines. In Figure 5.5 the direct-drive E40 has a higher failure rate than its geared-drive partners of the same size, whereas the direct-drive E66 has a lower failure rate than its partners, although the E66 data are rather limited in a number of wind turbines.
- From Figure 4.3 the aggregate failure rates of generators and converters in direct-drive wind turbines are generally greater than the aggregate failure rate of gearboxes, generators and converters in geared wind turbines. Therefore, the price paid by direct-drive wind turbines for the reduction of failure rate by the elimination of the gearbox is a substantial increase in failure rate of electrical-related sub-assemblies.
- On the other hand from Figure 5.7, it can be seen that the MTTR of electronic sub-assemblies is lower than the MTTR of gearboxes.
- From Figure 5.5, the failure intensities of larger direct-drive generators are up to double that of the geared-drive generators of a similar size. The following explanation is offered. The direct-drive machines in these machines were wound rotor, synchronous generators with high pole pair number, incorporating a large number of rotor and stator coils. On the other hand, the geared-drive

machines are four or six switchable pole, high-slip, induction generators or doubly fed, induction generators, with far fewer coils. It is suggested that the disparity in failure intensities is because of the following:

- The much larger number of coils in the direct-drive machine. The failure rate could be improved by replacing field coils by permanent magnets, but this would introduce other, reactive control issues.
- The larger diameter of the direct-drive generator making it difficult to seal the more numerous coils from the environment, exposing coil insulation to damage because of the air contaminants and environmental humidity.
- Insufficient standardisation in the manufacture of the large direct-drive machines, as a consequence of smaller production runs, compared to the more common doubly fed induction generator. From a general consideration of direct-drive or geared-concept wind turbines, the following issues arise associated with the design:
    - The reliability of these wind turbine generators, from Figure 5.5 and Table 5.4, is worse during early operational life than that achieved by generators in other industries.
    - In Figure 5.4 the reliabilities of these wind turbine gearboxes are seen to be that of a mature technology, constant or slightly deteriorating with time. The reliabilities are comparable with those obtained by gearboxes in other industries. Therefore, substantial improvements in the designed reliability of these gearboxes are unlikely in the future, although design improvements in gearboxes for newer, larger designs of wind turbines are being actively pursued, and it appears that maybe a greater onus is being placed on wind turbine gearbox reliability by the stochastically varying torque to which it is subjected.
    - In Figure 5.7 the reliabilities of these wind turbine converters are considerably worse throughout their operation than achieved by converters in other industries.
    - The MTTR of electrical components is relatively low, and industrial experience suggests that electrical sub-assemblies are more amenable to reliability improvement than mechanical sub-assemblies, e.g. the gearbox. Therefore, an all-electric direct-drive wind turbine may ultimately have an intrinsically higher availability than a geared-drive wind turbine.
    - From the observations earlier, improvements in generator and converter reliability design will be crucial to improving the reliability of both direct-drive and geared-concept wind turbines, and this design information is exceptionally important for off-shore wind turbines.

This chapter will go on to show that there is more that can be done to promote reliability during the design stage of off-shore wind farms, than simply changing the overall wind turbine configuration or concept.

### *6.2.3 Testing*

Testing of sub-assemblies, particularly converters and generators, can encourage the achievement of higher wind turbine reliability at the start of operational life by eliminating early failures. A suggestion is that off-shore wind turbines nacelles could be tested complete, at full or varying load, at elevated temperature, to accelerate the occurrence of early failures. This is a standard practice in the electrical machine and gearbox industry where prolonged heat runs at elevated temperatures are done as type tests on new products. These type tests are then repeated on individual machines from batch sizes specified, for example by IEC Standards 60034 and 61852. It is also a standard practice, in the volume production of low-rating power converters, <100 kW, to routinely age key converter sub-assemblies and then carry out extended load tests on assembled converters from batch sizes specified, for example by IEC Standard 60700, to identify generic weaknesses before despatch.

The issues of testing are discussed further in Section 6.4.

### *6.2.4 Monitoring and O&M*

The improving reliability of generators and converters in Figures 5.5 and 5.7 indicates that O&M activities are already having a reliability effect. Condition monitoring measures machine performance indicating the need for remedial action when performance deteriorates. The wind industry has applied SCADA and CMS systems to wind turbines and most wind farms now have a SCADA system providing data to remote control rooms. However, agreement has not yet been reached on processing the large quantities of data generated to indicate incipient failures. O&M methods need to use this information to predict failure and thereby schedule maintenance, although work is currently going on in this area of O&M, Caselitz *et al.* (2005). If the previous design and testing suggestions are developed and the monitoring techniques are resolved, the O&M approach will require the following:

- Maintenance based on the measured condition of the wind turbine so that failures of vital sub-assemblies like the generator, gearbox and converter can be pre-empted.
- The provision of adequate spares to reduce downtime when maintenance on the basis of condition takes place.

These issues will be raised in the following sections of this chapter and in Chapters 7 and 10.

## 6.3 Design techniques

### *6.3.1 Wind farm design and configuration*

Reliability does not only depend upon the wind turbine but also on the design of the wind farm in which the wind turbine is situated and its configuration, which contains not only the wind turbine but also collector cable arrays, sub-stations, cable

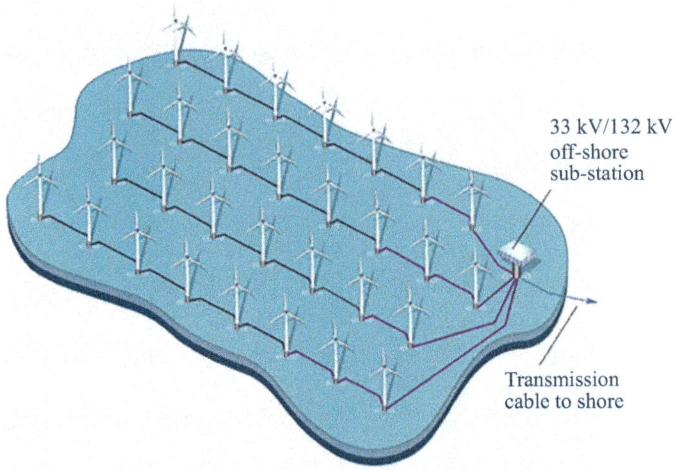


*Figure 6.1 Radial configuration of an off-shore wind farm with 33-kV collector voltage and 132-kV grid connection*

connection to shore and a shore sub-station. Figure 6.1 shows a typical radial cable configuration for a large wind farm.

As yet there has been no published failure modes and effects analysis (FMEA) for a wind farm array but the key issues to consider are as follows:

- Individual wind turbine transformer and switchgear arrangements for connection to the 20–33-kV cable collector array.
- Configuration of the 20–33-kV cable collector array itself, including any switchgear to allow sub-group isolation within that array.
- Configuration of the collector sub-station, including the collector array electrical protection.
- Switchgear, transformers for the cable connection to shore.
- Shore sub-station and protection for the cable connection to shore.

An important issue here is the degree of redundancy incorporated into the collector cable array and off-shore substation. Early wind off-shore wind farms had radial collector arrays, as in Figure 6.1, meaning that a single fault in a radial spur would interrupt power flow from the whole spur. However, by introducing some ring capability, with additional cable routes and switchgear, there can be an improvement in overall wind farm reliability and availability, by providing alternative power flow routes in the event of a failure in the collector network, but this adds to the project cost. Cable arrays have been investigated in Hou *et al.* (2019).

### *6.3.2 Design review*

A procedural method for raising prospective off-shore wind farm and off-shore wind turbine reliability is to apply design review procedures in the development phase. A process for off-shore wind turbine design, recommended by the draft standard, is shown in Figure 6.2.

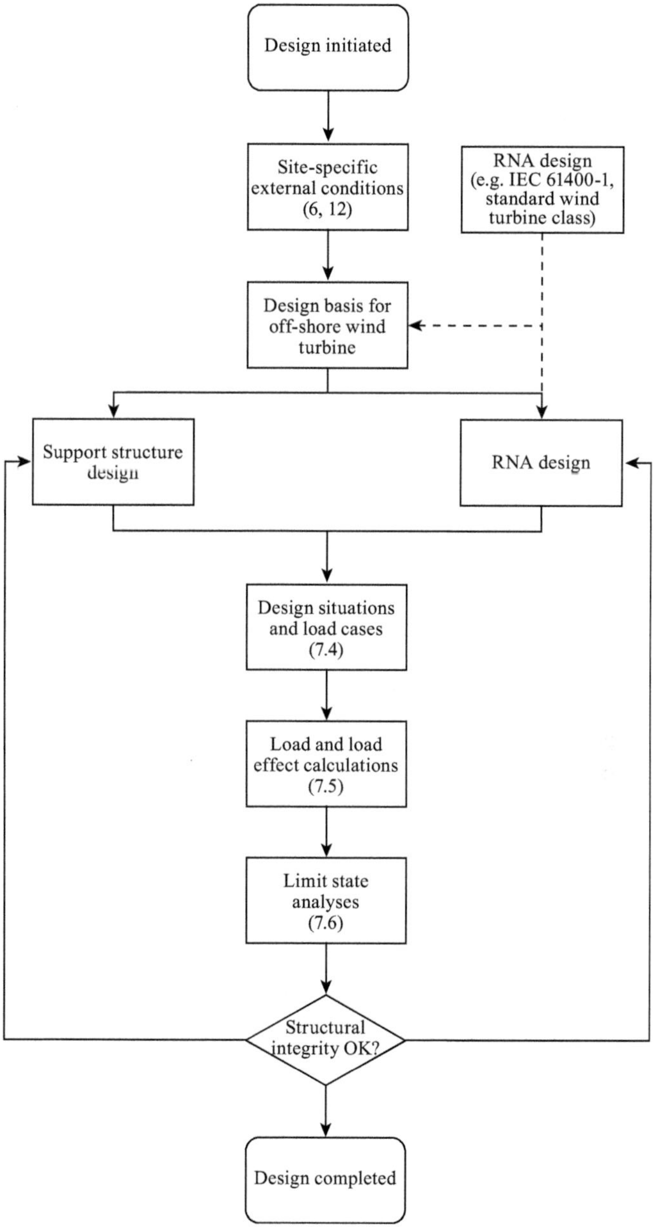


*Figure 6.2 Description of the design process for an off-shore wind turbine. Source: IEC 61400-3*

Off-shore wind turbines will be qualified for the rated wind speed and wind class, see Table 6.1, with a design lifetime for wind turbine classes I–III at least 20 years.

*Table 6.1 Basic parameters for wind turbine classes*

| Wind turbine class | I | II | III | S |
|---|---|---|---|---|
| $V_{ref}$ (m/s) | 50 | 42.5 | 37.5 | Values specified by the designer |
| $A\ I_{ref}$ (−) | 0.16 | | | |
| $B\ I_{ref}$ (−) | 0.14 | | | |
| $C\ I_{ref}$ (−) | 0.12 | | | |

Source: IEC 61400-1.

In Table 6.1 the parameter values apply at hub height, and $V_{ref}$ is the reference wind speed average over 10 min. *A* designates the wind turbine category for higher turbulence characteristics, *B* for medium turbulence characteristics, *C* for lower turbulence characteristics and $I_{ref}$ is the expected value of the turbulence intensity at 15 m/s.

An important issue here is that wind turbine reliability, described briefly at the start of Chapter 2, can be considered to consist of

- structural reliability;
- electro-mechanical reliability;
- control system reliability.

The process of certified design in Figure 6.2 is mainly directed towards structural survivability, when the off-shore wind turbine is subjected to the extreme events during its planned life.

Such analysis is made more complex because the turbine is also subject to aleatory uncertainty due to the stochastic effects of the weather itself, the wind from which the machine extracts energy and, in the case of off-shore wind turbines, the combined effects of wind and waves on the structure and of corrosion.

The impact of these extreme events is of primary importance to the off-shore wind turbines structure for its survival but does not impact the day-to-day operation and normal life, which depends upon the electro-mechanical and control reliability.

The reliability aspects of design must therefore concentrate upon these electro-mechanical and control issues and be wrapped around the process of Figure 6.2 as part of the design review process, and it will be essential that process is based upon genuine reliability data, either obtained from earlier developments of the off-shore wind turbine, or from surrogate data from the off-shore industry or from wind turbines of similar design, such as that described in Chapter 4.

It would also be advantageous to combine the design review process with an FMEA/FMECA (failure modes, effects and criticality analysis) process, such as that described in Section 6.3.3.

### *6.3.3 FMEA and FMECA*

FMEAs, or FMECAs, where failure rates are considered, are the best candidates for design stage reliability analyses as part of RMP. The process is well-defined IEC 60812 (2006) and has been used for many power generation engineering systems, although perhaps less with an emphasis on availability than concern for safety, design assurance or the avoidance of specific observed ones in service failure modes.

The FMEA is a powerful design tool that provides a means, from a risk point of view, of comparing alternative machine configurations, and it is also useful for considering designs improvements for a technology which is changing or increasing in rating, as wind turbine configurations are.

The FMEA is a formalised but subjective analysis for the systematic identification of possible root causes and failure modes and the estimation of their relative risks.

The main goal is to identify and then limit or avoid risk within a design. Hence, the FMEA drives towards higher reliability, higher quality and enhanced safety.

Since FMEA is used by various industries, including automotive, aeronautical, military, nuclear and electro-technical, specific standards have been developed for its application. A typical standard will outline severity, occurrence and detection rating scales as well as examples of an FMEA spreadsheet layout. Also a glossary will be included that defines all the terms used in the FMEA. The rating scales and the layout of the data can differ between standards, but the processes and definitions remain similar, for example, as follows:

- SMC Regulation 800-31 developed for aerospace.
- MIL-STD-1629 (1980) drafted by the US Department of Defense is the most widely used FMECA standard with over 30 years development and usage, having been employed in many different industries for general failure analysis. Due to the complexity and criticality of military systems, it provides a reliable foundation on which to perform FMEAs on a variety of systems. It also contains formulae for predicting the failure rates of electrical and electronic systems, coefficients of which are based on accelerated life tests.
- SAE J 1739 was developed as an automotive design tool and Ford has used it as a design review process.
- IEC 60812 (2006) is now considered a general standard.

It can also be used to assess and optimise maintenance plans. An FMEA is usually carried out by a team consisting of design and maintenance personnel whose experience includes all the factors to be considered in the analysis. The causes of failure are root causes and may be defined as mechanisms that lead to the occurrence of a failure. While the term failure has been defined, it does not describe the mechanism by which the component has failed. Failure modes are the different ways in which a component may fail. It is vitally important to realise that a failure mode is not the root cause of failure, but the way in which a failure has occurred. The effects of one failure can frequently be linked to the root causes of another failure.

The FMEA procedure assigns a numerical value to each risk associated with causing a failure, using severity, occurrence and detection as metrics. As the risk increases, the values of the risk rise. These are then combined in a risk priority number (RPN), which can be used to analyse the system, where RPN is calculated by multiplying the severity, occurrence and detection of the risk:

$$RPN = occurrence \times severity \times detectability \tag{6.1}$$

By targeting high RPN values, the most risky elements of a design can be addressed.

Severity refers to the magnitude of the end effect of a system failure mode. The more severe the consequence, the higher the value of severity will be assigned to the effect.

Occurrence refers to the frequency that a root cause is likely to occur, described in a qualitative way. That is not in the form of a period of time but rather in terms such as remote or occasional.

Detection refers to the likelihood of detecting a root cause before a failure can occur.

In conventional FMEA the severity, occurrence and detection factors are individually rated using a numerical scale, typically ranging from 1 to 10. These scales, however, can vary in range depending on the FMEA standard being applied. However, for all standards, a high value represents a poor score, for example catastrophically severe, very regular occurrence or impossible to detect. Once a standard is selected, it must be used throughout the FMEA. In this section, IEC 60812 (2006) will be used but with some amendment, principally to change the severity, occurrence and detection criteria by which the RPN is calculated. These modifications were necessary to make the FMEA methodology more appropriate to wind turbine systems.

The modified severity scale and criteria are shown in Table 6.2. The scale of 1–4 in IEC 60812 (2006) was maintained but changes were made to the category criteria definitions to emphasise their implications for a wind turbine.

An occurrence scale and criteria modified from IEC 60812 (2006) are tabulated in Table 6.3. Arabian-Hoseynabadi *et al.* (2010) *Int J Electrical Power & Energy Systems* have shown that severity can be related to $1/\mu = MTTR$.

Finally, the number of detection levels was reduced, according to Arabian-Hoseynabadi *et al.* (2010) *Int J Electrical Power & Energy Systems*, to 2 as shown in the modified detection scale and the criteria are tabulated in Table 6.4, and again it has been shown that occurrence can be related to $\lambda = 1/MTBF$.

*Table 6.2 Severity rating scale for a wind turbine FMEA*

| Scale no. | Description | Criteria |
|---|---|---|
| 1 | Category IV (minor) | Electricity can be generated but urgent repair is required |
| 2 | Category III (marginal) | Reduction in ability to generate electricity |
| 3 | Category II (critical) | Loss of ability to generate electricity |
| 4 | Category I (catastrophic) | Major damage to the turbine as a capital installation |

*Table 6.3 Occurrence rating scale for a wind turbine FMEA*

| Scale no. | Description | Criteria |
|---|---|---|
| 1 | Level E (extremely unlikely) | A single failure mode probability of occurrence is less than 0.001 |
| 2 | Level D (remote) | A single failure mode probability of occurrence is more than 0.001 but less than 0.01 |
| 3 | Level C (occasional) | A single failure mode probability of occurrence is more than 0.01 but less than 0.10 |
| 4 | Level A (frequent) | A single failure mode probability greater than 0.10 |

*Table 6.4 Detection scale for a wind turbine FMEA*

| Scale no. | Description | Criteria |
|---|---|---|
| 1 | Almost certain | Current monitoring methods almost always will detect the failure |
| 2 | Almost impossible | No known monitoring methods available to detect the failure |

It can be concluded from Tables 6.2 to 6.4 that with these gradations, the minimum RPN for any root cause is 1 and the maximum is 32. As long as the rating scales of a selected FMEA procedure remain fixed between alternative wind turbine designs, they can be used for the comparison of those alternatives and identification of critical assemblies. Defining these three criteria tables based on the standard MIL-STD-1629 (1980) is the first step in performing an FMEA. As mentioned before, the basic principles of an FMEA using different standards are similar and simple in the following:

- The system to be studied must then be broken down into its sub-systems, assemblies, sub-assemblies and components.
- Then for each sub-system, assembly, sub-assembly and components all possible failure modes must be determined.
- The root causes of each failure mode must be determined for each sub-system, assembly, sub-assembly and component.
- The end effects of each failure mode must be assigned a level of severity, and every root cause must be assigned a level of occurrence and detection.
- Levels of severity, occurrence and detection are multiplied to produce the RPN.

Therefore, the first stage in the FMEA procedure is obtaining a comprehensive understanding of the wind turbine system and its main assemblies. This is set out in Appendix B of this book based upon the experiences of ReliaWind.

The FMECA will require the designer to define failure modes and root causes for each sub-assembly in the wind turbine. Experience has shown that individual designers can generate a very wide spread of idiosyncratic failure modes and root causes, depending on their individual expertise and knowledge of the wind turbines field operation. The author's experience suggests that it makes the FMECA more meaningful if generic failure modes and root causes are adopted, at least initially, and that these form a standard for the designers to use across the sub-assemblies. A table of generic failure modes and root causes is shown in Table 6.5, which has been used in the author's paper and can be the basis for future development in specific FMECAs.

Software can be used to facilitate the FMECA and other system reliability studies. The author has had experience of the following software packages:

- ReliaSoft XFMEA
- Isograph, Reliability Workbench
- PTC-Relex, Reliability Studio 2007 V2

Users will need to evaluate these packages individually for their own needs. The more sophisticated aspects of the packages allow various forms of reliability modelling to be used, allowing access to database reliability information and discipline for the analysis structure. However, for an FMEA it is perfectly possible to assemble a professional analysis on the basis described earlier solely using an Excel spreadsheet.

There has been one published account of an FMECA applied to a wind turbine Arabian-Hoseynabadi *et al.* (2010) *Int J Electrical Power & Energy Systems*, and an EU FP7 project ReliaWind (2010) proposed the application to individual wind turbine and wind turbine sub-assembly OEMs with preliminary results reported in Wilkinson *et al.* (2011), a full report in ReliaWind (2011) and a detailed application

*Table 6.5 Suggested generic failure modes and root causes for a wind turbine FMEA*

| Failure modes | Failure root causes |
|---|---|
| Structural failure | Design defect |
| Electrical failure | Material defect |
| Mechanical failure | Installation defect |
| Software or control failure | Maintenance defect |
| Insulation failure | Software defect |
| Thermal failure | Corrosion |
| Mechanical seizure | Misalignment |
| Bearing failure | Low-cycle fatigue |
| Component fracture or material failure | High-cycle fatigue |
| Seal failure | Mechanical wear |
| Contamination | Lack of lubrication |
| Blockage | Thermal overload |
| – | Electrical overload |
| – | Weather incident |
| – | Grid incident |

to a common wind turbine type with three different drive trains (Tavner *et al.*, 2010). Other relevant reliability studies on drive trains and their electrical subassemblies are Arabian-Hoseynabadi *et al.* (2010) *Int J Electrical Power & Energy Systems* and Arabian-Hoseynabadi *et al.* (2010) *Renewable Energy*.

A useful analysis from the FMECA results is the occurrence frequency of the different failure modes and root causes. The repetition rate of these limited numbers of failure modes and root causes can be analysed for the wind turbine being considered, and this gives a good ranking for the key root causes to be mitigated and failure modes to be detected. Counting these failure modes and root causes over the whole FMECA can give histograms for each, which has been done identifying the top 10 failure modes and root causes in Figure 6.3.

As can be seen from Figure 6.3, the most significant failure mode was material failure, so improved material quality in wind turbines must be key point for reliability enhancement. It is worth mentioning that these failure mode frequencies are based upon FMEA results not chronological data of wind turbine performance.

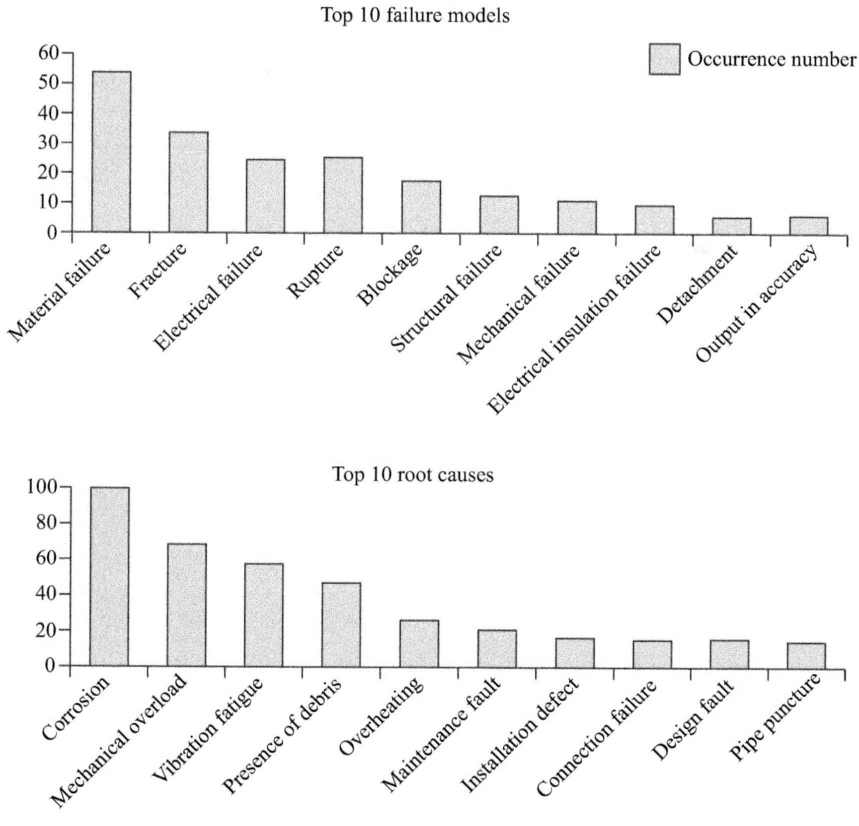


*Figure 6.3 Top 10 failure modes and root causes from the FMEA in Arabian-Hoseynabadi et al. (2010) Renewable Energy using generic examples as in Table 6.5*

Similarly, the most frequent root cause is corrosion, which affects the material quality. This will be more important in future off-shore wind turbines, so remedial design actions in this regard must be considered.

Identifying the most frequent failure modes and root causes will assist design improvement and maintenance optimisation. A cost–benefit analysis for reducing wind turbine failures could be conducted based on a priority list of the most frequent failure modes. A similar analysis could also be considered based on failure modes severity, for example by summating the severity of each failure mode in the FMEA, ranking the results and considering the costs incurred to alter the ranking.

### *6.3.4 Integrating design techniques*

The author's proposed method of integrating these previously mentioned design techniques during design, pre-production and production tests for an off-shore wind turbine is shown in Figure 6.4.

This is based upon the construction of a pre-production prototype off-shore wind turbine, development of an integrated SCADA/CMS system for the off-shore wind turbine, construction of a production prototype and construction of production machines. The design process needs to be integrated by a process of testing, data

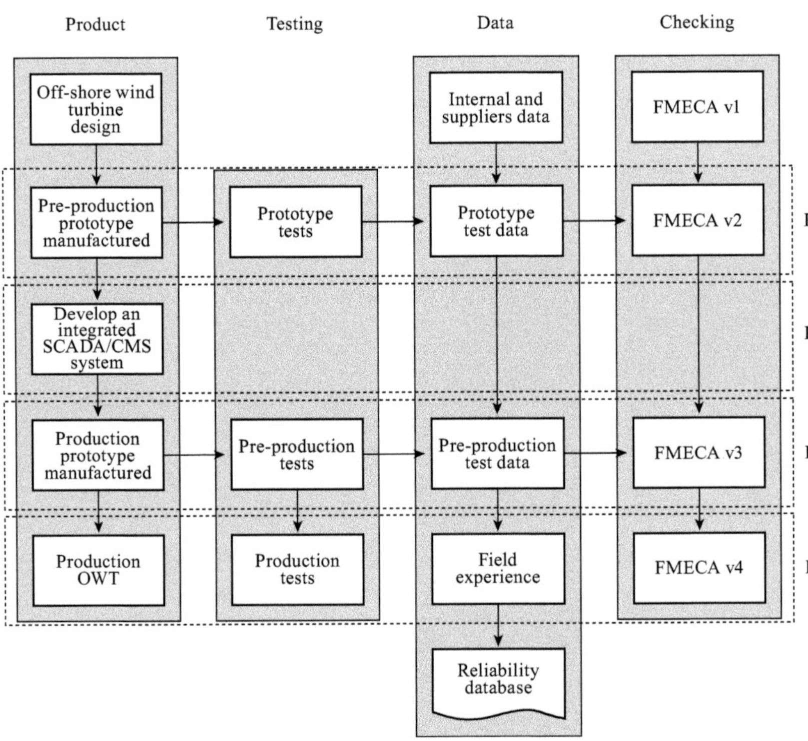


*Figure 6.4 Proposal for using FMEA as an off-shore wind turbine design review tool during design and manufacture*

collection and checking. In this case the FMECA document is used as a means to check progress.

This process would need to be extended after the design and prototype building phase to include commissioning and operations as shown in the next chapter.

## 6.4 Testing techniques

### *6.4.1 Introduction*

Section 6.2.3 emphasised the importance of testing as a further means to raise wind turbine reliability. Whole testing is intended to raise the reliability of components by lowering failure rate, $\lambda$, or increasing MTBF, see the bathtub curve, Figure 2.4, repeated here in Figure 6.5(a).

Pre-production testing would eliminate the early failures seen on the left of Figure 6.5(b). However, this testing can be broken down into a number of different activities at different stages of the off-shore wind turbine design, as described in the following sections.

### *6.4.2 Accelerated life testing*

Accelerated life testing (ALT) is aimed at measuring component and sub-assembly failure rates in a controlled test environment, whose conditions can be varied in such a way that the ageing process is accelerated. The acceleration of the test is achieved by applying a stress which is greater than that encountered in service but not beyond technological limits. This shortens the time to failure but without altering the failure mechanism which is assumed to be activated selectively by the increased stress producing an acceleration factor, *Acc* (Birolini, 2010).

ALT aims to collect reliability data for individual components or sub-assemblies to be used in reliability analysis to lower the intrinsic failure rate, $\lambda$, of a whole system, along the lines shown in Figure 6.5(b). ALT was originally developed for electronic components, where many of the ageing processes are driven by temperature, so acceleration is achieved simply by raising temperature and evaluating *Acc* using Arrhenius' rate law but has been expanded for use with electro-mechanical sub-assemblies.

The object is to derive detailed life reliability curves for individual components or sub-assemblies, as shown in Figure 2.5, in the environmental conditions they are likely to encounter in service, and for off-shore wind turbines, this should cover

- high or low temperatures;
- high humidity;
- saliferous atmosphere.

ALT can provide core data for reasonable component and sub-assembly reliability predictions to be used in an FMECA for a prospective off-shore wind turbine design.

Without ALT, designers need to obtain data from free or commercial data-bases, sometimes available from wind turbine OEM sub-suppliers. Release of such data can be part of the sub-suppliers procurement contracts.

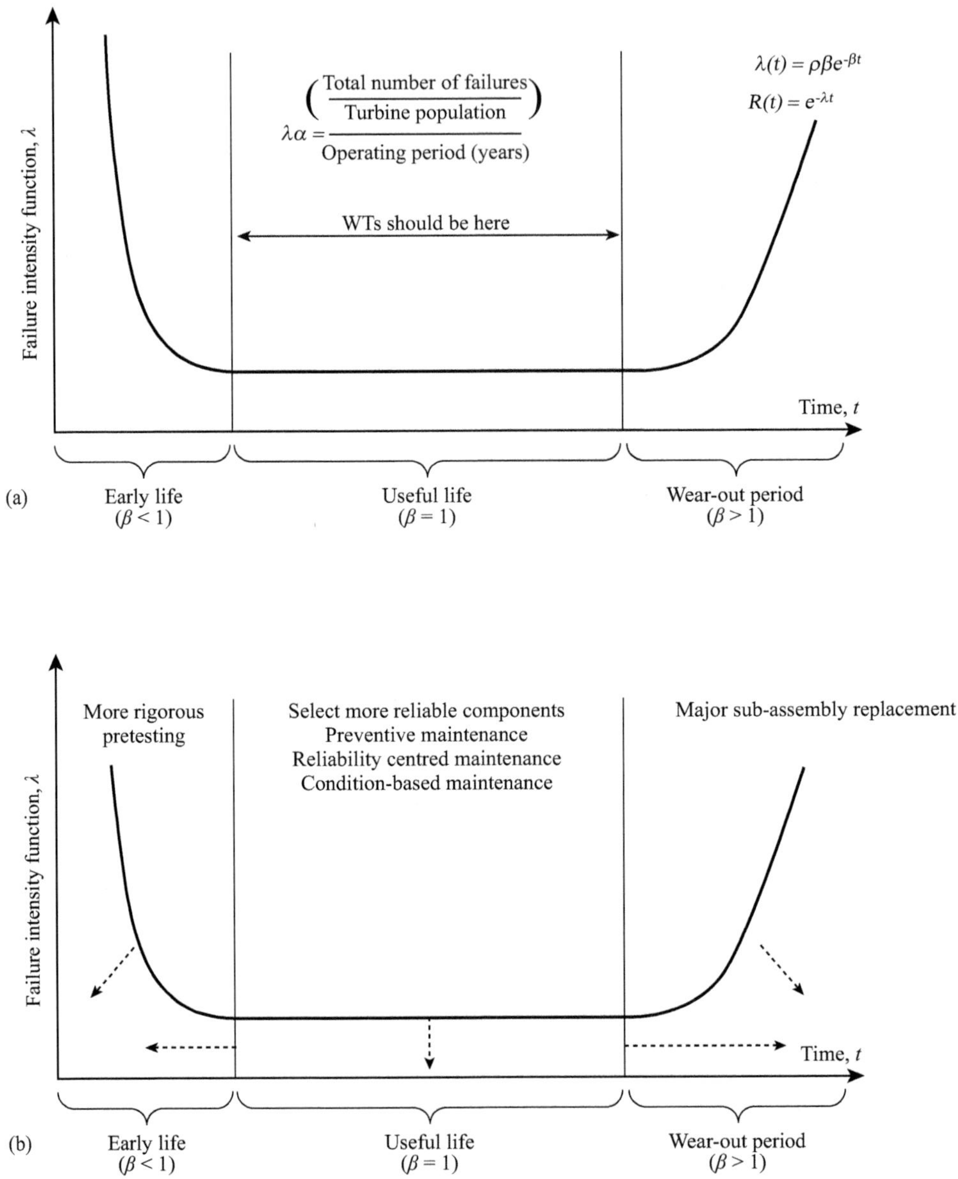


*Figure 6.5 (a) Bathtub curve of failure intensity showing (b) effects on the curve of testing and maintenance*

### *6.4.3 Sub-assembly testing*

In the absence of ALT data, or the ability to do ALT testing, off-shore wind turbine assemblies and sub-assemblies need to be thoroughly pretested in a low cost, benign test bed environment, perhaps at elevated load or temperature, to secure more reliable off-shore deployment.

This will reduce the early life failures at the start of the bathtub, Figure 6.5(b), reducing early failures in service. That process must start with the sub-assemblies most at risk, identified from public domain data, Figures 4.6–4.8, or data available to the off-shore wind turbine OEM and his sub-suppliers from operational experience with previous models. These sub-assemblies may be

- blades programmed for flap-wise and edgewise tests, as can be done at UK's ORE Catapult, see Greaves (2013), Doctoral Thesis;
- mechanical components, such as pitch motion units, lubricating oil systems or hydraulic power-packs; or
- power electronic modules, such as generator- and grid-side Inverters; or even
- controller sub-assemblies, such as the yaw, pitch and generator controllers.

The issue of sub-assembly testing has been shown in other industries, for example in power electronic variable speed drives, to be of particular importance for highly complex electrical and electronic sub-assemblies with high failure rates and low MTTR, with the potential for great reliability improvement. Considerable efforts have been made in the electronics industry to improve sub-assembly reliability through systematic testing and supplier quality control, see Birolini (2007).

Another benefit of controlled sub-assembly testing is that it generates the numerical data, which, when added to that produced from ALT, progressively builds an objective reliability model for a prototype wind turbine and provides the basis for future procurement quality control.

### *6.4.4 Prototype and drive train testing*

Despite the accumulation of data from ALT and sub-assembly testing, there will still be a need to prototype test the off-shore wind turbine itself or at the least major sub-systems. Chapter 5 has shown that that drive train reliability is a major cause for concern, less because of its failure rate than for the excessive MTTR and consequent drive train failure costs.

This is therefore becoming a major development area for off-shore wind turbine OEMs and their drive-train sub-assembly suppliers who are conducting a number of such tests, see, for example Figure 6.6 showing a 2.5-MW drive train under test, exemplifying this trend.

The process is particularly important for off-shore operations, where high off-shore installation and access costs must encourage wind turbine OEMs and developers to reduce subsequent interventions.

Figure 6.7 shows an example of the world's largest planned drive train test rig, valued at >£30M, which is intended to apply the torque and force components expected on modern large off-shore wind turbines to prototype drive trains. Again a major motivation for pursuing this kind of test exercise, which will be costly, is to

- test novel arrangements under known off-shore torque and lateral force transients;
- accumulate test information to inform drive-train sub-assembly testing;

*Figure 6.6* *Drive train test rig mounting a Samsung 2.5 MW drive train at the National Wind Technology Center. Source: National Renewable Energy Laboratory, USA*

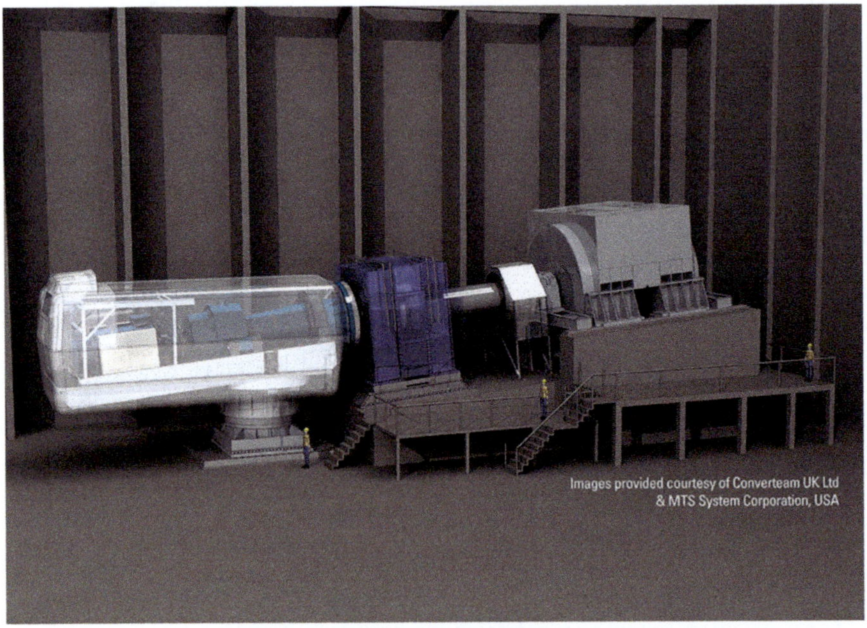


*Figure 6.7* *15 MW drive train test rig. Source: ORE Catapult, UK*

- involve gearbox, generator, converter or hydraulic sub-assembly OEMs in the development of a robust drive train;
- de-risk new drive train concepts.

### *6.4.5 Off-shore environmental testing*

Part of the off-shore situation is ensuring the reliability of parts to exposure to the more difficult ambient environment, from the point of view of temperature, humidity and saliferous atmosphere. This will be mitigated in most new off-shore wind turbine designs by the following:

- Nacelle and tower sealing, particularly at doors, hatches, nacelle and blade motion sites.
- Applying pressurising air treatment to the nacelle and tower.
- All control cabinets and electrical equipment to have heaters operational when shut-down.
- Dehumidifier in nacelle and tower to lower moisture content and protect power electronics and converter technology.

Part of the testing process should include exposure to those conditions, which can be achieved most cheaply at the sub-assembly stage, even if it does not contain the detailed ALT testing referred to in Section 6.4.2.

However, the exposure of combined systems in some pre-production off-shore test sites will generate a degree of experience and data to control the procurement of those sub-assemblies.

The off-shore oil and gas industry has shown that an important factor in achieving reliability in the harsh off-shore is by ensuring that the interfaces between pretested sub-assemblies, wiring, pipe work and junction boxes, are of the highest physical quality using stainless steel enclosures and high-quality armoured wiring as shown in Figure 6.8.

### *6.4.6 Production testing*

Considerable attention has focused on gearbox reliability and some gearbox OEMs are routinely back-to-back production testing their products as shown in Figure 6.9.

But Chapter 5 has shown that the converter too is a high-risk sub-assembly, even if its MTTR is lower and Figure 6.10 shows the routine production testing of a large converter prior to despatch.

An important innovation for off-shore wind turbines may be the back-to-back testing of complete nacelles at variable and full power before despatch from the factory, as was suggested in Spinato *et al.* (2009).

However, off-shore wind turbine OEMs will need to devise efficient means to achieve these processes cost-effectively in a timely way.

### *6.4.7 Commissioning*

Once an off-shore wind turbine has been installed, testing is not complete until commissioning testing is finished, Figure 6.11. Commissioning testing has an

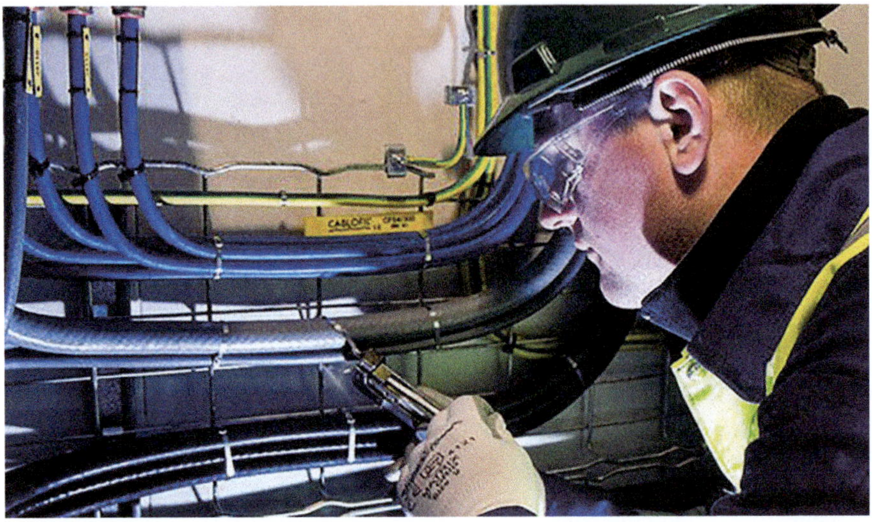

*Figure 6.8 High-quality off-shore wiring and cabling. Source: Cablofil*

*Figure 6.9 Back-to-back testing of two 3-MW wind turbine gearboxes. Source: Hansen Transmissions*

*Figure 6.10 Production testing of a large wind turbine converter. Source: ABB Drives*

*Figure 6.11 Off-shore wind turbine commissioning Horns Rev 1, showing crew landing baskets. Source: ABB Drives*

important further part to play in identifying early failures and resolving them early in operational life. High-quality commissioning also plays a major part in the accurate setting of alarms of SCADA and CMS systems, which is crucial for the reliable operation of the off-shore wind turbine but requires considerably more resource to execute effectively in the off-shore environment.

## 6.5 From high reliability to high availability

### *6.5.1 Relation between reliability and availability*

The relationship between reliability and availability is shown in Figure 1.11, and the relationship of reliability and availability to cost of energy is shown in Figure 1.15. The processes described in this chapter are designed to ensure the prospective availability of an off-shore wind turbine in a wind farm. But these processes cannot deliver high wind farm availability without additional support to maintain reliability in service. High availability in service depends upon installing a high reliability off-shore wind turbine, as described earlier, and then

- the off-shore environment itself, including access to the asset;
- the ability to detect and interpret low reliability in service;
- planned preventative and corrective maintenance in response to that detection and interpretation;
- a programme of asset management based on the previously mentioned ones to consider the through life performance of the asset.

### *6.5.2 Off-shore environment*

The environment plays an enormous part in our achievement of good performance. Off-shore wind resource is strong but can also adversely affect performance because gusts and turbulence can damage the wind turbines, and higher wind speeds lead to raised wave height limiting access.

### *6.5.3 Detection and interpretation*

An off-shore wind turbine is a remote unmanned robotic power generation unit. Good availability performance cannot be achieved unless we can remotely and accurately detect performance deterioration and interpret it prior to action. Therefore, the installation of reliable and accurate SCADA and CMS systems will be essential to achieve this part of the off-shore mission. It will be vitally important that the data arising from detection and interpretation are fed back into the off-shore wind farm management system.

### *6.5.4 Preventive or corrective maintenance*

The action needed from operation and detection of performance deterioration is an organised programme of maintenance.

This must include OPEX actions for preventive maintenance, based upon the off-shore wind turbine design, and corrective maintenance, driven by SCADA and CMS detection.

The results of maintenance must also be fed back into the database of reliability information for the off-shore wind farm.

### 6.5.5 Asset management through life

Finally the whole wind farm asset will need to be managed holistically against the energy produced, not only to justify the ongoing OPEX needed to maintain performance, but also to plan for the large-scale CAPEX to maintain the asset over its planned life, including the longer term deterioration and planned replacement of larger sub-assemblies, such as blades, gearboxes and generators.

## 6.6 Summary

This chapter has described techniques for improving the reliability of wind turbines during the design and manufacturing processes, including design review, FMEA/FMECA and testing. It has shown how these can be coupled together during the prototype process through testing of sub-assemblies and prototype turbines leading to the development of an RCM plan for full deployment of the product. The key to this is the availability and the generation of reliability data from these processes in a database, including

- prior design field reliability data;
- accelerated lifetime testing;
- sub-assembly suppliers data;
- prototype test data;
- pre-production test data;
- commissioning test data;
- maintenance logs;
- SCADA/CMS in-service data.

Finally this chapter has shown the connection between design for reliability and pretesting and real operational availability, demonstrating what is needed to deliver low cost of energy from operational off-shore wind farms. This chapter will also demonstrate our early experience with off-shore wind farms, and the later chapters will address individually the points raised in Section 6.5 to put these lessons to work to improve our performance in the field.

# *Chapter 7*
# Early off-shore unreliability and availability

## 7.1 Early European off-shore wind farm experience

### *7.1.1 Horns Rev 1 wind farm, Denmark*

The first large off-shore wind farm in the world, consisting of 80 Vestas V80-2 MW wind turbines each of 5,027-$m^2$ swept area, was completed in 2002 in 6–14 m of water in the North Sea at Horns Rev, 14–20 km off the West Jutland coast of Denmark.

The project was managed by the West Danish utility Elsam, then DONG Energy, now Oersted and the wind farm is connected via an off-shore substation using 30-kV AC collector cables and to shore from the substation via a 150-kV AC export cable. Maintenance of the wind farm was conceived on the basis of using helicopters' access to individual wind turbine nacelles via specially designed access platforms built onto the nacelles to accommodate drops and lifts from the Eurocopter EC135.

There were many difficulties in the early operation of the wind farm arising from the process of installing and commissioning the wind turbines and from the use of the V80 wind turbine, which had previously been used largely on-shore. Table 1.2 demonstrates the scale of the challenge undertaken by the industry at Horns Rev.

The problems arose from some aspects of the wind turbine and wind farm design, as follows:

- Wind turbine dry-type transformers, installed in the wind turbine nacelles, experienced winding failures, due to seismic vibration fretting.
- Vibration and other damages to the DFIG generators.
- Gear and bearing damages to the wind turbine gearboxes.
- Problems with the wind turbine pitch control systems.
- Subsequent problems with the collector and export cable arrays.

These difficulties lead to a large number of commissioning and post-commissioning visits to individual wind turbines and the replacement of some wind turbine gearboxes and generators.

The situation worsened and the entire fleet of 80 V80 nacelles were returned to shore for full refurbishment, although this drastic decision was almost certainly facilitated by the proximity of Horns Rev 1 to the Vestas manufacturing plants and the fact that the manufacturer, developer and operator were of the same nationality.

However, many lessons were learned from this early large off-shore wind farm experience.

### *7.1.2 Round 1 wind farms, United Kingdom*

The following five United Kingdom (UK) Rounds 1 and 2 wind farms are considered:

- North Hoyle, operational in July 2004, 30 Vestas V80-2 MW wind turbines of a 5,027-$m^2$ swept area, in 7–11-m water depth, 9.2-km off-shore in the Irish Sea, operated by RWE Npower Renewables.
- Scroby Sands, operational in January 2005, 30 Vestas V80-2 MW wind turbines of a 5,027-$m^2$ swept area, in 5–10-m water depth, 3.6-km off-shore in the North Sea, operated by E.ON Climate & Renewables GmbH.
- Kentish Flats, operational in January 2006, 30 Vestas V90-3 MW wind turbines of a 6,362-$m^2$ swept area, in 5-m water depth, 9.8-km off-shore in the English Channel, operated by Vattenfall.
- Barrow, operational in July 2006, 30 Vestas V90-3 MW wind turbines of a 6,362-$m^2$ swept area, in 15–20-m water depth, 12.8-km off-shore in the Irish Sea, operated by Centrica/DONG Energy.
- Teesside, operational in 2013, 27 Siemens SWT-2.3-93 2.3 MW wind turbines of a 6,800-$m^2$ swept area, in 35-m water depth, 1.5-km off-shore at Redcar, operated by EdF.

It is clear from Table 1.2 and the previous information that these five, smaller off-shore wind farms were less challenging in location than the larger Horns Rev 1, although the distance off-shore and water depth at Barrow were similar to Horns Rev 1.

It is also clear that considerable experience had been gained between 2002 and 2006 in deploying, commissioning and successfully operating off-shore wind farms.

All the difficulties recorded at the UK Round 1 wind farms, with the possible exception of V90 wind turbine pitch system problems, replicate those Horns Rev 1 experiences, although they seem to be of lesser magnitude and no complete nacelles had to be replaced, although there was a learning curve operating the Vestas V90 and SWT-2.3-93 wind turbines.

The operational problems at four of the five wind farms were set out in Feng *et al.* (2010) and are summarised from the published operational reports, while the results for the last wind farm were summarised by Papatzimos *et al.* (2019) and Papatzimos (2020), Doctoral Thesis.

#### 7.1.2.1 Scroby Sands (V80s)

In 2005, there was substantial unplanned work attributed to minor commissioning issues, corrected by remote turbine resets, local turbine resets or minor maintenance work, mostly resolved within a day. A smaller number of unplanned works involved larger scale plant problems with more serious implications, the primary cause being gearbox bearings.

In 2005, 27 generator side intermediate speed shaft bearings and 12 high-speed shaft bearings were replaced. A number of reasons for the gearbox bearing damage were identified related to the bearing designs.

In 2005, four generators were replaced with generators of alternative design.

In 2006, unplanned work involved three outboard intermediate speed shaft gearbox bearings, nine high-speed shaft gearbox bearings and eight generator failures. Generating capacity was also significantly reduced for 2 months when one of the three transition joints in the cable to the beach failed.

In 2007, problems experienced with the generators were resolved by replacing all original generators with a generator of proven design. The gearbox bearing issue was managed in the short term by a proactive replacement of the outboard intermediate speed bearings; in addition, 12 high-speed shaft bearings were identified as worn during routine internal inspections and proactively replaced before failure. Three gearboxes were also identified as requiring replacement. Capacity was also affected by a transition joint failure in another cable to the beach; commissioning tests also identified a fault in the subsea portion of the cable, for which replacement was planned for spring 2008.

Scroby Sands wind farm is mounted on large sand banks, North of Great Yarmouth, formed by longshore drift. As a result, the wind farm experienced some scour around its monopile foundations, initially detected by changes in the tower natural vibration frequencies. This has been mitigated by dropping rock ballast around the foundations.

#### 7.1.2.2 North Hoyle (V80s)

In 2004–05, unplanned work involved an HV cable fault, generator faults associated with cable connections and supervisory control and data acquisition (SCADA) electrical faults.

In 2006 the following issues arose:

- Two generator bearing faults.
- Six gearbox faults.
- An unplanned grid outage.
- Preparation and return of turbines to service further extended downtime.
- Downtime owing to routine maintenance and difficulties in the means of access to the turbines.

In 2007, the following issues arose:

- Four gearbox bearing faults and chipped teeth resulting in gearbox replacements delayed by the lack of a suitable maintenance vessel.
- Two generator rotor cable faults.
- Two circuit breaker failures.
- One cracked hub strut.
- One turbine outage for yaw motor failures.
- An unplanned grid outage.
- Again downtime owing to difficulties in the means of access.

#### 7.1.2.3 Kentish Flats (V90s)

In 2006, there was substantial initial unplanned work attributed to minor commissioning issues corrected by remote turbine resets, local turbine resets or minor maintenance work.

Other unplanned work involved larger scale plant problems and included the following:

- Main gearbox.
- Generator bearings.
- Generator rotor cable connections from the slip-ring unit.
- Pitch system.

  The generator bearing and rotor cable problems were prolonged as the generator sub-supplier undertook the repairs to avoid jeopardising the warranty.

  The first main gearbox damage was detected in late 2006 and an intensive endoscope campaign revealed that 12 gearboxes required exchange. In 2007, all 30 gearboxes were exchanged owing to incipient bearing failures in the planetary gear. The exchange programme was scattered over the years, and due to waiting time and the lack of a crane ship, the outages were longer than the repair time. About half of the generators were refurbished owing to the following:

- Damage on internal generator rotor cable connections.
- Shaft tolerances.
- Grounding of bearings to avoid current passage.

Other unplanned tasks included the following:

- Pitch system repair.
- Blade repair on one turbine due to crane impact during gearbox exchange.

#### 7.1.2.4 Barrow (V90s)

In 2006–07, unplanned work on the turbines was substantial although some issues were minor, solved by a local reset or minor work to the turbine. Other larger issues were as follows:

- Generator bearings failed and replaced with a new type.
- Generator rotor cables replaced with a new type.
- Pitch systems modified.

Owing to gearbox problems seen on other turbines of the same type, an inspection process commenced in 2007 showing a few gearboxes beginning to show similar problems. It was decided proactively to replace gearboxes before failure and this started in July 2007 completing in October 2007.

#### 7.1.2.5 Teesside (SWT-2.3-93s)

Papatzimos *et al.* (2019) considered the reliability of this site, showing failure distribution of the 27 Siemens SWT-2.3-93 2.3 MW wind turbines over a 5-year period but not commenting on the layout or connection reliability.

### 7.1.2.6 Egmond aan Zee, the Netherlands

The Egmond aan Zee wind farm in the Netherlands consists of 36 Vestas V90-3 MW wind turbines of a 6,362-m$^2$ swept area, in 17–23-m water depth, 10–18-km off-shore in the North Sea, went operational in April 2007. The wind farm is operated by NoordzeeWind, a joint venture between utility company Nuon, now part of Vattenfall, and Shell.

The location of Egmond aan Zee can be considered to be as challenging as Horns Rev 1; however, Egmond aan Zee has an operational advantage in being so close to a maritime centre at the mouth of the River Ij at Ijmuiden.

Reliability analysis for the Egmond aan Zee wind farm used information taken from operational reports in its first 3 years of operation recording 'stops' and not 'failures' (Feng *et al.*, 2010), and the results are shown in Figure 7.1.

It can be seen from Figure 7.1(a) that there were a significant number of stops associated with the turbine control system; however, the average downtime per stop data, Figure 7.1(b), shows that the control system stops must be easily reset as the downtime was short.

Figure 7.1(b) shows high gearbox and generator downtimes, although these components have a relatively low stop rate. This combination leads to very high

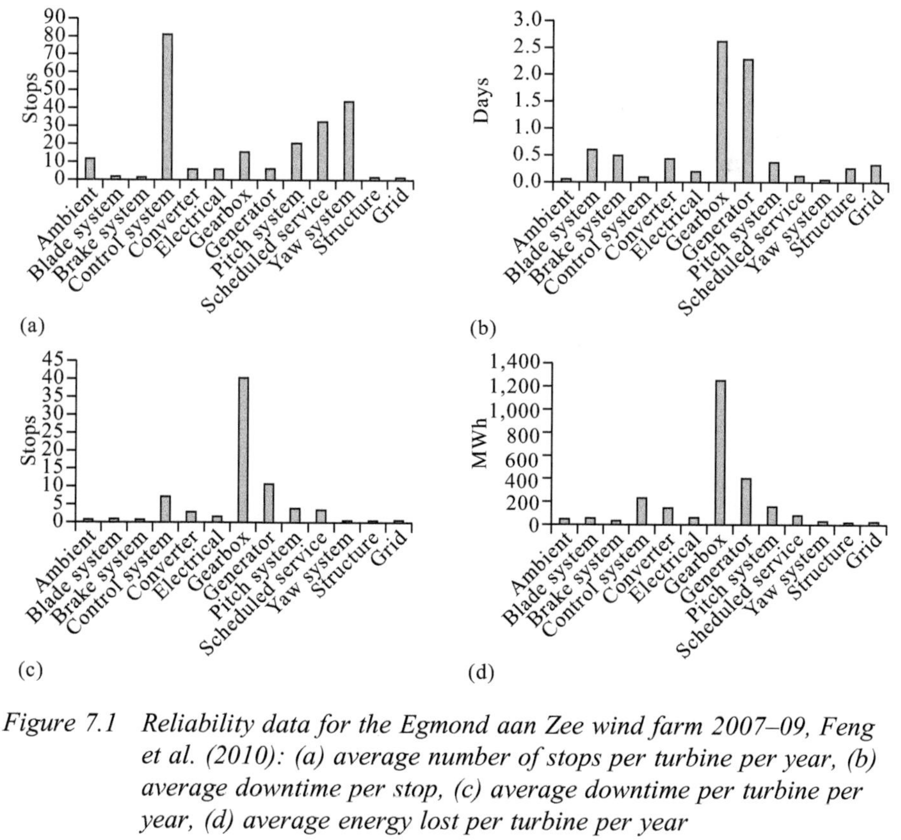


*Figure 7.1 Reliability data for the Egmond aan Zee wind farm 2007–09, Feng et al. (2010): (a) average number of stops per turbine per year, (b) average downtime per stop, (c) average downtime per turbine per year, (d) average energy lost per turbine per year*

average downtime per turbine per year. This is confirmed in Figure 7.1(c) where the gearbox is shown to have a significant effect on the turbine availability. The V90-3 MW turbines have been subject to an extensive gearbox replacement programme due to type faults and some generator replacements, reflecting the experience at Kentish Flats and Barrow in the UK, which also operated the V90-3 MW wind turbines.

These programmes will have a significant effect on the downtime figures.

As expected, the average energy lost per turbine per year, Figure 7.1(d), is closely correlated with the average downtime per turbine per year.

## 7.2 Lessons learnt

### *7.2.1 General*

The information on early experience was available from off-shore wind farms operating Vestas wind turbines although Siemens turbines were used at Teesside, whereas information on other makes of wind turbines, particularly Siemens, is now also becoming more available, see Section 7.2.5.

The author makes a side comment that it is surprising, how little information manufacturers and operators gave away to the future industry in these reports, considering the size of government subsidies that facilitated the building of these wind farms.

But an interesting conclusion from Horns Rev 1, the UK Round 1 and Egmond aan Zee off-shore experiences is that the failure modes do not seem markedly different to those described in Chapters 4 and 5 from on-shore wind turbines. There seem to be a few, new and unexpected failure modes associated with the off-shore environment, except those due to the off-shore AC connector cable arrays and the AC export cables.

There were few problems with blades and a large number of problems associated with gearboxes, generators, pitch systems and the turbine control, almost certainly aggravated by the relatively low off-shore operational experiences with these two makes of wind turbines.

However, it is clear that the root causes of failure were exacerbated by off-shore operations and conditions, for example due to the following:

- High wind resource.
- Consequent drive train transient torques arising from that resource variability.
- Wind turbine control system operation.
- Seismic vibration of drive trains.

In addition, it is clear that the most pressing issue for all these wind farms was that of access.

### *7.2.2 Environment*

The effect of the off-shore environment can be most clearly seen by comparison with the effect of wind speed on the availability of a large on-shore wind farm in the United States over a period of 2 years (Figure 7.2).

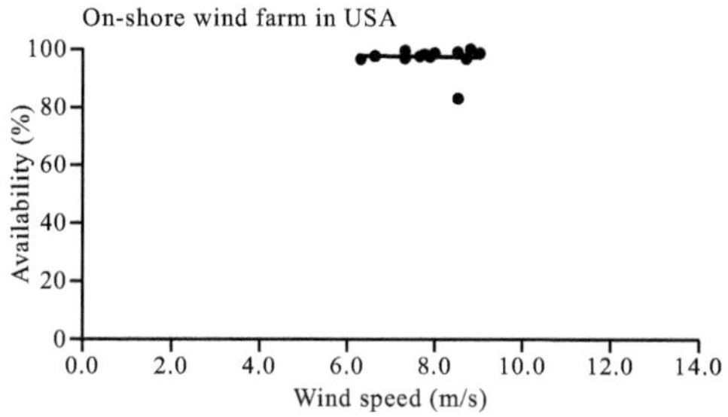


*Figure 7.2 Effect of wind speed on wind turbine capacity factor from a large on-shore wind farm over a period of 1–2 years*

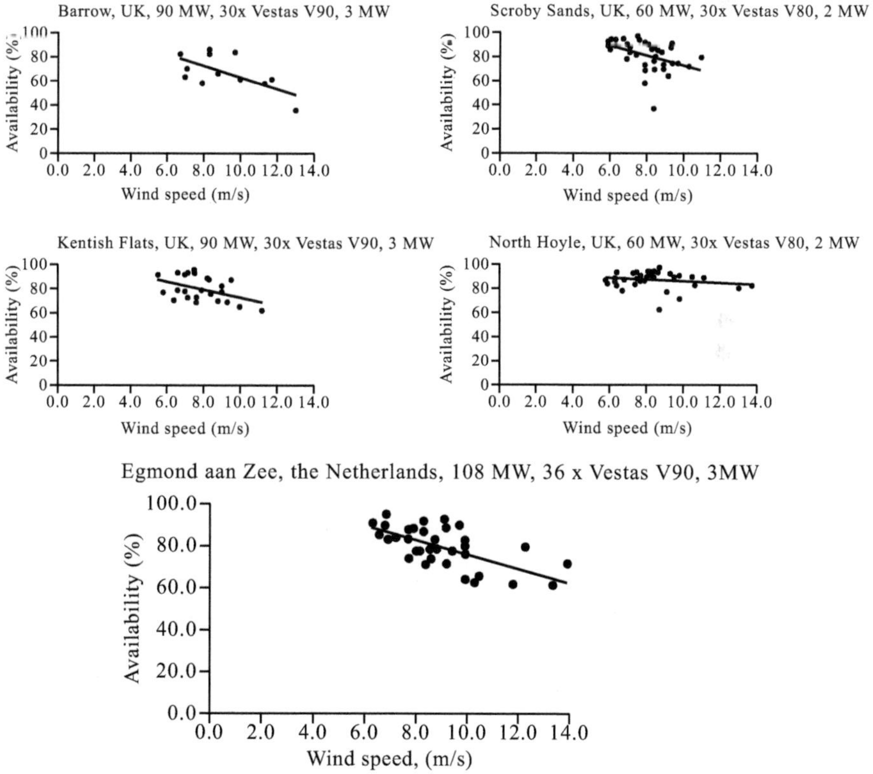


*Figure 7.3 Effect of wind speed on wind turbine capacity factor from six off-shore wind farms over a period of 1–2 years*

Then consider in Figure 7.3 the effect of the off-shore environment, in particular wind speed to the same scale as Figure 7.2, on the capacity factors of five of the off-shore wind farms considered earlier. Figure 7.3 shows the much increased range of wind speed available to the off-shore wind farms but a drop in capacity factor as wind speed rises, although this is less marked for one wind farm.

The importance of achieving high availability at high wind speeds off-shore is exemplified by Figure 7.4, taken from Feng *et al.* (2010) and Harman *et al.* (2008), which shows availabilities from a large database of on-shore wind turbines and confirms from the energy curve that 40% is available at wind speeds >11 m/s. Feng *et al.* (2010) and Castro Sayas *et al.* (1996) also consider wind farm availability.

It is not clear whether the drop in capacity factor is due to increased outages from higher wind speeds or the fact that already defective wind turbines cannot be repaired at higher wind speeds due to limited access. It is likely to be a combination of both.

These data from five of the off-shore wind farms described earlier can then be compared in Figure 7.5 with the predicted capacity factors for the V80 and V90

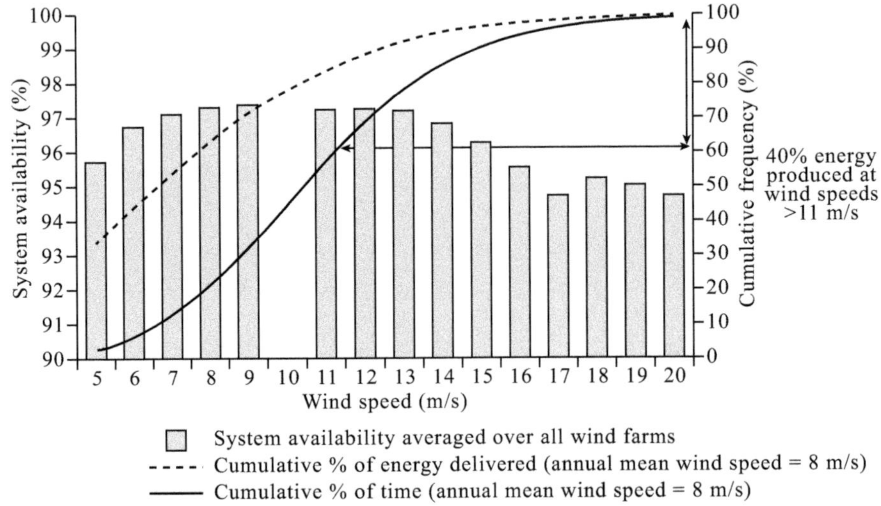


*Figure 7.4 Availability of on-shore wind turbines showing the fall in availability at higher wind speeds. Source: GL Garrad Hassan, Feng et al. (2010) and Harman et al. (2008)*

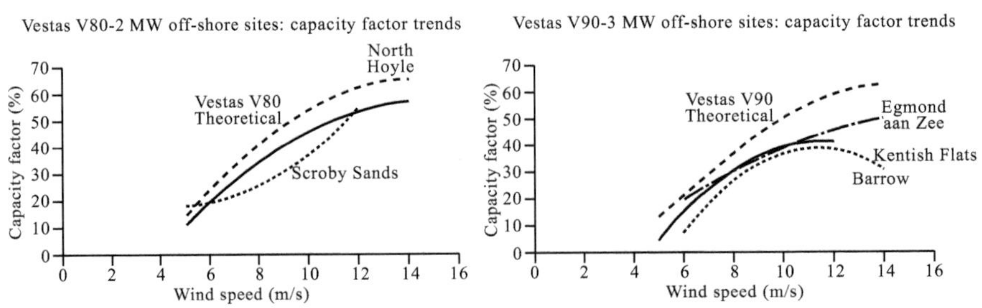


*Figure 7.5 Predicted capacity factors for Vestas V80 and V90 wind turbines compared with measured values achieved at five off-shore wind farms*

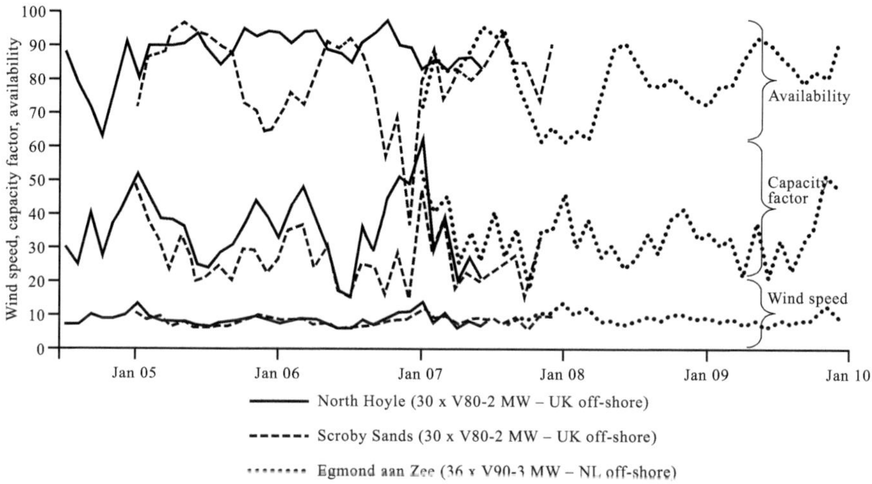


*Figure 7.6 Summary of three off-shore wind farms with Vestas wind turbines in early operation*

wind turbines. It is clear that both types of wind turbines are performing reasonably well compared to their theoretical capability, with some fall-off at higher wind speeds, which deserves further investigation.

The overall environmental effects are summarised from 2004 to 2009 in Figure 7.6, showing the average wind speed, capacity factor and availability for North Hoyle, Scroby Sands and Egmond aan Zee off-shore wind farms, situated, respectively, in the Irish Sea, North Sea west coast and North Sea east coast. The grey bands highlight the operational winter seasons, from October to March.

Figure 7.6 shows that the average monthly wind speeds at the three sites were similar, but capacity factors peaked in the highlighted winter seasons, due to higher wind speeds, while availability dipped during those periods. However, a close observation of Figure 7.6 shows that at North Hoyle, where Feng *et al.* (2010) and perhaps Figures 7.3 and 7.5 suggested an emphasis on operations and maintenance (O&M), those winter dips in availability were not so severe. This implies that if maintenance and repair are appropriately managed, it is possible to profit from the strong winter capacity factors without a drastic fall in availability. However, these issues depend upon good access and planned maintenance.

There are few records of environmental failure modes, due to corrosion or marine fouling, from the wind farms considered previously, although these were well known from earlier, older off-shore wind farms. Perhaps these issues will arise as the wind farm life progresses but do not appear to be root causes in the first 3 or 4 years of wind farm life.

### *7.2.3 Access*

The access problems recorded, exemplified in Figure 7.5 in the winter periods by the lower availabilities, were all especially severe when dealing with major repairs, such as the changing of generators and gearboxes, leading to large delays and consequent loss of generated energy.

These access issues were not necessarily mitigated by the use of helicopter access at Horns Rev 1, which proved costly and involved personnel difficulties. However, it is believed that these difficulties can be resolved and many recent off-shore wind turbines are now fitted with helicopter drop and lift access platforms, see Figure 6.11, like those used at Horns Rev.

### *7.2.4 Off-shore LV, MV and HV networks*

#### 7.2.4.1 Substations

The off-shore substation at Horns Rev 1 proved to be a success.

#### 7.2.4.2 Collector cables

One UK Round 1 site experienced problems with a collector cable array and other sites experienced problems with some collector cables.

Some of these difficulties arose because buried cables became exposed and then subject to damage from fishing or anchor activity. But some difficulties have arisen because of damage to collector cables due to subsequent construction activity on the wind farm site due to the activity of jack-up vessels.

#### 7.2.4.3 Export cables

Only Horns Rev 1 had an HV export cable and this did experience some problems.

One UK Round 1 site experienced problems with a transition joint in cables coming to shore.

### *7.2.5 Other Round 1 wind farms, United Kingdom*

Whilst Figure 7.6 shows the performance of three wind farms with Vestas wind turbines, there are relatively little data in the public domain about other wind turbine makes. However, the UK Rounds 1 and 2 are now operating with Siemens SWT-3.6 wind turbines with an induction generator drive and fully rated converter, Type D in Figure 4.2. Capacity factor information is available for at least four of these wind farms, summarised as follows, with results shown in Figure 7.7:

- Burbo Bank, operational in July 2007, 25 Siemens SWT-3.6-107-3.6 MW wind turbines of a 9,000-m$^2$ swept area, in 0–8-m water depth, 7-km off-shore in the Irish Sea, operated by DONG Energy.
- Lynn, operational in April 2008, 27 Siemens SWT-3.6-107-3.6 MW wind turbines of a 9,000 m$^2$ swept area, in 5–10-m water depth, 5.2-km off-shore in the North Sea, operated by Centrica.
- Lynn and Inner Dowsing, operational in June 2008, 27 Siemens SWT-3.6-107-3.6 MW wind turbines of a 9,000-m$^2$ swept area, in 5-m water depth, 5.2-km off-shore in the North Sea, operated by Centrica.
- Rhyl Flats, operational in July 2009, 25 Siemens SWT-3.6-107-3.6 MW wind turbines of a 9,000-m$^2$ swept area, in 4–15-m water depth, 8-km off-shore in the Irish Sea, operated by RWE Npower Renewables.

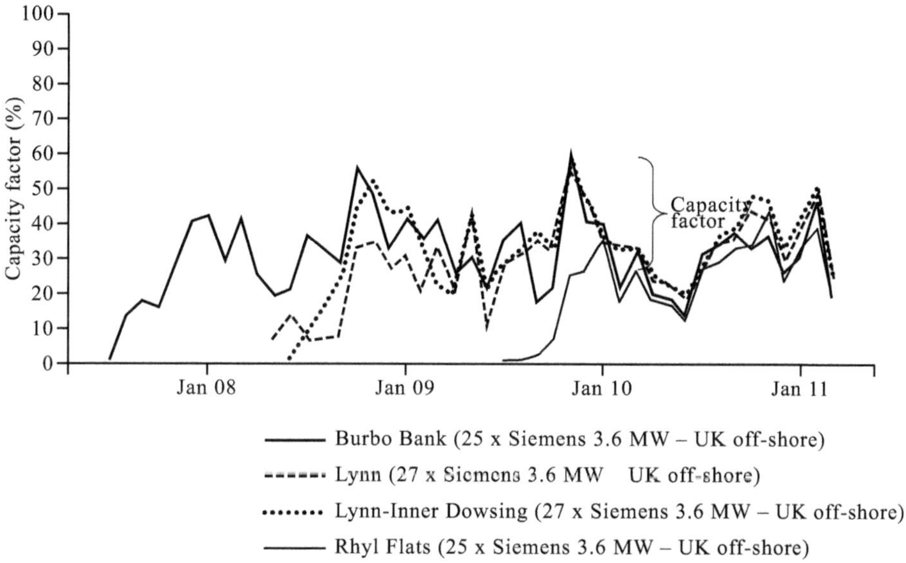


*Figure 7.7 Summary of four UK off-shore wind farms with Siemens 3.6 MW wind turbines in early operation*

Figure 7.7 shows capacity factors rising to high values in the winter seasons, similar to Figure 7.6. The most interesting feature is how the performance of the wind farms improves progressively in the first 3 years of operation, exemplified by all four off-shore wind farms in this case. This is the result of effective early commissioning.

## *7.2.6 Commissioning*

High-quality commissioning proved an important early experience lesson from off-shore wind farms for Horns Rev 1, UK Round 1 and at Egmond aan Zee and is exemplified by the improving capacity factors during early operation seen in Figure 7.7.

A notable feature mentioned in a number of the early operating reports was that many early faults were corrected by remote or local turbine resets and that there were a number of SCADA electrical faults, and remote and local turbine resets also had to be adjusted to ensure reliable operation.

## *7.2.7 Planning off-shore operations*

To plan off-shore O&M and avoid the issues that arose in our early European experience, an approach needs to be adopted for O&M similar to that proposed in Figure 6.4 for design and manufacture. This is shown in Figure 7.8 and the basis is to link that work to the planning of maintenance through the reliability-centred maintenance plan. Again the key to success for Figure 7.8 is that O&M must be soundly based upon data generated from the wind farm in operation, just as Figure 6.4 was based upon data generated during design and manufacture.

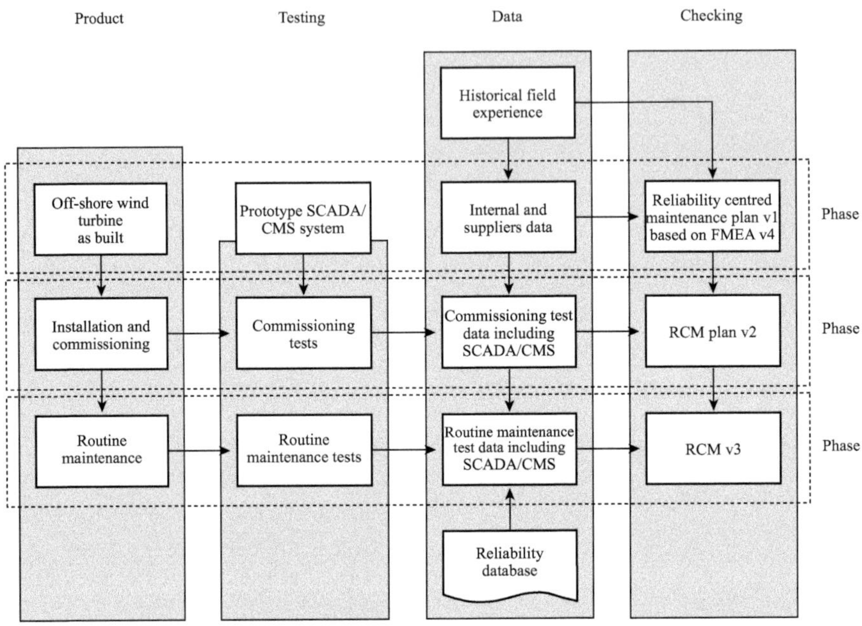


*Figure 7.8 Proposal for using FMEA and RCM as off-shore wind turbine review tool for O&M*

## 7.3 Summary

This chapter has shown Europe's early experience of operating off-shore wind farms. The experience available in the public domain has been limited to wind farms operating Vestas V80 and V90 wind turbines but has identified some important general lessons, the major ones being as follows:

- On-shore wind turbines do experience problems in the off-shore environment; however, many of the failure modes off-shore were similar to those that arose in on-shore experience.
- Thorough pretesting of the subassemblies and of the wind turbine, designed for off-shore operation, are necessary preliminaries to de-risking off-shore operation.
- Thorough and efficient commissioning of the wind turbines in the off-shore wind farm lowers the risk of subsequent problems.
- Thorough preparations of off-shore access facilities, both at the shore base and wind farm, are essential de-risking preliminaries to off-shore O&M.

The next chapter will show how layout and grid connections affect this picture.

Later chapters will show how SCADA and CMS monitoring can assist in producing the data to solve the problems of this early experience and turn wind turbine reliability into wind farm availability and a lower cost of energy.

# Chapter 8
# Off-shore wind farm layouts and grid connection

## 8.1 Introduction

As the size of off-shore wind farms increase and we move from fixed to floating structures, the reliability of individual off-shore wind turbines becomes less important than the overall reliability and availability of the off-shore power stations that they form.

In turn, this depends on their layouts and grid connections, the subject of this chapter.

We have seen that agreed designs for off-shore wind, wave and tidal arrays, layouts and grid connections have been proposed and are being deployed but we have not yet achieved consensus on ideal arrangements because of the following:

- We need to resolve when devices should be fixed to the seabed and when floating and moored is a better option.
- There has been a lack of consensus on the most applicable marine off-shore renewables extraction technologies, which includes off-shore wind.
- These all are dependent on local marine resources, orography, geography and power network considerations plus the availability of the skills and assets to deploy them.

The most developed installed marine power off-shore arrays have been deployed at test facilities around the world. These arrangements to date have been aimed at technology demonstration, rather than maximising power generation and transmission, fulfilling the demonstration mission. However, these test facilities, listed in Figure 8.1, do demonstrate some prospective off-shore wind site arrangements with the least understood reliabilities and availabilities:

- Denmark:
  - LORC wind turbine nacelle test facilities:
    - 25-MW test;
    - 16-MW test;
    - 14-MW test.
  - Aalborg University – Wave and Current Basin
- France:
  - Lir NOTF – Deep Ocean Basin;

(a)

(b)

(c)

(d)

*Figure 8.1* *Testing Facilities around the World: (a) United Kingdom, ORE Catapult Testing Facilities, Blyth, Northumberland; (b) United Kingdom, ORE Catapult Levenmouth Demonstration Turbine, Fife, Scotland; (c) United Kingdom, Billia Croo Wave Energy Test Site, Orkney, Scotland; (d) United States, National Wind Testing Centre at NREL, Boulder, Colorado*

  - Lir – Ocean Emulator;
  - Ifremer – Basin of Brest;
  - Sem-Rev off-shore test site, Pays de Loire coast.
- United Kingdom:
  - ORE Catapult, Blyth, Northumberland, facilities:
    - blade test;
    - large drive train test;
    - smaller drive train test;
    - electrical system test facility;
    - off-shore wind test site;
    - 7-MW Levenmouth Demonstration Turbine;
    - grid connection.
  - Vattenfall European Off-shore Wind Deployment Centre, Aberdeen, Scotland:
    - Deploying innovative next generation technology, 11×8.8 MW turbines;
    - Grid connection.
  - European Marine Energy Centre, Orkney, Scotland, facilities:
    - Fall of Warness Tidal Energy Test Site, off the island of Eday;
    - Billia Croo Wave Energy Test Site, Stromness;
    - Both designed with electrical connection points available for a variety of devices;
    - Grid connection.
  - Wave Hub, Hayle, Cornwall, facilities:
    - Deployment of wave and tidal energy devices;
    - Grid connection.
- The USA:
  - National Wind Technology Center (NWTC) at the National Renewable Laboratory (NREL), Boulder, Colorado, see Figure 8.1. NWTC is accredited to perform the following tests in accordance with international standards:
    - acoustic noise to IEC 61400-11 and MEASNET;
    - power performance to IEC 61400-12-1 and MEASNET;
    - mechanical loads to IEC 61400-13;
    - power quality to IEC 61400-21 and MEASNET;
    - duration testing to IEC 61400-2;
    - safety and function to IEC 61400-2 and IEC 61400-22.

These locations were based upon their proximity to scientific resources, access to industrial expertise and readily available existing grid connections and can teach us the necessary power collection arrays for off-shore wind.

However, off-shore wind now has a significant number of different and extensive off-shore arrays, some of the largest being the London Array off the Thames Estuary, and the BorWind and DolWind sites in North Germany, and these arrays could be used to explain some of the issues involved regarding reliability and availability.

## 8.2 Arrangements

Figure 8.2 shows mounting and foundation arrangements for a range of off-shore fixed or floating wind turbines, based on Musial *et al.* (2004), which helps understanding of the basis for off-shore wind farm and foundation or mooring architectures.

The differences between the six arrangements shown in Figure 8.2 affect the size of the off-shore wind farm and will dictate the cable collection arrangement for the following reasons:

- Fixed off-shore wind turbines can be arranged in a tighter wind farm as there is no wind turbine movement between turbines to be accommodated.
- Floating and moored wind turbines need to be in a wider arrangement to allow for the spread of their mooring anchors and for the floating structure to adjust itself in the seaway; however, the tension leg arrangement allows closer arrangements.
- Floating and moored wind turbines also require greater shipping clearance, because of their movement and to avoid the risk of damage to mooring lines and flexible cables from other maritime actions, as ships come to anchor near to a wind farm.

## 8.3 Device and collector arrays

Device and collector array layouts depend upon the resource and the location, see Lundberg (2006), PhD Thesis.

(a) (b)

*Figure 8.2 Examples of off-shore wind turbine foundation and mooring arrangements, on the left fixed structures, monopile, jacket or twisted jacket (a), on the right floating structures, semi-submersible, tension leg or spar buoy (b), US Office of Energy Efficiency (2017)*

Some prospective off-shore arrangements, developed for wave or tidal marine energy converters, which could be applied to fixed or floating and moored wind turbines, are illustrated in Figure 8.3 showing the following arrangements:

- Full string cluster: all the wind turbines are strung on one cable. This offers no redundancy but is cheap and is being adopted for early small marine renewable arrays but not for off-shore wind farms.

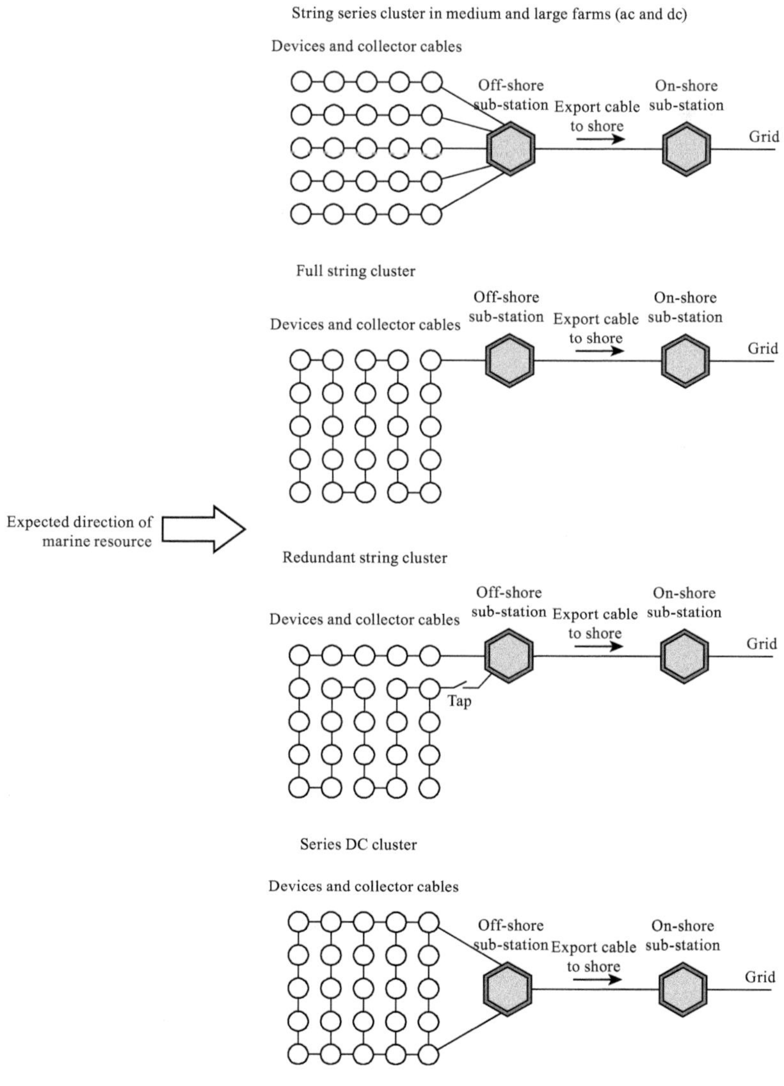


*Figure 8.3 Typical marine energy converter farm layouts, Alcorn et al. (2014)*

- String series cluster: for larger arrays as it offers some redundancy in that not all wind turbines are dependent on one cable.
- Redundant string cluster: a single loop with a switch. This offers some redundancy as it allows wind turbines to feed on-shore via redundant cable routes. It is cheap but with the expense of a switch, mounted in a wind turbine or sub-station. This is being adopted for some off-shore wind farms.
- Series DC cluster: offers extensive redundancy but is designed for devices generating DC, where generation devices are in series. This is being contemplated for the very large off-shore wind farms planned for the middle of the North Sea and the German BorWind sites are adopting this approach, see Figure 8.4.

A typical operational off-shore wind farm array layout is shown in Figure 8.5, which is the London Array in United Kingdom, a large off-shore wind farm with an AC radial structure, incorporating separate redundant string clusters, Figure 8.3, and

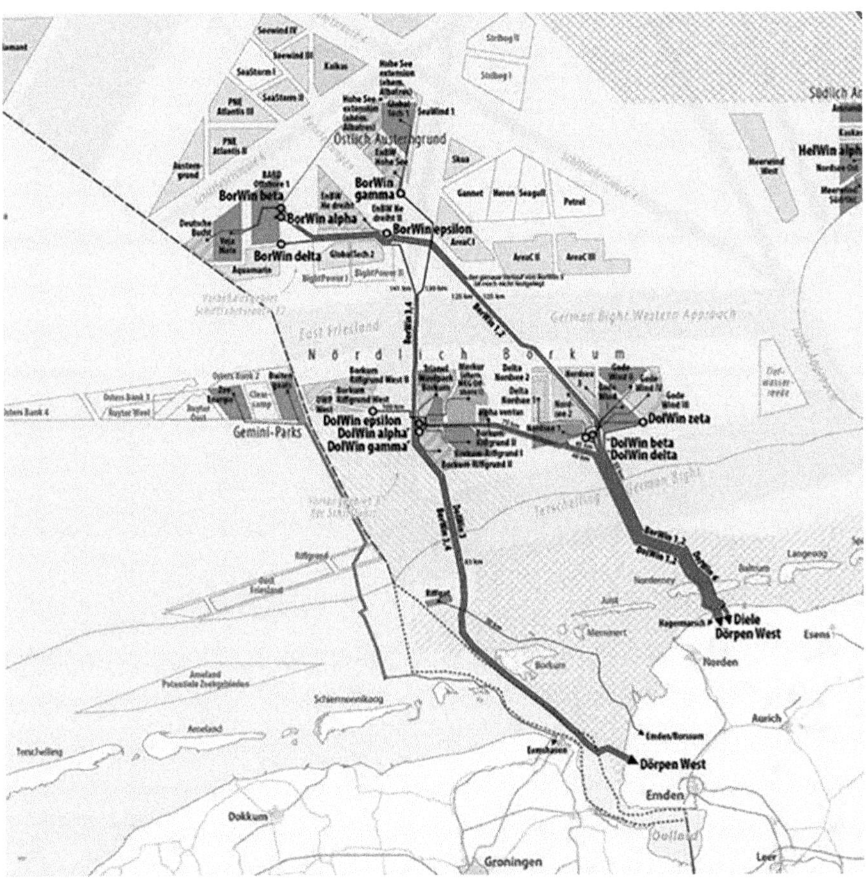


*Figure 8.4 Layout of the North German BorWind and DolWind array sites, with DC export cable systems*

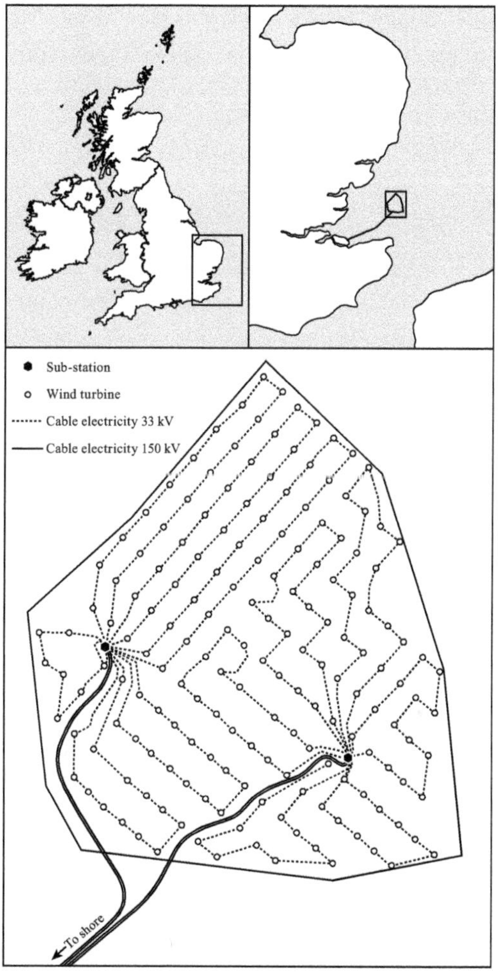


*Figure 8.5 Layout of an existing off-shore WT farm London Array*

AC export cables. Each radial spur is a loop with switches at the end of the loop and at the collector sub-station to provide redundancy in the event of a failure in an array cable. This is again a larger example of the third system in Figure 8.3. The array has the following:

- 175 3.6-MW off-shore WT devices.
- 200 km of fixed, buried, 33-kV XLPE AC cable collector array.
- Collector array connected to two off-shore sub-stations, stepping voltage up from 33/150 kV AC.
- Export from sub-stations via two 53-km double lengths of fixed, buried, 150-kV XLPE AC export cables, thus ensuring some redundancy for each sub-station export connection.

- The export cables supply a shore sub-station, on the Isle of Sheppey, where connection is made to the 400-kV AC grid via Static VAr Compensators, to balance the reactive VArs produced by the wind farm and filters to ensure 400-kV waveform compliance.

## 8.4 AC versus DC collectors and export connection

Most on-shore wind turbines to date generate 900 V, AC, LV with voltage amplitude and frequency regulated by power electronic technology. This voltage was then transformed in the wind turbine tower from that voltage up to 20–33 kV.

However, conventional gas, oil and coal-fired power stations generate at higher voltages of 15–25 kV, AC, MV because generator MW ratings favour a higher voltage and industry has developed high-quality MV insulation materials and standards to support them with higher reliability at acceptable cost, reducing connection system current ratings and losses.

Hitherto the wind industry has adopted lower voltages because smaller generators, with LV windings and insulation, were cheap but less reliable and carried higher currents.

As we move towards 12–20-MW wind turbine ratings, the industry is raising generation voltages to 1.5–4.0 kV, AC, MV also implying the use of MV converters, maybe more complex than LV converters, but because of improved switching module package sealing are also more reliable. MV generators and converters therefore use higher quality materials and have learnt valuable reliability lessons from other parts of the electrical industry. Some larger off-shore wind turbines are now generating at MV voltages higher than 4 kV.

Turbine voltages will still be stepped up by wind turbine transformers from the generation of voltage to 20–33 kV, AC, for trenched collector cable transmission under the seabed from the wind turbines within the wind farm to the sub-stations. These collector cables in turn allow wind turbine groups to be paralleled and voltages raised further in sub-station transformers, for transmission to shore at HV AC voltages of 110–300 kV, see Figure 8.6(a).

It is desirable to export power to shore in HV AC cable, to reduce current flow, cheapen cable costs and reduce loss, Figure 8.5. Such shore cables must also be trenched and buried but they are also extra-armoured near to shore for protection and they carry fibre-optic cables for off-shore wind farm control and SCADA data collection.

However, when transmitting to shore in export cables over long distances, more than 90 km, HV AC is less efficient than HV DC, so larger wind farms are using off-shore HV AC-to-DC converter stations and long-distance HV or even EHV DC cables, as shown in Figure 8.4.

The off-shore HV AC sub-station Figure 8.6(a) is then replaced by a larger HV AC sub-station and HV DC converter station, such as Figure 8.6(b).

A relevant reference for control of these assets is Sanchez *et al.* (2020).

Figure 8.6 compares the cross-sections of typical AC and DC, MV export cables.

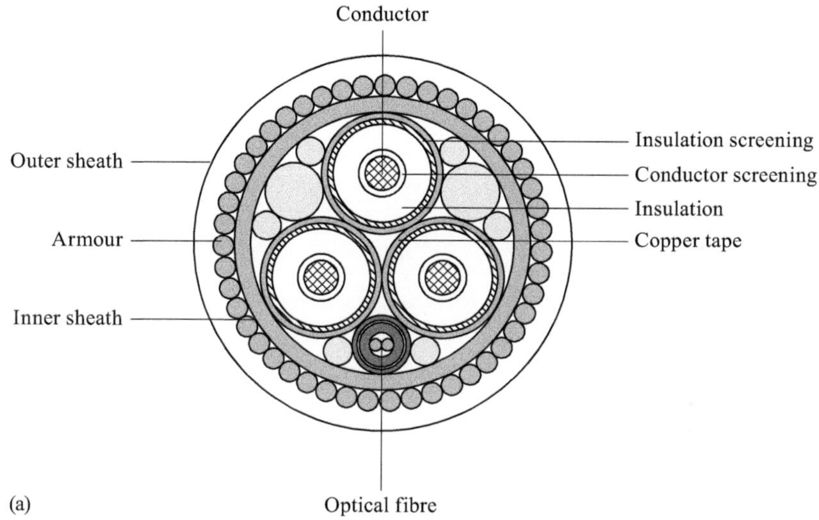


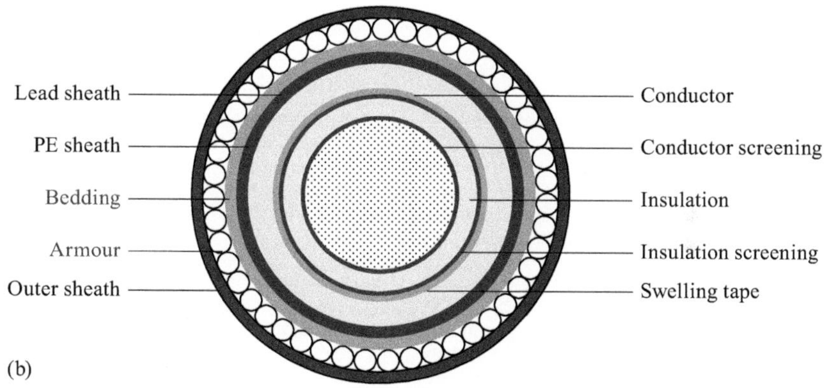


*Figure 8.6 Cross-sections of cables from an off-shore wind farm: (a) three-phase MV, AC export cable and (b) single-phase MV, AC export cable*

As made clear earlier most off-shore wind turbines generate at LV AC, but some larger turbines generate at HV. The majority of current off-shore wind farms collect energy from individual wind turbines at MV voltages after step-up in a wind turbine mounted three-phase transformer. These have the advantage that voltages can simply be raised and power collected using MV AC collector cables, such as shown in Figure 8.6(a). These collector cables feed an off-shore sub-station, such as shown in Figure 8.7(a), where collector circuits are protected by switchgear and voltages can be raised using transformers and power exported to shore in large HV AC export cables.

However, at longer distances to shore, >90 km the use of HV DC cables, Figure 8.6(b) are desirable. These avoid the need to rate electrical equipment for higher currents to cater for reduced AC power factors and the converter station can

(a)

(b)

*Figure 8.7 Examples of off-shore connection infrastructures: (a) United Kingdom, London Array off-shore HV AC–AC sub-station, note supply vessel at the sub-station and (b) Germany, BorWind off-shore array, HV AC–AC sub-station and AC–DC converter station note sub-sea structure, fixing points and collector cables* allow an increased level of control over the export power. In this case an off-shore HV AC–AC sub-station and AC–DC converter station is required, Figure 8.7(b), which will be considerably larger size for a given rating than an off-shore HV AC–AC sub-station, Figure 8.7(a).

Calculations show that DC transmission over export cables becomes more economic than AC transmission at distances to shore >40 km, Alcorn *et al.* (2014). But DC operation is still in relative infancy for off-shore wind, where control strategies, voltage management and cabling technology are not yet fully agreed, but off-shore wind farms at locations >90 km from the shore are now under consideration. However, in Germany the BorWind and DolWind projects, Figure 8.4 has overcome a number of issues and provided valuable experience, whereas AC parallel operation has a long and well-developed experience base, both off-shore and on-shore, enhanced by recent off-shore wind farm operations.

## 8.5 Sub-stations and converter stations

On-shore sub-stations are entirely suitable for small off-shore wind farms of up to 60 MW, close to shore, with MV AC export transmission, as it avoided the need for large off-shore sub-station structures.

However, as off-shore wind farm sizes have increased, it has become essential to adopt large off-shore sub-station structures, as shown in Figure 8.7. A description of the issues in AC and DC sub-station design and control is given in van Hertem *et al.* (2019). There are also issues of frequency control for the grid into which off-shore wind energy is injected, which is discussed in Shah *et al.* (2019).

## 8.6 Off-shore wind farms

### *8.6.1 One line diagrams*

The reliability of off-shore collection layouts, sub-stations and converter stations can best be assessed using the architecture shown by their electrical one-line diagrams. Billinton *et al.* (1996) described the methodology to be used, and this reference gives a clear guide to their application to transmission power systems, relevant to off-shore wind collection systems.

Figure 8.8 shows typical one-line diagrams for some different off-shore arrangements.

These diagrams are valuable for itemising the non-turbine components in an off-shore wind farm so that a reliability assessment could be methodically made, along the lines adopted for generators and converters in Tables 5.4, 5.5 and 5.7, using metadata such as listed in Appendix E, see Section 8.7.

### *8.6.2 Reliability*

Chapter 7 gave valuable wind turbine reliability information but we must now add that to the reliability of the off-shore wind farm layout.

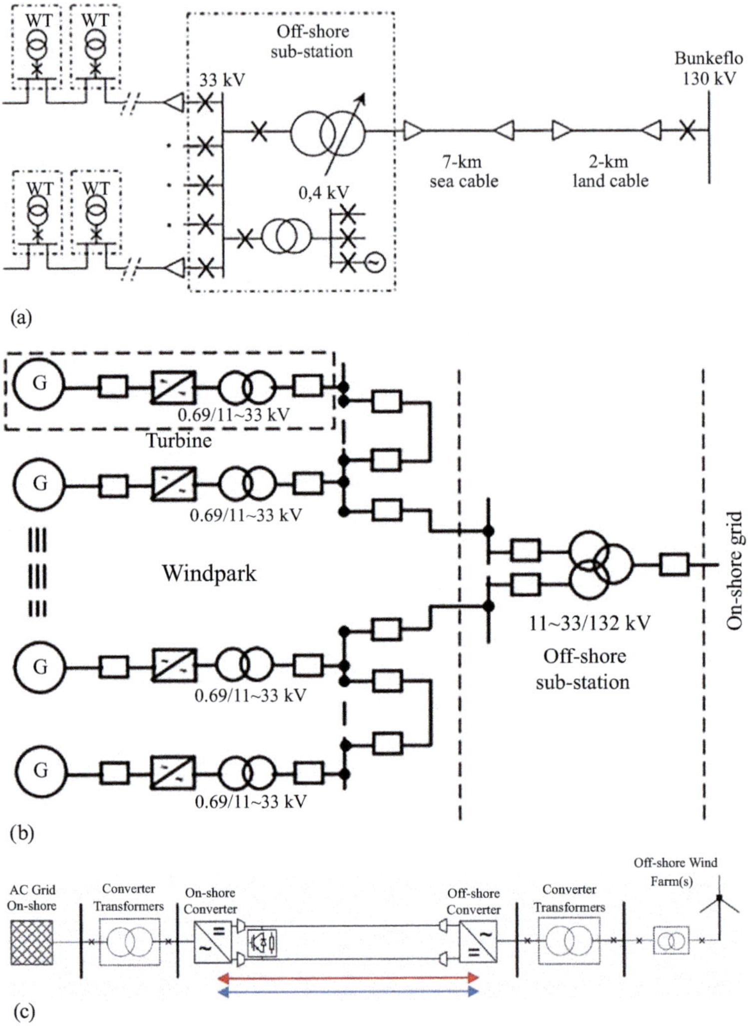


*Figure 8.8 One-line diagram of near or far off-shore wind farms: (a) Showing: (i) five parallel series WT MV off-shore collector arrays; (ii) one bus-bar sub-station with two-winding tapped MV/HV transformer; (iii) one LV standby generator and LV/MV transformer in the off-shore sub-station; (iv) a short HV export cable connection to an on-shore sub-station. (b) Showing: (i) two separate paralleled series MV off-shore collector arrays; (ii) combined at a split two bus-bar off-shore sub-station with three-winding MV/HV transformer for collector cable connection. (c) HVDC export cable array showing: (i) an off-shore collector array; (ii) an off-shore sub- and converter-station with an HV/MV transformer and an AC/DC converter; (iii) DC cable with DC link control and snubber; (iv) an on-shore sub- and converter-station with a DC/AC converter and an MV/EHV transformer*

The reliability of off-shore wind MV AC collector cables, off-shore AC–AC sub-stations and HV AC export cables has received some attention but few reliability results are available in the public domain.

For the MV AC collector cables, off-shore AC–AC sub-stations and HV AC export cables parts of off-shore wind systems, IEEE Std 493 (1997) provides detailed failure results for switchgear, transformers and cables, albeit from the 1990s, which could be incorporated into any FMEA performed on the off-shore and on-shore electrical assets of an off-shore wind farm. One-line diagrams, such as Figure 8.7, are invaluable in this respect.

Delorm (2013) gave a useful tabulation of failure rates for a study of tidal distributed generation systems, quoted in Appendix E, this method being based upon IEEE Std 493 (1997) following the Billinton *et al.* (1996) methodology.

However, there is less documented information available for HV AC–AC sub-stations or AC–DC converter stations with their HV DC export cables.

The most up-to-date papers on the subject of off-shore wind farms are as follows, with their conclusions listed.

The first two papers, sponsored by EdF, concern respectively an installed wind farm in United Kingdom and a proposed wind farm in France:

- Papatzimos *et al.* (2019) considered the reliability of the United Kingdom Round 2 Teesside off-shore wind farm in the United Kingdom, with 27 2.3-MW wind turbines, showing the turbine failure distribution over a 5-year period but not commenting on the layout or connection reliability, so this work is simple, like Chapter 7, giving very similar results.
- Dahmani *et al.* (2017) considered the architecture and reliability optimisation of the planned Banc de Guerande off-shore wind farm in France, consisting of 80 6-MW wind turbines with 33-kV collector cables and a 220-kV AC export cable from one or two sub-stations.

The four optimisations, see Figure 8.9 and Table 8.1, were the following:

- A reference design, collector cable in full string cluster with one sub-station;
- And three alternative designs:
  - A collector cable in string series cluster with two sub-stations;
  - B collector cable in ring cluster with one sub-station;
  - C collector cable in ring cluster with two sub-stations.

The results demonstrated that the reference design had a low investment cost but Case A had a lower investment cost, whilst Cases B and C had higher investment costs but lower through-life costs with Case C being the cheapest total cost.

The next three papers, on off-shore wind farms in China, are respectively a literature review and two computer modelling simulations in the following:

- Hou *et al.* (2019) sponsored by China State Grid but writing from Aalborg University, Denmark is a literature review of off-shore wind farm layout optimisation methods, including their electrical system design methodology, giving valuable references.

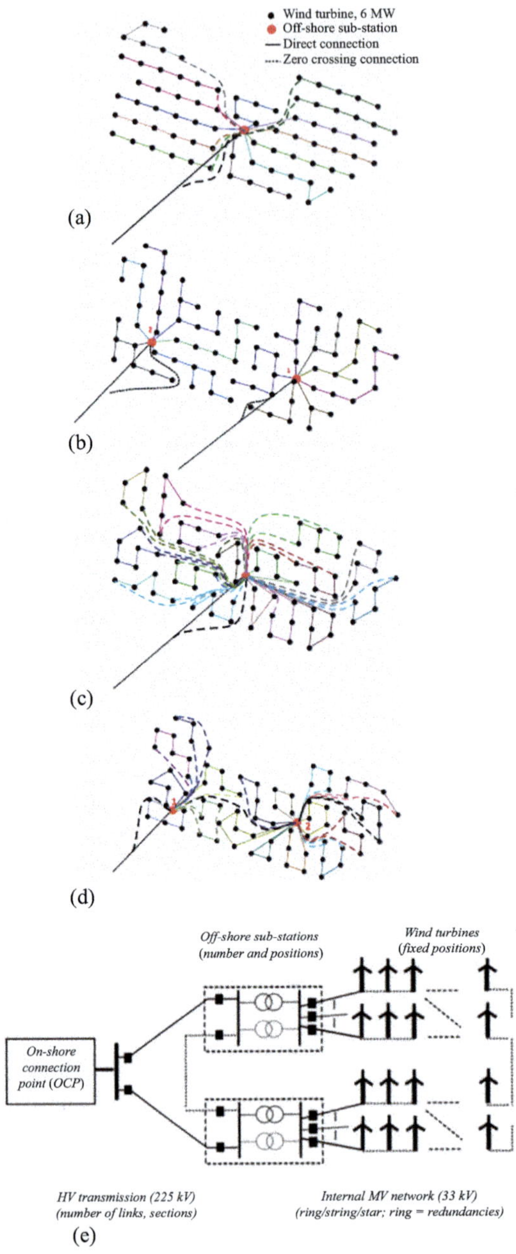


*Figure 8.9 Optimisation of Banc de Guerande off-shore wind farm connection infrastructures, Dahmani et al. (2017); (a) reference case, WT collector cable in full string cluster with one sub-station; (b) Case A, WT collector cable in string series cluster with two sub-stations; (c) Case B, WT collector cable in ring with one sub-station; (d) Case C, WT collector cable in ring with two sub-stations; (e) collector schematic*

*Table 8.1 Component reliability data for Banc de Guerande wind farm, Dahmani et al. (2017), compare reliability figures with Table E.1 from Appendix E, Delorm (2013) and Tavner (2016)*

| Equipment | Failure rate, $\lambda$ (failure/year) | MTTR (h) |
|---|---|---|
| AC generator | 1.500 | 490 |
| LV AC contactor | 0.067 | 240 |
| WT tower AC cable, 70 m | 0.015 | 240 |
| WT transformer | 0.013 | 240 |
| MV AC switch | 0.025 | 240 |
| MV AC circuit breaker | 0.025 | 72 |
| MV AC cable | 0.015/km | 1,440 |

*Table 8.2 Component reliability data, Huang, Wang, Fan et al. (2019), compare reliability figures with Table E.1 from Appendix E, Delorm (2013) and Tavner (2016)*

| Equipment | Failure rate, $\lambda$ (failure/year) | MTTR (h) |
|---|---|---|
| AC collector cable | 0.008/km | 2,160 |
| Sub-station transformer | 0.013 | 240 |
| Sub-station AC bus bar | 0.004 | 720 |
| Sub-station AC circuit breakers | 0.024 | 2,160 |
| Converter transformer | 0.020 | 4,320 |
| Sub-station AC/DC converter | 0.050 | 720 |
| DC collector cable | 0.071/km | 1,440 |

*Table 8.3 Reliability comparison for various off-shore wind farm connection configurations, Huang, Wang, Fan et al. (2019)*

| Equipment | Collector system unavailability | Total system unavailability | Average available output power (MW) |
|---|---|---|---|
| Single AC ring | 0.0039 | 0.1455 | 376 |
| Double AC ring | 0.0039 | 0.1455 | 361 |
| AC parallel | 0.0039 | 0.1455 | 368 |
| DC parallel | 0.0026 | 0.1360 | 369 |
| DC series–parallel | 0.0046 | 0.1285 | 365 |

- Huang, Wang, Fan *et al.* (2019) give a theoretical reliability and economic assessment of off-shore wind farms, with reliability data shown in Table 8.2 comparing either AC or DC collector cables. This shows, for a modest 380-MW modelled off-shore wind farm, that the highest available energy comes from a simple single-AC ring configuration, whereas the greatest availability comes from a DC series–parallel configuration, Table 8.3.

- Huang, Wang, Qian *et al.* (2019) give a theoretical reliability assessment of off-shore wind farm DC turbine and collector series–parallel connections, see Figure 8.10, avoiding the need for larger capacity DC/DC converters for the export cable, lowering up-front capital costs.

Table 8.4 shows the component data and the resultant reliability parameters, Table 8.5, again compares data with that previously provided in this book, implying that DC collector and export systems could dominate distant off-shore wind farm unreliability.

An idea of the scale of the wind farms being described in the previous papers can be seen in Figure 8.11, which shows the first turbines of the new Dogger Bank A wind farm, located in the North Sea, United Kingdom, being installed.

This wind farm will comprise three 1.2-GW off-shore wind farms, Creyke Beck A, Creyke Beck B and Teesside A, each with 10-GE Haliade turbines, located more than 80-mi off-shore.

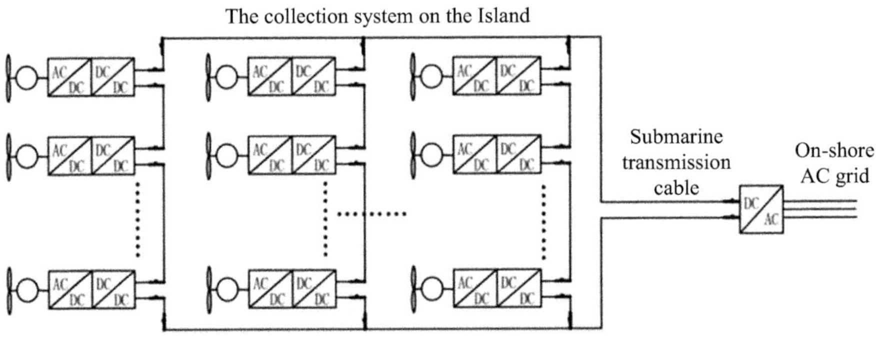


*Figure 8.10 Example of a DC series–parallel export off-shore wind farm connection system, Huang, Wang, Qian et al. (2019)*

*Table 8.4 Component data for reliability calculation, Huang, Wang, Qian et al. (2019), compare reliability figures with Table 18.7, Delorm (2013) and Tavner (2016)*

| **Equipment** | **Failure rate, λ (failure/year)** | **MTTR (h)** |
|---|---|---|
| AC collector cable | 0.008/km | 2,160 |
| 70-kV off-shore platform AC circuit breakers | 0.032 | 720 |
| AC/DC off-shore platform converter | 0.05 | 720 |
| 150-kV off-shore platform DC circuit breakers | 0.032 | 720 |
| DC transmission/export cable | 0.008/km | 720 |
| On-shore DC/AC converter | 0.05 | 50 |
| 150-kV on-shore AC circuit breakers | 0.032 | 4 |

*Table 8.5 Parameter values used, Huang, Wang, Qian et al. (2019), compare with Table 18.7, Delorm (2013) and Tavner (2016)*

| **Parameters** | **Turbine string collector system** | **Sub-station/ converter** | **Transmission/ export system** | **Total system** |
|---|---|---|---|---|
| Failure rate, $\lambda$ (failure/year) | 1.29 | 0.056 | 1.71 | 1.77 |
| Unavailability, $U$ | 0.19 | 0.005 | 0.12 | 0.13 |
| Instantaneous repair rate, $\mu$, $1/MTTR$ | 5.57 | 12.18 | 12.06 | 12.01 |

*Figure 8.11 Dogger Bank A wind farm, 2 GE 12-MW Haliade Wind Turbines mounted on jackets with on the left an associated off-shore transmission module and on the right installation ship. © NS Energy*

This wind farm arrangement will ultimately be connected to power grids on either side of the North Sea by HVDC cables, see Figure 8.4 for an idea of the complexity of such arrangements.

## 8.7 Floating technology, effect on reliability

Despite the small number of operational floating off-shore wind production arrays, Figure 8.12, and the lack of operational experience to date, certain key reliability conclusions can be reached about these devices and their arrays, based upon the work of this book:

- Redundancy in off-shore wind cable arrays will raise operational reliability and availability, reducing operational expense (OPEX) and lowering cost of energy

*Figure 8.12 Off-shore floating wind turbine, giving an impression of the scale of these huge structures. © The Energyst*

(CoE), see Section 8.6. However, capital expenditure (CAPEX) will be higher and there will be a CAPEX/OPEX trade-off for these reliability and availability benefits. Device array foot-print and grid connection arrangements will affect this trade-off.

- There will be a significant difference in the CAPEX/OPEX ratio between large floating and moored off-shore wind arrays, which will have a larger CAPEX/OPEX, compared to smaller foot-print fixed device arrays, which will have a smaller CAPEX/OPEX.
- The foot-print of an off-shore floating and moored array will be larger than fixed device arrays of the same rating, because of the larger floating devices and their mooring arrays. Therefore, the annual energy capture in $GWh/km^2$ and consequent payback could be higher for fixed than for a floating and moored array of the same rating.

  However, the development of much larger wind turbines for off-shore floating arrays will mitigate this and we should eventually see lower CAPEX costs per GW and greater $GWh/km^2$ capture from off-shore floating and moored arrays.
- Off-shore fixed device arrays may eventually have higher CAPEX costs/MW than off-shore floating and moored arrays, because of their limited ratings.
- Off-shore floating and moored arrays with non-detachable connections to fixed cable arrays will have lower array CAPEX costs than arrays with detachable connections, because of higher detachment costs and the smaller array area. However, detachable arrays will in the long term have a lower OPEX cost

because of the possibility of towing large devices to harbour for repair than repair in situ.

Finally some of the off-shore wind turbine array issues and their effect on reliability and availability have been described, in particular:

- The lack of thorough analyses of off-shore wind layout reliabilities and the need to develop a methodology to determine optimum technology and layouts, although the works by Dahmani *et al.* (2017), Huang, Wang, Fan *et al.* (2019) and Huang, Wang, Qian (2019) are steps in the right direction, demonstrating the prospective balance between CAPEX and OPEX.
- The larger size of prospective off-shore wind turbines and potentially even a larger size of floating turbines and their sub-stations, which affects the wind farm foot-print. The consequent altered issues for human access to such large structures in feasible sea states to permit maintenance and the consequent effects on maintainability.
- The different clearances between fixed and floating wind turbines, because of their disparate prospective sizes to
  - maximise energy extraction;
  - allow movement between floating structures and provide adequate sea-ways for access.
- The importance of wind turbine and collection cable layout to provide redundancy to raise reliability and availability.
- The differences between fixed and floating off-shore wind farm configurations and their relative economic performance, reliability, availability and maintenance, including the issue of stresses on dynamic cables for floating off-shore wind, Young *et al.* (2018).

Therefore, detachable, floating and moored off-shore wind turbines will probably deliver an OPEX benefit, because there will be fewer, larger devices, but will carry a CAPEX penalty because of their large device sizes, but this will increase turbine capacity factors and energy yield, eventually lowering CAPEX/MW and improving their investability.

Floating off-shore wind systems may be the best future solution to reduce project risk, deliver project operational reliability and CAPEX pay-off in an acceptable period. Key publications include Musial *et al.* (2004), Leimeister *et al.* (2018), Leimeister, Kolios (2018), Musial *et al.* (2018, 2019, 2020).

The latter was demonstrated for other off-shore renewable sources in Tavner (2017) by simple comparison between the costs of a fixed off-shore tidal device and floating and moored wave device of the same rating, Figure 8.13. The installation costs and prospective CoE were compared, using data from Hardisty (2009) and O'Connor *et al.* (2012).

More recently researchers have also been considering the collocation of off-shore wind and wave assets, Gideon *et al.* (2021), and conclude that that for the US West Coast and Scottish North Sea, collocated wave and wind generation systems have the potential to reduce power variability and mitigate costs. However, for an

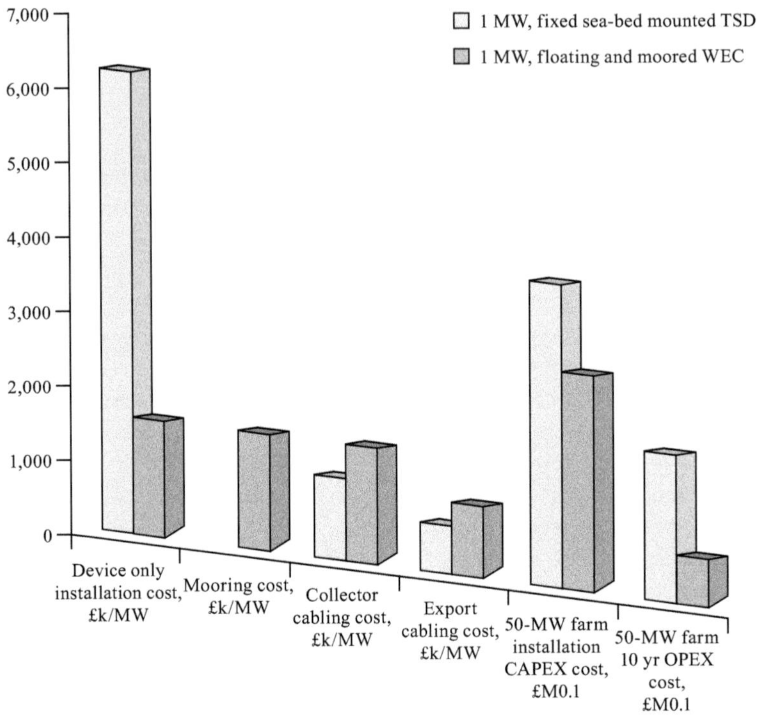


*Figure 8.13 Comparison between installation costs, CoE, CAPEX and OPEX costs, for a fixed off-shore tidal device compared to a floating and moored off-shore wave device*

immature ocean energy industry, where installation and operational financial data are in extremely short supply, such conclusions must be treated with extreme caution.

This will change as the energy available from floating off-shore wind, particularly in the USA and Japan, becomes appreciated, see Musial *et al.* (2018, 2019, 2020). For a thorough analysis of methods for predicting the effect of off-shore wind turbine reliability on the long-term CoE, see Dao, Kazemtabrizi *et al.* (2019) and Dao, Kazemtabrizi *et al.* (2020).

## 8.8 Summary

This chapter has shown that there is a balance to be made, for availability, reliability and maintenance, in the device layout and grid connection of off-shore wind turbines and associated sub-station or converter assets, whether their sub-stations are fixed to the seabed or are detachable, floating and moored.

Fixed off-shore wind turbine arrays will incorporate devices which are up to 5× larger than on-shore devices but will incorporate larger more reliable infrastructure.

Floating and moored off-shore wind turbine arrays will contain fewer, larger devices, perhaps 10× larger than on-shore devices spread over a larger area. Such larger devices maybe more difficult to access for maintenance but could provide crew accommodation to allow them to undertake longer maintenance tasks.

There will be an optimisation of the redundancy incorporated into arrays, but experience from fixed off-shore wind farms suggests that the practicable degree of array redundancy will be modest.

There is interest in the development of DC collection and export networks, because of device architecture and possible savings in energy transmission from arrays far off-shore.

Experience from current off-shore wind farms suggests that long-distance DC connections will become important because of the larger available off-shore energy resource; however, the technical issues associated with interconnection between off-shore DC networks need to be resolved so that large, diverse assets can manage the flow of energy between different arrays and shore.

There is insufficient installed experience and relatively few published papers on floating off-shore wind farms to draw specific conclusions about the best way to specify turbine layouts and grid connections. However, some broad conclusions can be reached about array details:

- Fixed off-shore wind turbines are likely to have a higher unit CAPEX costs and lower array CAPEX costs but higher OPEX costs, than detachable, floating and moored off-shore wind turbines.
- Whereas detachable, floating and moored off-shore wind turbines are likely to have a lower unit CAPEX cost and higher array CAPEX costs but lower OPEX costs, than fixed off-shore wind turbines.
- Therefore, detachable, floating and moored off-shore wind turbines should deliver an OPEX benefit, enhanced by higher energy take in $GWh/GW_{rated}$ but could carry an array CAPEX penalty. However, they may be the best solution to reduce project risk, deliver adequate project operational reliability and pay-off CAPEX costs in an acceptable period.
- AC collection and export from off-shore wind may be the preferred initial solution, before deployment at greater off-shore distances gives advantage to DC devices and DC export, because of the following:
  - Fixed off-shore wind turbines can be arranged in a tighter off-shore wind farm as there is no requirement for positional motion of the turbine.
  - Floating and moored off-shore wind turbines will need to be in a wider arrangement to allow for the spread of mooring anchors and the off-shore wind turbines to adjust themselves in the seaway.
  - Floating and moored off-shore wind turbines also require greater clearance for shipping, to avoid the risk of damage to mooring lines and flexible cables.

There is an important balance between CAPEX and OPEX expenditures for fixed or floating off-shore arrays, but in general judicious extra CAPEX-spending can reduce many years of excessive OPEX-spending.

# Chapter 9
# Monitoring for off-shore wind farms

## 9.1 Introduction

The monitoring of modern wind turbines and wind farms may include a variety of systems that can be generally classified as follows:

- Supervisory control, or signal conditioning, and data acquisition (SCADA) system, to provide low-resolution monitoring to supervise wind turbine and farms operations and provide a channel for data and alarms.
- Condition monitoring system (CMS), to provide high-resolution monitoring of high-risk sub-assemblies in wind turbines for the diagnosis and prognosis of faults, included in this area are blade monitoring systems, aimed at the early detection of blade defects.
- Structural health monitoring (SHM), to provide low-resolution signals for the monitoring of key structural items of fixed or floating wind turbines, sub-stations and converter stations structures.

These systems each have different data rates as summarised in Figure 9.1, but as the wind industry develops they are slowly being integrated together.

## 9.2 Supervisory control and data acquisition

### *9.2.1 Why SCADA?*

Supervisory control and data acquisition (SCADA) systems originated in the process, oil and gas industries where large, physically distributed processes could only be controlled by accurate measurements of the states of valves, pumps and storage vessels and of the consequent temperatures, pressures and flows.

These data acquisition systems were originally developed independent of the controls. However, where their measurements were needed to control the plant, they evolved into industrial control systems. More recently, where plant control was distributed throughout the plant and embedded into the data acquisition system, some SCADA systems evolved further into distributed control systems (DCS).

The power generation industry has been using SCADA for 35 years and DCS has been used to control modern power station units in the United Kingdom since about 1985. Therefore, it was natural for the wind industry to apply these techniques to wind turbines, unmanned remote robotic power generation units. However, the emphasis in

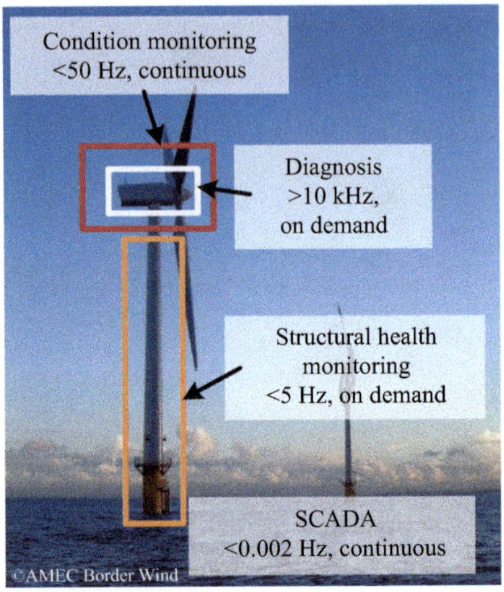


*Figure 9.1 SCADA, CMS and SHM for a wind turbine*

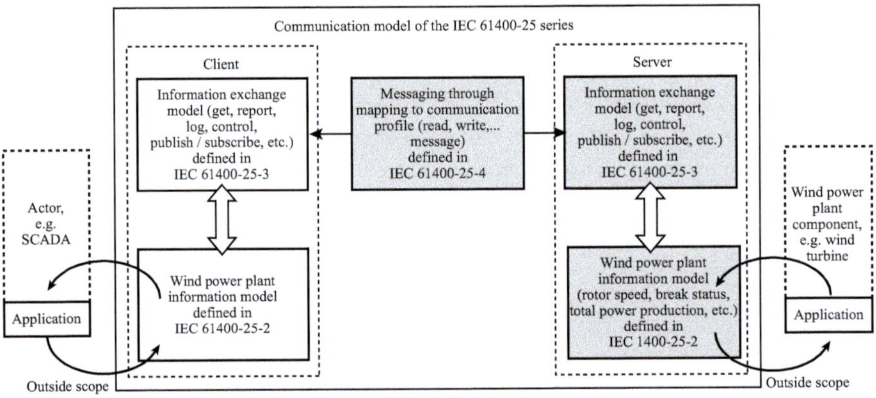


*Figure 9.2 Conceptual communication model for a wind turbine, IEC 61400-25 series*

the wind industry has been on monitoring rather than control, which in a wind turbine is exercised primarily by the wind turbine controller mounted in the nacelle, although that can be overrode by external signals from the operator via SCADA.

In fact the majority of SCADA signals and alarms in wind turbines originally derived from within the wind turbine controller, generally an industrial programmable logic controller that ensures that the wind turbine remains within its safe operating envelope supervising cut-in, synchronisation, adherence to the power curve, cut-out and emergency stop action in the event of excessive gusts or untoward operation.

The international standard which prescribes the layout for wind turbine communication systems, including SCADA, is IEC 61400-25 series, see Figure 9.2.

However, the evolution in wind turbine size, the number of units and designs has encouraged the wind industry to apply SCADA more widely than elsewhere in the power generation industry. This may have been because of the growth and cheapness of measurement, IT and communications technology but is also because of the prototype nature of early large wind turbines, exemplified by the error codes in the latter part of the table in Appendix A.

A survey of the SCADA systems available to the wind industry is given in Appendix C.

### *9.2.2 Signals and alarms*

The SCADA system handles both input/output (I/O) signals and alarms and usually samples at 10-min intervals, although emergency alarms are handled immediately. Fast changing or commercially important signals, such as wind speed or power output, the system transmits and records maximum, mean, minimum and standard deviations of those signals, so that they can be reconstructed in arrears.

The majority of data are output, flowing from the wind turbine to the control room, but some signals and commands are input, fed from the control room to the wind turbine.

An example of the growth of modern wind industry SCADA is as follows:

- An operational 500-MW fossil-fired generation unit may have 1–2,000 SCADA I/O channels.
- Whereas a modern 5-MW off-shore wind turbine may have 4–500 I/O channels, including signals and alarms, emphasising the unmanned, remote, robotic and developmental nature of some modern, large wind turbine units.

### *9.2.3 Value and cost of SCADA*

The value of SCADA is that it gives the wind turbine original equipment manufacturer (OEM), or operator, at their operational base, on-line data about the functioning and alarms of remote wind turbines. This allows the generation of graphical information so that operations can be optimised and maintenance planned.

However, the volume of data generated by SCADA require careful organisation, for example an off-shore wind farm with 100 wind turbines each generates 40,000 data items every 10 min, that is 96 Mbytes of data per day, requiring considerable analysis for on-line interpretation.

In general wind turbine OEMs have developed these techniques, in order to manage wind turbines during the warranty period and using SCADA they are able to compare the performance of different wind farms and of individual wind turbines against the whole populations of the same type.

A great benefit of SCADA is that it can provide an overview of a whole wind turbine, looking at production measures, such as wind speed and energy output, monitoring signals, such as lubrication oil and bearing temperatures and control system alarms from the pitch and power electronics systems, for example. Therefore, it allows an operator to compare signals widely across the wind turbine system giving

*Figure 9.3 Analysing SCADA data to detect wind turbine problems*

confidence in indicated results. The weakness of SCADA is that its low data rate does not allow the depth of analysis usually associated with accurate condition monitoring (CM) diagnosis. However, as the next section will show this weakness in-depth is more than compensated by the breadth of SCADA scope, which can produce easy to interpret graphical images, such as the power curves shown in Figure 9.3.

On the other hand, wind turbine operators generally do not have these facilities, except by access permission from the wind turbine OEM, and may face difficult decisions at the end of the warranty period, whether to extend an OEM maintenance contract or attempt managing the wind turbines themselves.

SCADA is generally a low-cost monitoring system, integrating cheap, high-volume measurement, IT and communications technology into the wind turbine controller, designed by the OEM during original manufacture. A typical cost of SCADA provision depends upon the size of a wind farm but can be typically £5–10k/ wind turbine.

## 9.3 Condition monitoring systems

### *9.3.1 Why CMS?*

Wind turbine CM first appeared in the industry in the 1990s, following pressure from insurance companies, as a reaction to a large number of high profile claims due to wind turbine gearbox failures. Wind turbine CM technology was largely adapted from other rotating machine vibration experience. Wind turbine CMS initially came from reputable CM OEMs, such as Bruel & Kjaer, Bently Nevada and National Instruments with systems largely based upon experience from traditional rotating machine vibration CM experience. CMS system selection also became part of the wind turbine certification process DNV GL (2013).

However, wind turbine CM throws up a number of issues which are uncommon in traditional rotating machines, because of the stochastic nature of the wind resource that is that modern large wind turbines operate at continuous and rapidly varying power, torque and speed and are usually remote from technical support. CMS has proven successful in on-shore wind turbines when used by experienced operators and is now installed on new wind turbines $\geq$1.5 MW almost as standard and are being fitted to all off-shore wind turbines.

Wind turbine OEMs make considerable use of CMS technology during the warranty period but despite the continued installation of wind turbine CMS, little attention is paid by wind farm operators to system's alarms or data generated. This stems primarily from the fact that operators may not have the specialist knowledge required to interpret complex CM data. As a result, many operators, particularly those with less experience, subcontract wind turbine CM to specialist companies or maintain a monitoring contract with the wind turbine OEM. This can be a costly exercise and, as a result, CMS may be neglected and non-reactive maintenance strategies may be adopted.

A survey of the CMS systems available to the wind industry is given in Appendix D.

## *9.3.2 Different CMS techniques*

### 9.3.2.1 Vibration

Vibration techniques were the first to be used in wind turbine CMS for monitoring the generator, gearbox and main turbine bearings. A variety of techniques have been used, including low-frequency accelerometers for the main bearings and higher frequency accelerometers for the gearbox and generator bearings and in some cases proximeters. Figure 9.4 shows the frequency range appropriate for vibration displacement, velocity and acceleration measurements.

A particular issue for wind turbine vibration analysis is that vibration periods and amplitudes change with time, because of the continuous and rapidly changing drive train torque, and care is essential during analysis.

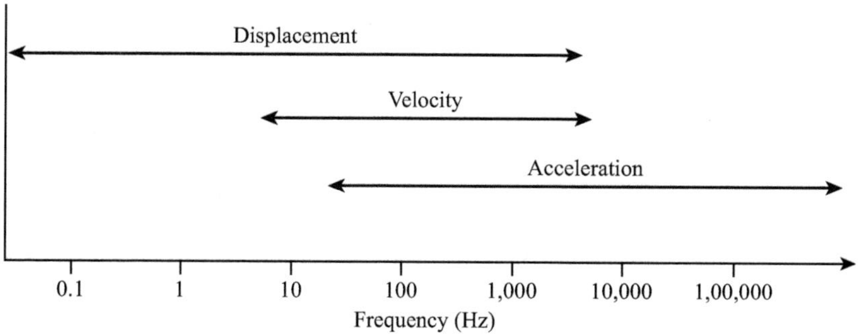


*Figure 9.4 Frequency range for displacement, velocity and acceleration vibration measurement*

A feature of wind turbine CM is that the majority of bearings within the drive train are rolling element and that, combined with the use of high-ratio gearboxes, means that when faults are present vibration signals contain a high impulsive content. This and the continuously changing drive train torque has encouraged some to advocate the use of wavelets DNV GL (2013) and Watson *et al.* (2010) to analyse wind turbine CMS signals to deal with their time-varying, impulsive nature. However, this is computationally expensive.

The majority of current CMS vibration analysis methods used by the wind industry, described in Appendix C, capitalise upon the periodic origin of the vibration signals and use of conventional Fourier transforms, but with the signal collected within a limited, pre-defined speed and power range.

The most important issues to consider in analysing these vibration measurements are the following:

- peak and rms amplitude trends;
- signal time domain;
- signal-frequency content.

A rising rms vibration trend indicates that a worsening fault but low rms vibration with high peak value indicates impulsive energy in the signal and the need to observe the time domain to determine waveform content and identify that impulsive component. Finally if the time domain confirms an unusual impulsive component structure, vibration frequency analysis becomes necessary. This can identify the harmonic origins of the impulsive energy content, for example gear- or ball-passing frequencies, enabling location of a vibration source. Vibration signals are rich in harmonic information, which needs to be understood if diagnosis is to be performed with confidence. Some wind turbine CMS systems allow the mechanical parts of the drive train to be represented within the CMS to provide FFT spectral cursors to aid this interpretation.

#### 9.3.2.2 Oil

Because of the seriousness of gearbox failures, in terms of downtime, gearbox oil debris analysis has assumed more importance in the industry and has been encouraged by other machinery operators with more gearbox experience. The function of the oil in the gearbox is 3-fold:

- to provide gearbox cooling;
- to provide rolling element bearing lubrication;
- to provide meshing gear lubrication.

The lubrication oil itself will have base properties to ensure the proper formation of the lubrication film in the gear pairs and bearings and have additives to minimise wear. Maintenance of these good lubrication properties depends upon

- a high-quality charge of oil in the first place;
- regular removal of debris from the oil;
- maintenance of the oil properties at suitable temperature;
- oil cleaning at appropriate intervals or renewal with the same grade and quality.

Gears and bearings are all wearing components and inevitably produce ferrous and non-ferrous debris from their natural operation. Debris produced during the gearbox running-in process should have been removed during running-in tests during production, such as that shown in Figure 6.9. Most wind turbine gearboxes, in common with other large marine industry gearboxes, for example, utilise a spray lubrication system. That means oil is pumped from the gearbox sump, via a cooler and in-line filter, to the top of the gearbox, from whence it is sprayed onto the operating components from a number of nozzles. Oil is not therefore introduced directly to the bearings or gear pairs for lubrication via an oil port, as would be the case in an internal combustion engine, for example.

This means that the wind turbine gearbox oil stream is both universal and mixed, gaining heat and debris from all parts of the gearbox. An advantage of this for oil monitoring is that the process is global for the whole gearbox, making it attractive for CM.

Crucial to the value of oil debris detection is the length of the warning that it can give of impending failure, Figure 9.5, giving time to arrange for inspection and maintenance.

This latter point is at the heart of effective CM.

However, oil debris detection cannot then locate a fault, except by distinguishing between the types of debris produced. The arrangement of a three-stage gearbox shows the location of parts and the oil system diagrammatically in Figure 9.6, see Feng *et al.* (2013).

In-line filters can remove large debris items, >100 μm in diameter, but cannot remove smaller debris without excessive pressure drop. Studies have shown that

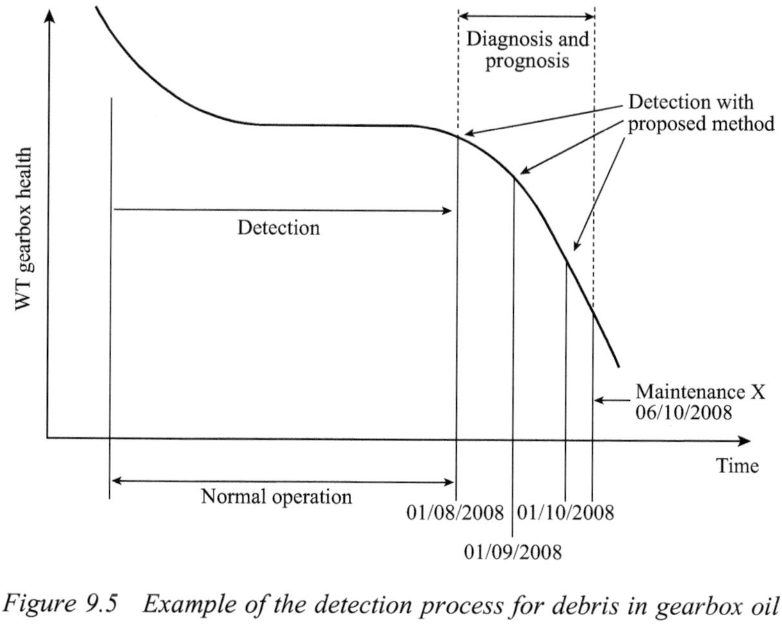


*Figure 9.5 Example of the detection process for debris in gearbox oil*

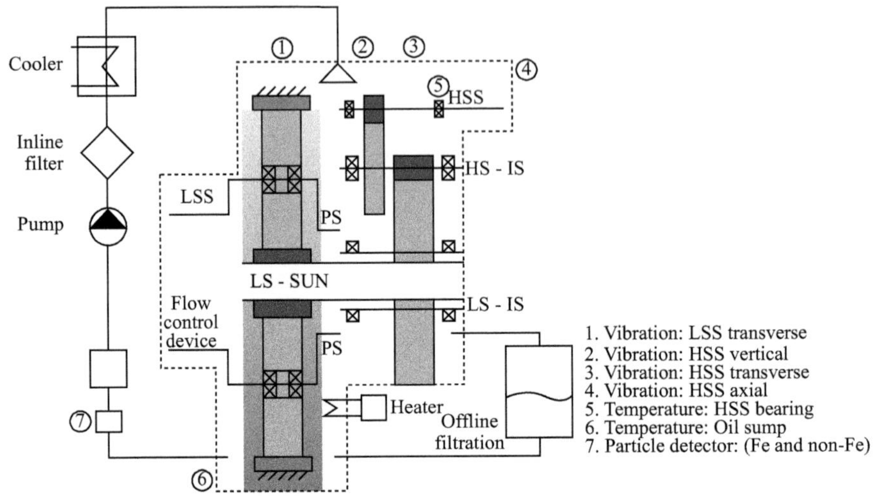


*Figure 9.6 Diagrammatic layout of a three-stage gearbox and lubrication system showing measurement points, Feng et al. (2013)*

gearbox oil should be maintained below 2 μm and there are life advantages if it is kept below 1 μm. However, few gearboxes achieve this level of cleanliness and therefore suffer the effects of large debris damage unless the oil is changed at frequent intervals.

Modern oil debris counters take a proportion of the lubrication oil stream from the downstream of the filter in Figure 9.6 and detect and count both ferrous and non-ferrous particles of varying sizes. The counts can be fed as on-line data to a CMS. Increasing measurement detail increases the cost of the on-line instrument.

There is a strong argument for modern wind turbine gearboxes to be fitted with powerful offline oil filtration units to remove debris down to 1 μm. This could either be done with an in-nacelle unit attached to the gearbox, see Figure 9.6, or a portable facility, brought to the wind turbine as needed, but more of this later in Section 9.5.3 and Chapter 10.

#### 9.3.2.3 Strain

To improve wind turbine performance, there has been a trend, over the past 20 years, towards variable pitch wind turbines. More recently attempts have been made to pitch control turbine blades independently. Independent pitch control reduces the torques and lateral forces on a wind turbine during turbulent wind, prolonging life. This is possible, through independent, blade root, bending moment measurements, made using circular fibre optic arrays incorporating fibre Bragg gratings (FBG) strain gauges, as shown in Figure 9.7.

This technique was originally developed by Insensys as an SHM method for measuring strain in the carbon fibre masts of high-performance yachts, then applied to strain measurement in glass-fibre wind turbine blades and finally applied to pitch control.

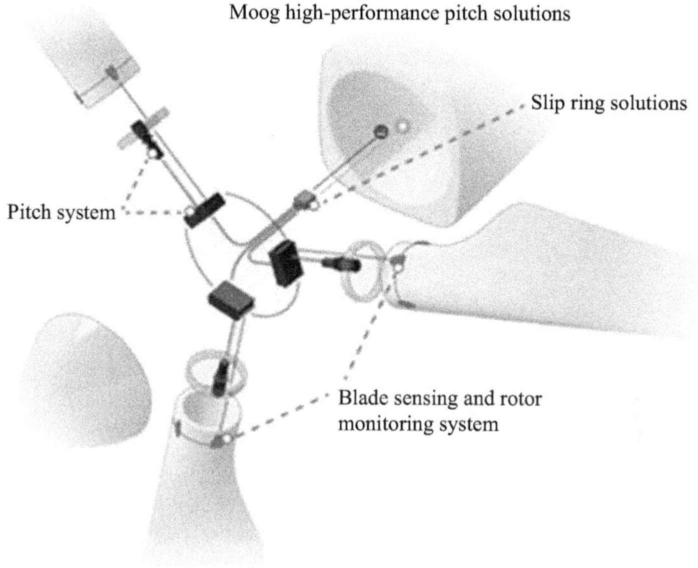


*Figure 9.7 Diagrammatic layout of a three-blade fibre FBG pitch control system*

Insensys was then bought by Moog, and the FBG pitch control technology was incorporated into Moog's wind turbine control offerings.

The measurements from these strain gauges were primarily intended for blade pitch control. However, these measurements can now also be used to condition monitor a wind turbine performance and such techniques are growing in the industry, see the survey in Appendix D.

#### 9.3.2.4 Electrical

Finally, the newest potential source of wind turbine CM information comes from the generator and converter electrical signals used to control speed and excitation, voltage, current and power. These signals have been used in other industries, for CM electrical machines and their coupled drive trains, for many years, Yang *et al.* (2009, 2010). They can now be used as wind turbine drive train global monitoring signals, particularly the power signal, as described by Zappalá (2014), Zappalá *et al.* (2019), Tavner *et al.* (2020) and Brigham *et al.* (2020).

These electrical signals are very close to the torque production process, but they are rich in harmonic information, which must be accurately understood if diagnosis is to be performed with confidence. A simple method to aid electrical interpretation, by generating CMS FFT spectral cursors, as has been used for vibration monitoring.

### *9.3.3 Value and cost of CMS*

The cost of the hardware and software for a mid-range wind turbine CMS is approximately £7k and would require approximately £7k to retrofit to an existing wind turbine, that is £14k/wind turbine, more expensive than SCADA, with less coverage and of less significance to turbine control.

These costs would fall if the CMS were installed in volume to a large number of wind turbines, by the OEM. Wind turbine OEMs will wish to fit their own specified CMS, which they have developed over time with their CMS OEM, as this is their main diagnosis tool during the warranty period. Operators may have their own preferences, because of experience with other wind turbine plant but cannot expect to retrofit their own choice without incurring costs similar to those described earlier.

Therefore, CMS is not as cheap as SCADA nor is it essential for turbine control. In addition, CMS data interpretation incurs additional costs, dependent upon the availability of skilled manpower. There has been considerable debate in the industry about the true value of CMS; however, experience from other industries, power, rail and aerospace, is that accurate CMS is essential for OPEX improvement.

Recent studies by the author showed for traditional power generation plant that CMS was justified solely on cost-saving from unplanned lost production, prevented by CMS.

For on-shore wind turbines CMS can be justified, at the cost levels described earlier, if the costs of replacement equipment, associated labour and lost production are taken into account, particularly if gearbox and generator failures are prevented.

For off-shore wind turbines CMS can be justified if the costs of site access, replacement equipment, associated labour and lost production are taken into account, again particularly if blade, gearbox, generator or other large sub-assembly failures are prevented.

However, in all these cases wind turbine CMS can only be justified if the system is capable of detecting a fault and giving sufficient warning, Figure 9.5, to avoid full sub-assembly replacement, the costliest aspect of failure, and if that CMS detection and warning can be acted upon by operators and wind turbine OEMs.

## 9.4 Structural health monitoring

SHM is specialist systems applied to the main structural components of the wind turbine and their support structures to monitor deflection and strain of key components including

- blades;
- turbine towers;
- monopile and jacket structures;
- floating structures;
- fixations, such as bolts, nuts and welds between structural components, for example wind turbine tower bolts and wind turbine blade attachment bolts.

In general, they involve measurements of components strain, usually at regular but low-frequency intervals, such as daily, or at occasional inspection intervals, monthly or

*Figure 9.8 Range of instrumented bolts used in wind turbines SHM*

annually. There are some proprietary products available such as strain measuring bolt assemblies to simplify the inspection and processing of SHM, Figure 9.8.

Some experimental or developmental wind turbines had extensive SHM measurement systems to de-risk the product.

From experience newer turbines have a smaller range of SHM measurements, dependent on their size. For off-shore floating wind it is to be expected that a significant application of SHM monitoring may be necessary in early deployment to reduce risk in later more numerous deployments.

SHM is especially important for larger turbines off-shore and their associated sub-station and converter station structural fixtures. SHM will be particularly important for floating off-shore wind turbines and structures.

Note that the Insensys pitch monitoring CMS system, referred to in Section 9.3.2, evolved from an SHM system and there may be future repetitions of such a progression in the future wind industry.

## 9.5 SCADA and CMS monitoring success

### *9.5.1 General*

The processes necessary for successful SCADA and CMS monitoring are set out in Figure 9.5, namely:

- Detection, that is the perception that something is faulty in part of the machinery and ideally a location for that fault.
- Diagnosis, that is the determination of the nature of the fault, including its more precise location.
- Prognosis, that is the determination of what needs to be done to remove the fault.
- Maintenance action, that is to remove the cause of the fault or to replace the faulty item.

A fault will take a certain time to develop before it can interrupt the operation of the wind turbine and monitoring must consider that time span if it is to be effective. For example, some faults take a short time to grow from inception to failure. A generator earth fault may take 10 s to grow from inception to failure. Such a fault may give sufficient time for detection but certainly not for diagnosis, prognosis and maintenance action. On the other hand Figure 9.5 shows that an oil debris detection process may give some weeks of warning, which if successfully detected by the monitoring system, will allow effective diagnosis, prognosis and potentially successful maintenance action. This period from detection to maintenance outcome has been called the prognostic horizon.

SCADA and CMS monitoring must concentrate on the measurements and detections which can extend this so-called prognostic horizon. The following examples of analysis methods being used to monitor on SCADA and CMS signals are shown to demonstrate this, including in ascending order of complexity:

- simple trending;
- physics of failure methods;
- narrow band spectral methods;
- Fourier transform methods;
- wavelet and non-stationary methods;
- artificial intelligence methods:
  - artificial neural networks,
  - Bayesian methods.
- multi-parameter monitoring.

The following successful SCADA and CMS signal or alarm detection of known incipient wind turbine or wind turbine CM test rig (CMTR) sub-assembly faults are taken from referenced examples made by the author's research workers and students using data supplied through our research contracts.

### *9.5.2 SCADA successes*

#### 9.5.2.1 General

The problems of wind turbine monitoring are clearly demonstrated in Figure 9.9 showing 18 days of SCADA power, wind speed, rotor speed and generator-bearing temperature data from a single wind turbine. During this period a significant storm and high winds were experienced for 2 days, whilst for the remaining days the wind turbine operated each day with a diurnal wind speed variation. The reader can see the wide and rapid changes of power and rotor speed, which must be accommodated in any analysis of the signals for detection, diagnosis and prognosis of faults.

#### 9.5.2.2 Gearboxes

The most common monitoring analysis applied to wind turbine SCADA data has been to study changes in the wind turbine power curve, as shown in Figure 9.3.

However, more detailed analysis is possible and is exemplified here by studying SCADA signals for two wind turbine sub-assemblies of concern, the gearbox and converter.

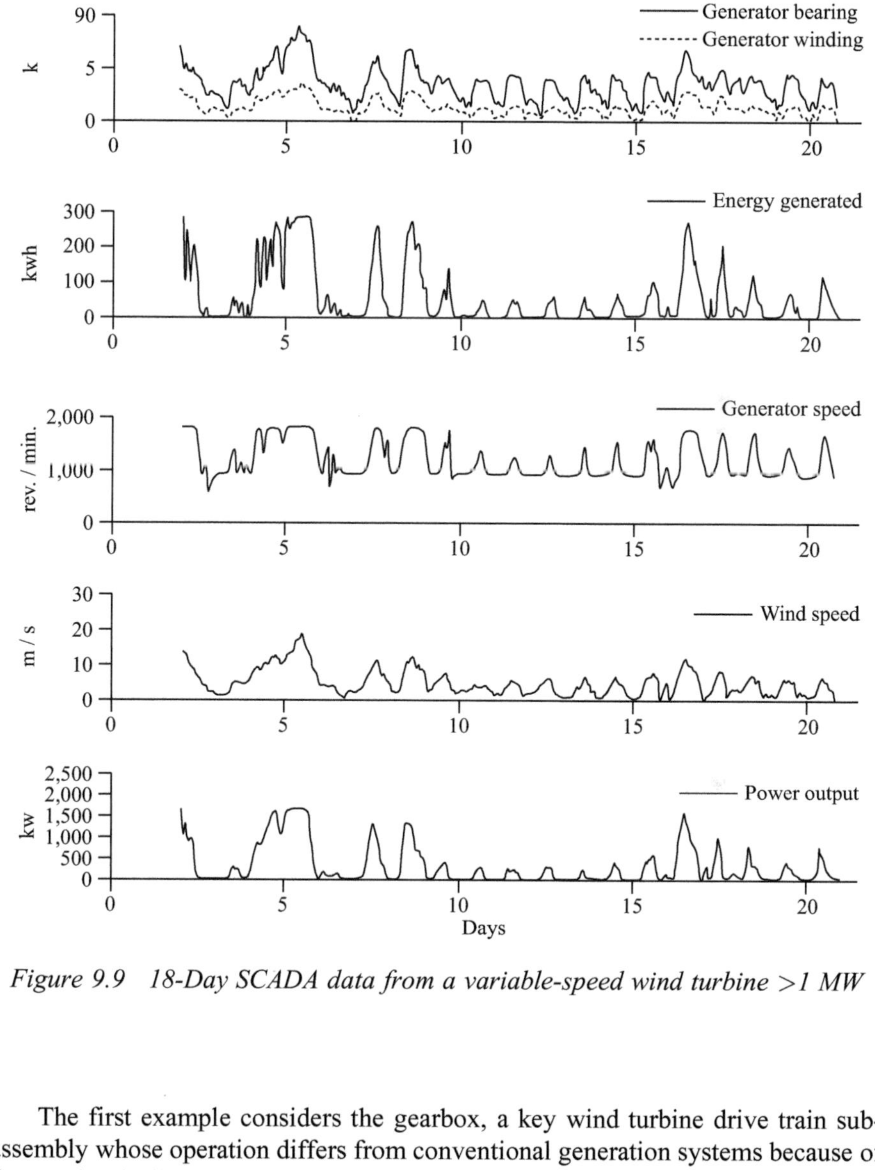


*Figure 9.9 18-Day SCADA data from a variable-speed wind turbine >1 MW*

The first example considers the gearbox, a key wind turbine drive train sub-assembly whose operation differs from conventional generation systems because of the stochastically varying torque experienced in the wind turbine, considered a major root cause for gear and bearing fatigue. Gearbox root cause analysis requires detailed understanding of the effects of the operating environment and cumulative high- and low-cycle fatigue damage, using information from the gearbox and its neighbouring sub-assemblies, the rotor and generator. This can simply be analysed from SCADA data, using a physics of failure approach, by monitoring the transmission efficiency and rotational speed, at different shaft stages and relating them to the gearbox temperature rise to detect and prognose fault development. The heat generated in a gear stage or bearing will be proportional to the work done on that component, which means

$$Q \alpha W \alpha \Delta T \tag{9.1}$$

where $Q$ is the heat generated from the gear stage or bearing, $W$ the work done upon it and $\Delta T$ is its temperature rise above nacelle temperature. The work done by a gear stage can be physically expressed as

$$W = \frac{1}{2} I \omega^2 \tag{9.2}$$

Supposing the gear efficiency is $\eta_{gear}$ and the bearing has efficiency $\eta_{brg}$, the energy dissipated will be transferred as heat onto the gear or the bearing. Therefore,

$$Q_{gear} = (1 - \eta_{gear}) \frac{1}{2} I_{gear} \omega_{gear}^2 = k_{gear} \Delta T_{gear} \tag{9.3}$$

or:

$$Q_{brg} = (1 - \eta_{brg}) \frac{1}{2} I_{brg} \omega_{brg}^2 = k_{brg} \Delta T_{brg} \tag{9.4}$$

also expressed as the inefficiency $1-\eta_{gear}$:

$$1 - \eta_{gear} = \frac{2k_{gear} \ \Delta T_{gear}}{I_{gear} \ \omega_{gear}^2} \tag{9.5}$$

or

$$1 - \eta_{brg} = \frac{2k_{brg} \ \Delta T_{brg}}{I_{brgr} \ \omega_{brg}^2} \tag{9.6}$$

Therefore, $2k/I$ is constant for any gear stage or bearing whose inefficiencies will then be proportional to $\Delta T_{gear}/\omega_{gear}{}^2$ or $\Delta T_{brg}/\omega_{brg}{}^2$, respectively. When a fault, leading to an efficiency reduction, occurs in a gear stage, (9.6) shows that $\Delta T_{gear}$ will increase for the same $\omega_{gear}{}^2$.

Assuming that the remainder of the kinetic energy transmitted through gearbox is converted into generator power output such that

$$P_{out} = W - Q_{gear} \tag{9.7}$$

then

$$P_{out} = \eta_{gear} \frac{1}{2} I_{gear} \omega_{gear}^2 \tag{9.8}$$

By comparing (9.3) and (9.8), we have

$$\frac{1 - \eta_{gear}}{\eta_{gear}} = k_{gear} \frac{\Delta T_{gear}}{P_{out}} \tag{9.9}$$

or

$$\Delta T_{gear} = P_{out} \frac{1}{k_{gear}} \left( \frac{1}{\eta_{gear}} - 1 \right) \tag{9.10}$$

Equation (9.10) shows that the temperature rise of the gear stage is proportional to the power output $P_{out}$, given unchanged gear stage efficiency. At a certain power output, the efficiency $\eta_{gear}$ for a healthy gearbox under ideal conditions will be fixed; therefore, $\Delta T_{gear}$ is proportional to power output $P_{out}$. When a fault occurs in a gear stage, leading to an efficiency reduction, (9.10) shows that $\Delta T_{gear}$ must increase for the same power output $P_{out}$.

In the following, this approach has been used retrospectively on the SCADA data of a variable-speed wind turbine of ~2 MW, in which the maintenance record showed a subsequent catastrophic wind turbine gearbox planetary gear failure, undetected by any wind turbine monitoring system. The SCADA analysis has been done on data for three successive identical length periods:

- 9 months before failure,
- 6 months before failure,
- 3 months before failure.

Figure 9.10 shows the average temperature rise $\Delta T_{gear}$ bands plotted against $\omega_{gear}^2$ and grouped into the separate periods. The data for the 3-month period preceding the failure clearly show the worsening situation predicted by (9.10).

Alternatively, the wind turbine output power can be normalised to the rated power $P_N$ and the gearbox oil temperature rise $\Delta T_{gear}$ assumed proportional to the

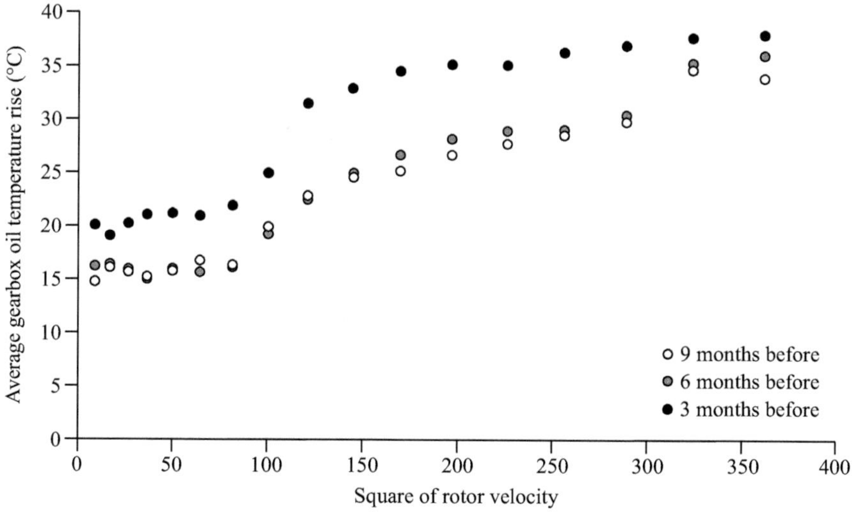


*Figure 9.10 Trends of gearbox oil temperature rise $\Delta T_{gear}$ versus square of rotor velocity $\omega_{gear}^2$*

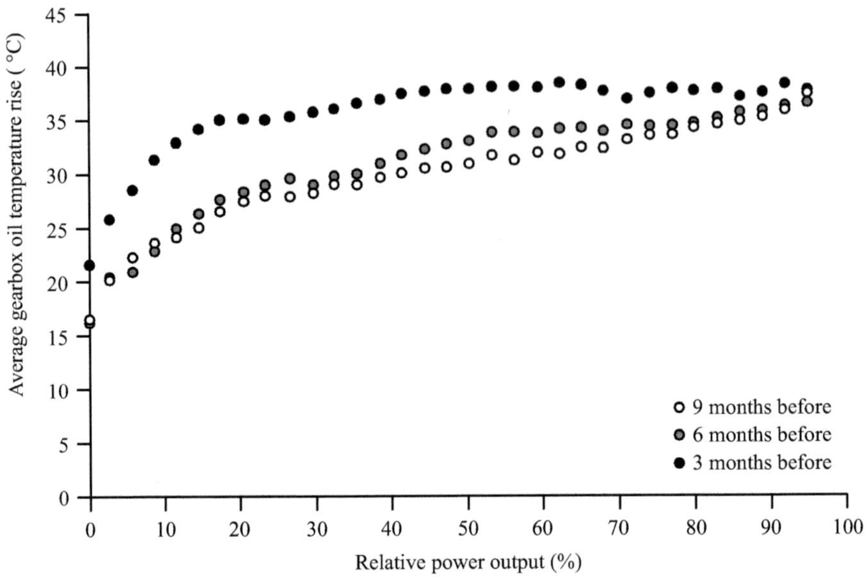


*Figure 9.11 Trends of gearbox oil temperature rise versus relative wind turbine power output*

power out, according to (9.10). Figure 9.11 shows this for the three periods. In this figure, the average gearbox oil temperature rise was binned for 50-kW power output increments in the three periods.

Figure 9.12 shows the histogram of frequency of gearbox oil temperature rises for the three periods.

Both Figures 9.10 and 9.11 clearly show rising gearbox inefficiency in 9 months before failure, with a worsening trend in the last 3 months predicting the failure.

The results of Figures 9.10–9.12, conforming to (9.5) and (9.10), demonstrate clearly that slow speed SCADA temperature data can provide long-term detection and prognosis of internal gearbox problems.

Probably, the simplest detection algorithm to adopt, based on Figure 9.12, would be to measure gearbox oil temperature rise and bin results into temperature rise bands, placing an alarm on bands above a 35°C rise.

#### 9.5.2.3 Alarm showers

Others have shown how SCADA monitoring can predict failures, including Zaher *et al.* (2009), Gray *et al.* (2009), Garcia *et al.* (2006).

Another SCADA data monitoring example, intending to predict wind turbine converter sub-assembly failures, investigated alarm showers from wind turbine controller alarms Qiu *et al.* (2012). This again adopted a physics of failure approach. To do this, the normalised cumulative percentage of selected generator,

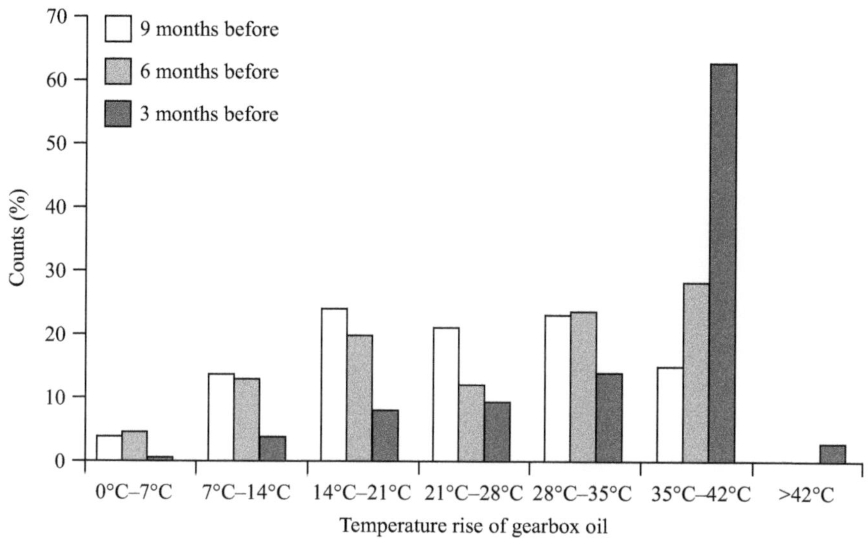


*Figure 9.12 Histogram of gearbox oil temperature rise frequencies*

grid and converter alarms was plotted against calendar time, Figure 9.13, during an extended period of operation for two specific variable-speed wind turbines of ~2 MW, chosen at random from the same wind farm.

Figure 9.13 shows the following:

- The impact of two-grid voltage dip incidents on days 39,200 and 39,500 in the two wind turbine alarm patterns, the same patterns were observed on other wind turbines in the WF during the same days.
- Serious grid voltage dips of >75% caused more than ten converter or inverter alarms during the period investigated.
- Converter general alarms were strongly correlated with grid voltage dip alarms, indicating grid voltage dip as a converter failure root cause.
- Wind turbine EFC pitch alarms also responded to these conditions and their alarms are also shown in Figure 9.13.
- The steps observed in the normalised cumulative alarm percentage indicate alarm triggers with long cumulative duration. The numerous alarms in these steps were accompanied by inverter IGBT failure alarms, suggesting the use of these steps to accumulate inverter sub-assembly stresses, giving advance converter fault warning.

In total 15–20 alarm triggers were observed for each wind turbine for each of these incidents. Therefore, for a WF with 30–35 wind turbines, 450–700 alarms could be triggered simultaneously by such incidents. With the probable repetition of some alarms, this suggests a possible alarm rate >1,000 per 10 min, suggesting the need to optimise wind turbine alarms.

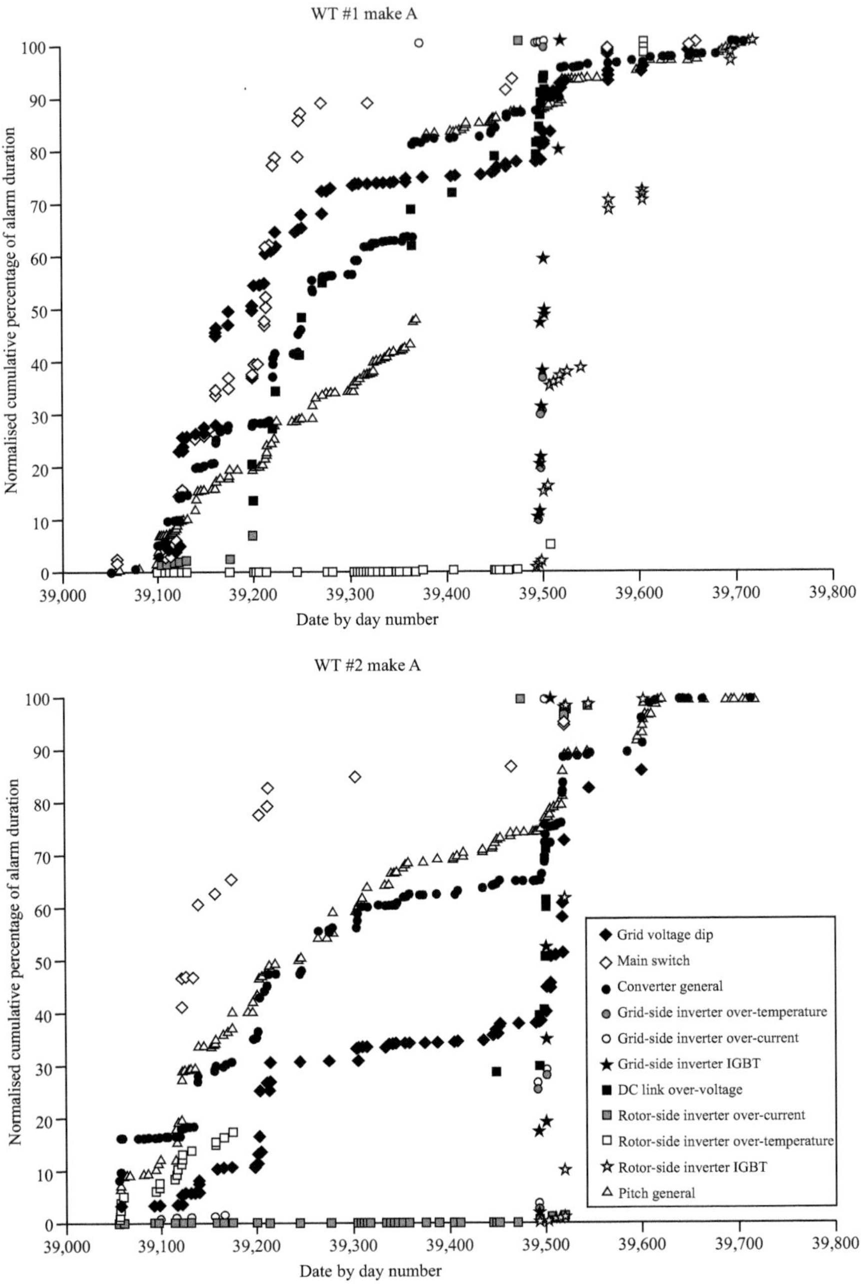


*Figure 9.13 Alarm showers from two wind turbines in the same on-shore wind farm: normalised cumulative alarm percentage duration versus calendar time, Qiu et al. (2012)*

Such simple algorithms could easily be implemented either in the wind turbine controller to give a graph and local alarm, or in a remote-control centre where the same algorithm could be applied to each wind farm turbine.

#### 9.5.2.4 Pitch systems

An important SCADA lesson came from turbine pitch systems, first indicated to be important wind turbine unreliability in Figure 3.9, Table 3.3 and Section 4.6. Defects in pitch systems affect potentially costly blade and main turbine structure defects. This work demonstrated that turbine reliability depends crucially on pitch system reliability so that early and ongoing monitoring could prevent more serious failures.

Chen, with a computer science background, described a pitch system analysis, using only SCADA data, from two wind turbine families:

- 22 Mitsubishi older, constant-speed, stall-controlled 1-MW wind turbines with 53 WT-years of SCADA data, installed in Brazos, USA.
- 26 Alstom newer, variable speed, pitch-controlled 1.67-MW wind turbines with 63 WT-years of SCADA data, installed in Spain.

The SCADA analysis employed two separate methodologies:

- Consideration of pitch system fault prognosis using an a-priori, knowledge-based, adaptive neuro-fuzzy inference system, Chen *et al.* (2013).
- Automated on-line fault pitch fault detection to provide prior fault prognosis, Chen *et al.* (2015).

The purpose of these two cases was to investigate the use of pitch control errors for giving an early warning of pitch control difficulties, thereby preventing later wind turbine damage. Figure 9.14 summarises the results.

The work is described more fully in Chen (2013), Doctoral Thesis, which has formed an important bridge between slow-speed, cheap SCADA and high-speed but more expensive CMS, demonstrating that slow-speed, SCADA, wind turbine pitch control signals can provide wind farm operators with early fault prognosis.

The work also hints at the future potential influence of artificial intelligence, and machine learning methods could have on SCADA and CMS to provide better, more valuable fault information to wind farm operators from scanning SCADA data, see Godwin (2015), Doctoral Thesis and Stetco *et al.* (2019).

#### 9.5.2.5 SCADA summary

These results show that SCADA is a rich source of data, which despite its slow data rate can provide clear indications of faults in pitch, gearboxes and converter systems.

Data have also shown that in certain conditions, it provides a reasonable prognosis horizon for upcoming faults, of up to 3 months.

However, SCADA data lack detail and require complex analysis, although they do hint at the potential for the application of artificial intelligence and machine learning, to provide a complete monitoring capability, see Stetco *et al.* (2019).

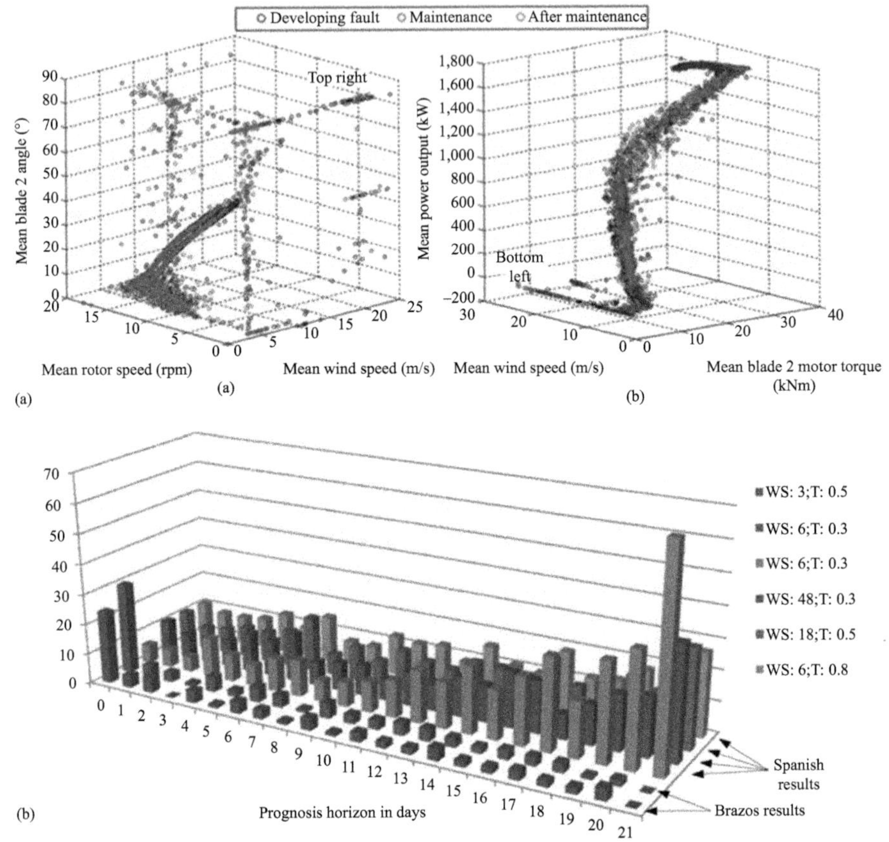


*Figure 9.14 Pitch defects and fault prognosis from pitch control SCADA data, Chen et al. (2015): (a) two different analyses of SCADA pitch signals, left blade angle, right turbine power; (b) defective pitch signal magnitudes from six sets of SCADA data from wind farms from 22 Mitsubishi (Brazos) and 27 Alstom (Spanish) WTs, showing fault prognosis horizon*

## *9.5.3 CMS successes*

### 9.5.3.1 Generators

The following are examples, using CMS signals, of successful detection of incipient faults in various sub-assemblies from operational wind turbines in the field and a wind turbine CMTR.

The first examples, using simple narrow band spectral analysis on electrical signals from a variable speed wind turbine CMTR, detected wind turbine DFIG generator rotor electrical unbalance in Figure 9.15 and mechanical unbalance in Figure 9.16. These results are taken from Crabtree (2011).

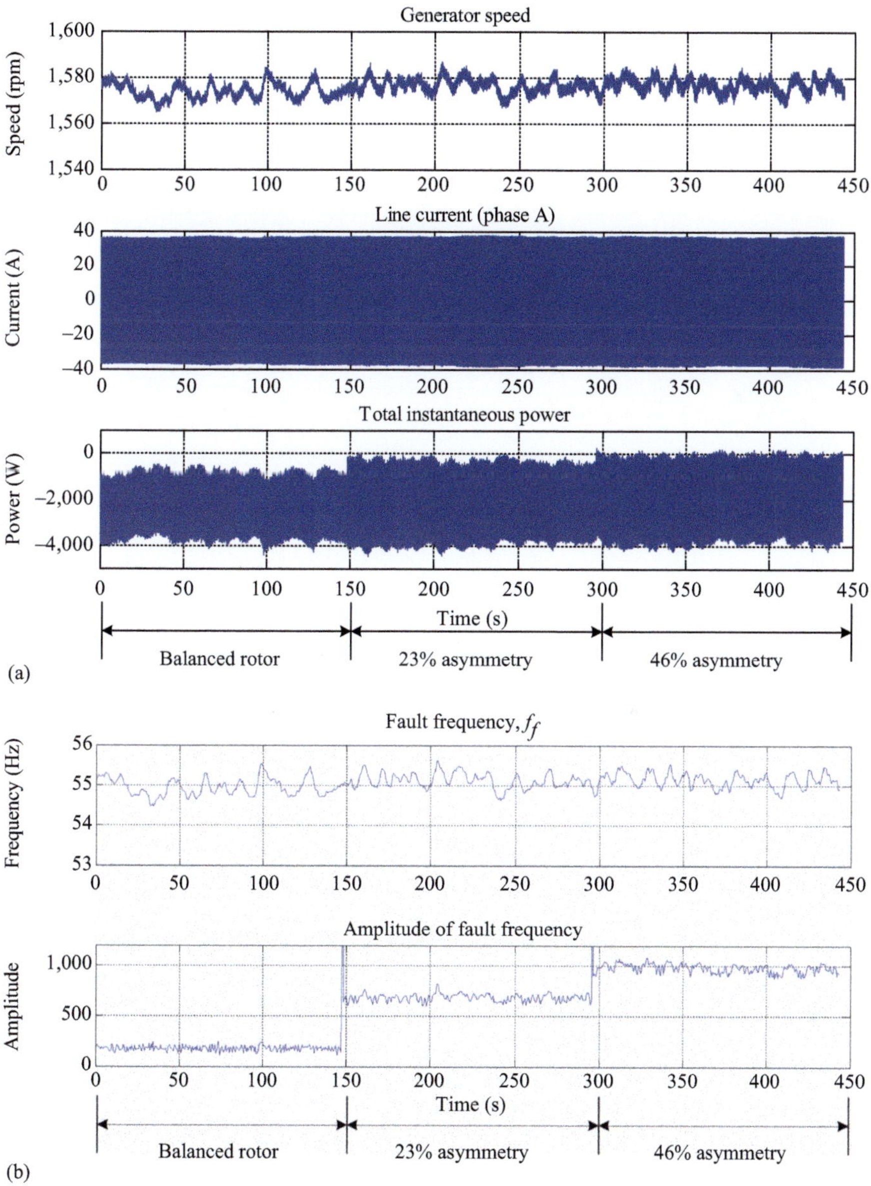


*Figure 9.15 From a variable speed wind turbine CMTR DFIG, spectral component analysis of rotor electrical asymmetry: (a) raw electrical signals monitored; (b) line current analysis $(1-2s)f_{se}$; (c) total power analysis $sf_{se}$*

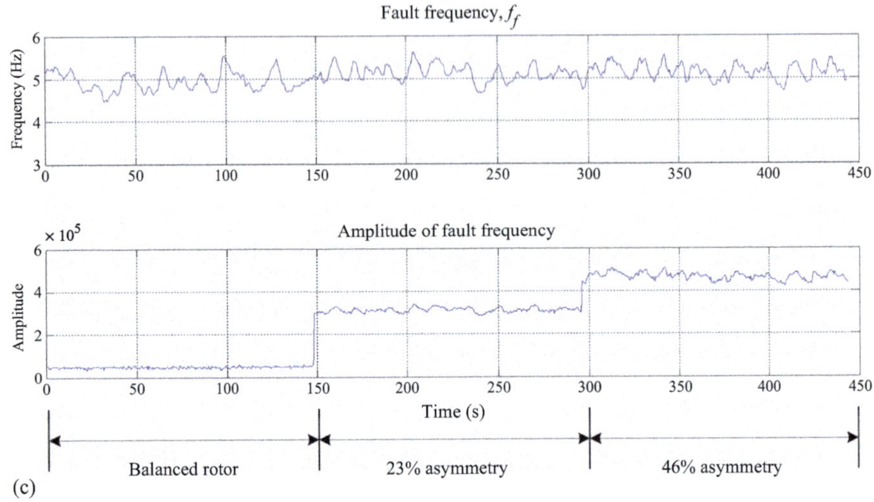


*Figure 9.15 (Continued)*

Figures 9.15 and 9.16 clearly show that wind turbine DFIG electrical or mechanical rotor faults can be detected by a simple narrow band spectral analysis of generator CMS electrical signals.

Parameter variations, accompanied by excessive currents, indicate an electrical machine's condition and such a method was first suggested by Ran *et al.* (1998) for investigating conducted EMC, then for machine monitoring by Bellini *et al.* (2000).

Although wind turbine generator phase current, under closed-loop control, is subject to control limits, this is intended to protect the converter rather than the generator. However, machine insulation breakdown, such as an inter-turn short circuits or core faults, may cause internal circulating currents in the generator which do not appear externally to the machine. CMS needs to detect rotor unbalance in such situations and VSD in-loop control allows us to do this.

Using the method of VSD control in-loop current detection proposed by Bellini *et al.* (2000), Zaggout (2013) developed current and power monitoring methods for detecting wind turbine DFIG rotor current unbalance. This was tested on a WTCMTR test rig, albeit with limited experimental results.

Under both fixed and variable speed conditions rotor winding unbalance was detected by measuring rotor current converter feedback signals, $i_d$ and $i_q$ from Figure 9.17, comparing the measured result against a control simulation. In so doing during variable speed operation, the method obtained benefit from the closed loop gain, as recommended by Bellini *et al.* (2000), increasing fault detectability and detection confidence, although the measured results proved less significant than the simulation, Figure 9.18.

This is a considerable step towards in-loop monitoring of VSD-fed machines, as needed for wind turbine generators as originally proposed by Bellini *et al.* (2000).

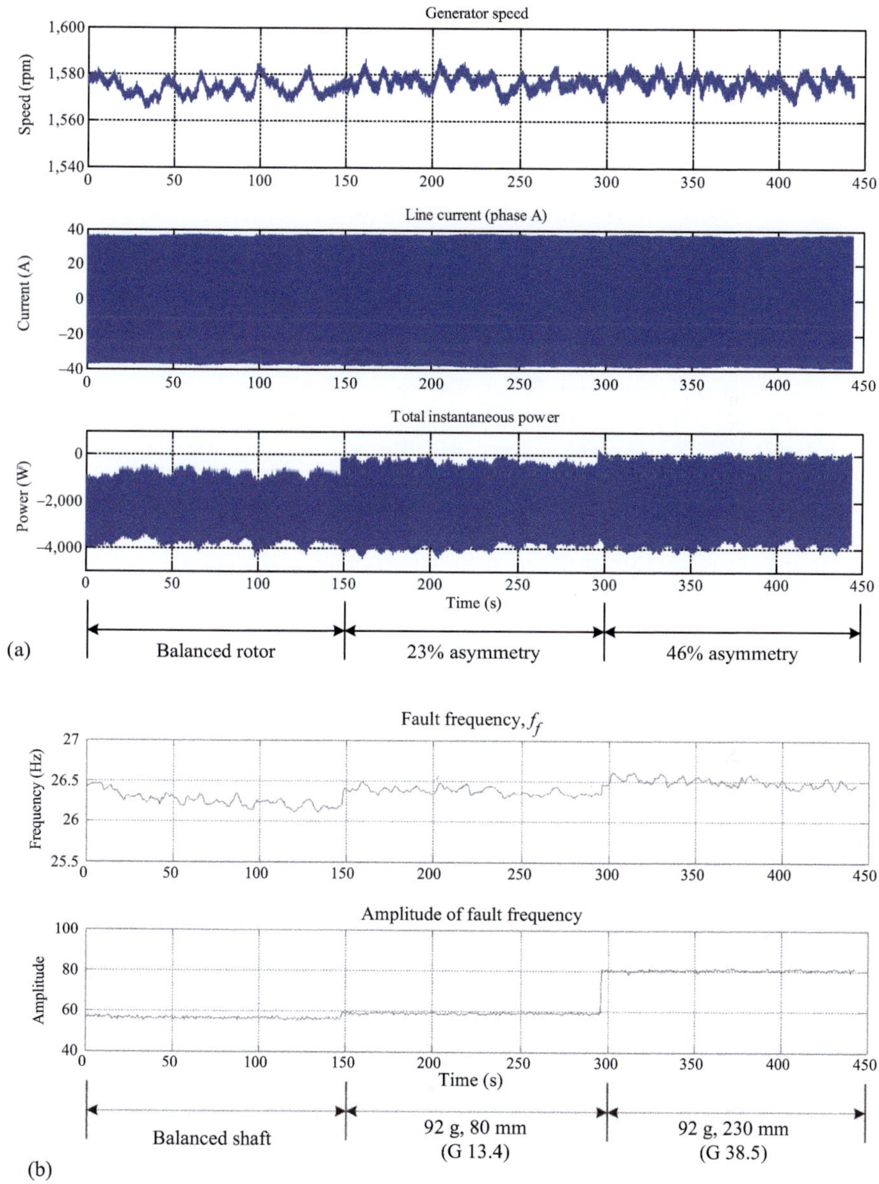


*Figure 9.16* *From a variable speed wind turbine CMTR DFIG, spectral component analysis of rotor mechanical asymmetry: (a) raw electrical signals monitored; (b) high-speed shaft displacement analysis $f_{rm}$, machine rotational speed; (c) narrow band analysis at machine rotational speed, $f_{rm}$, of a gearbox HSS accelerometer*

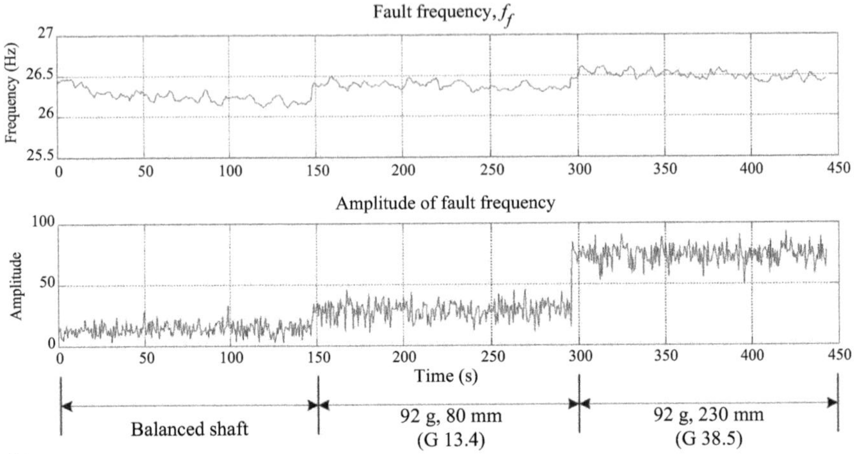


(c)

*Figure 9.16 (Continued)*

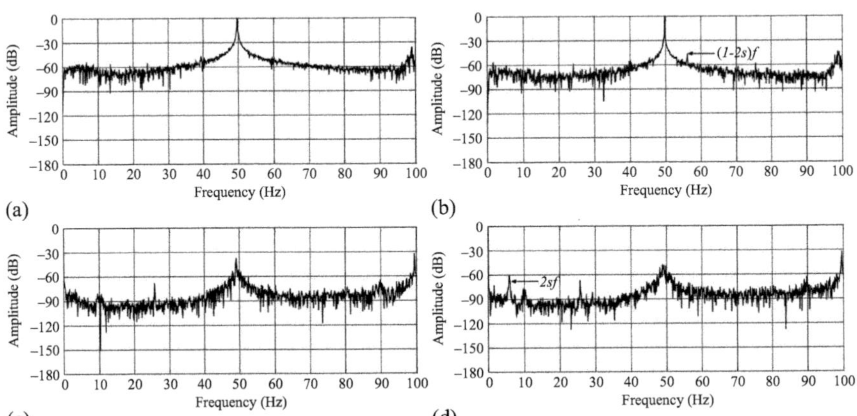


*Figure 9.17 Measured healthy and faulty stator current and power spectra from a DFIG with an unbalanced rotor fault, Zaggout et al. (2014): (a) current healthy; (b) current faulty; (c) power healthy; (d) power faulty*

Further work on a more sophisticated WTCMTR at Manchester University, Zappalá *et al.* (2019), showed how multiple signals, current power, vibration and speed, from a wind turbine generator can all be used to increase generator fault detectability.

### 9.5.3.2 Gearboxes

Turning now to the use of CMS on operational wind turbines in the field, Figure 9.19 shows rising gearbox high speed stage (HSS) vibration accompanied by

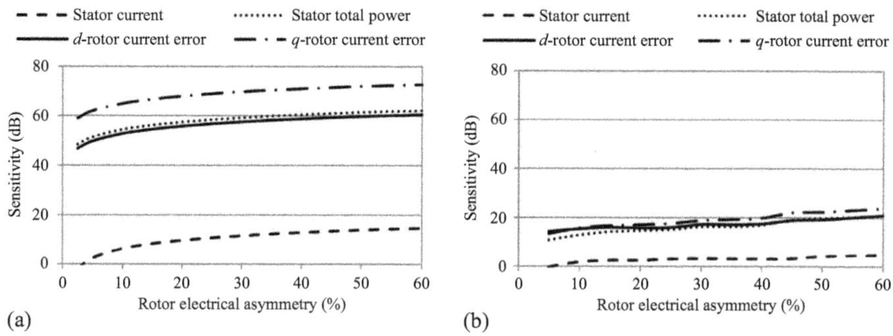


*Figure 9.18 Detectability of a DFIG unbalanced rotor fault from simulated or measured stator current and power signals, noting increased detectability with $i_d$ and $i_q$, Zaggout et al. (2014): (a) simulated; (b) measured*

a rising oil debris count, indicative of a deteriorating intermediate stage (IMS) bearing. The removed IMS bearing inner race is shown in Figure 9.19(c) to indicate the damage causing the indications earlier. The key points to note from Figure 9.19 are as follows:

- The combination of indications arising from two disparate sensors.
- The substantial warning obtainable from the measurements, in this case more than 120 days.;
- The opportunity to plot data against a variety of variables, see Chapter 2, Section 2.3, may improve detection visibility.
- In this case timely CMS detection of an incipient IMS bearing fault prevented bearing failure and potential failure of the whole gearbox.
- However, the most shocking revelation from this CM example, evident in the gearbox oil monitoring trace Figure 9.19(b), is that after changing the IMS bearing the gearbox oil was not cleaned, retaining all the metallic debris from the IMS bearing's collapse.

This kind of metallic debris will cause gear wheel damage, of the kind to be seen in Figure 9.21 in a 750-kW wind turbine gearbox.

In the case of Figure 9.19(b), the gearbox oil should have been cleaned by off-line filtration, see Section 9.3.2, or replaced. What the reader sees in this figure could never be permitted in a railway or aerospace industry maintenance regime.

The next example, Figure 9.20, uses CMS narrow band spectral analysis to detect progressive gear tooth failure in the 2-stage gearbox of a wind turbine CMTR. The sensitivity of detection shown in Figure 9.19(c) is relatively low but the fault is clearly visible.

The next example of CMS detection, Figure 9.21, looked at the same fault as Figure 9.20 but applied wideband spectral analysis and the result, Figure 9.21(b), is much more convincing than Figure 9.20(c).

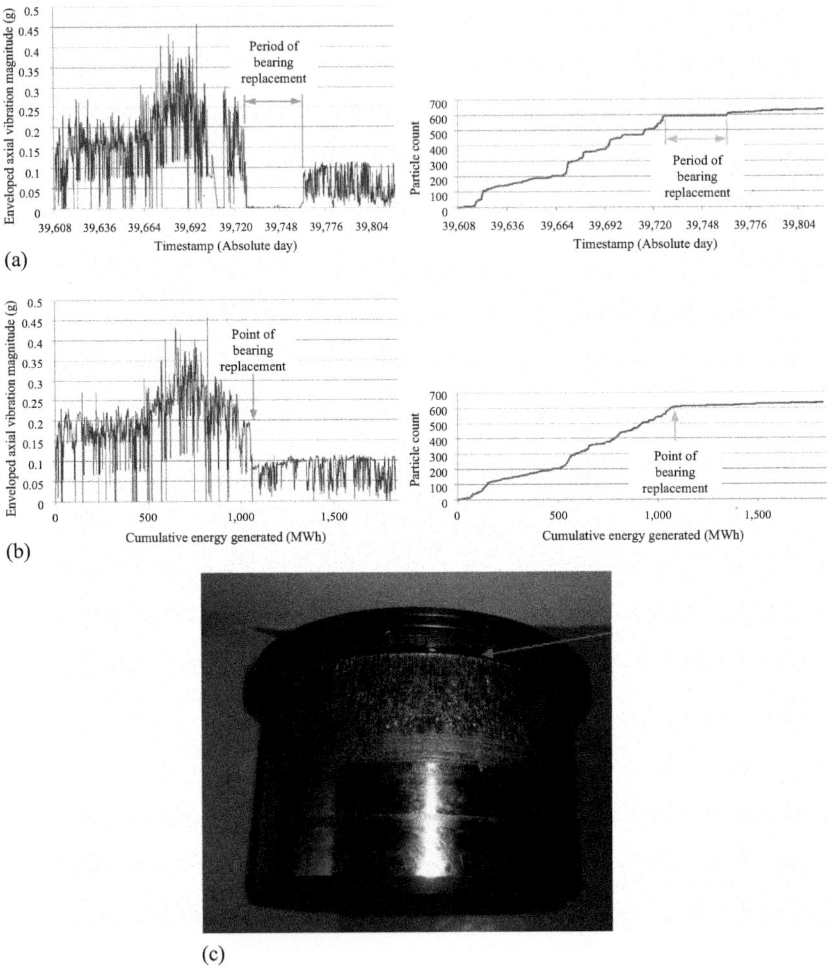


*Figure 9.19 From a ~1.5-MW fixed speed stall-regulated wind turbine gearbox, detection of incipient IMS bearing failure. HSS axial vibration amplitude envelope and simultaneous 100–200-μm oil debris, plotted against (a) absolute date stamp and (b) cumulative energy generated; (c) IMS bearing inner race under inspection following its replacement showing damage*

A final successful example of gearbox condition vibration monitoring, using sideband frequency analysis, is shown in Figure 9.22, which was reported by Zappalá, Tavner *et al.* (2014), comparing results from a small WTCMTR gearbox at Durham University with those from a full-size 750-kW test rig gearbox at NREL, US, Figure 9.22(a), each with damaged gearbox pinions.

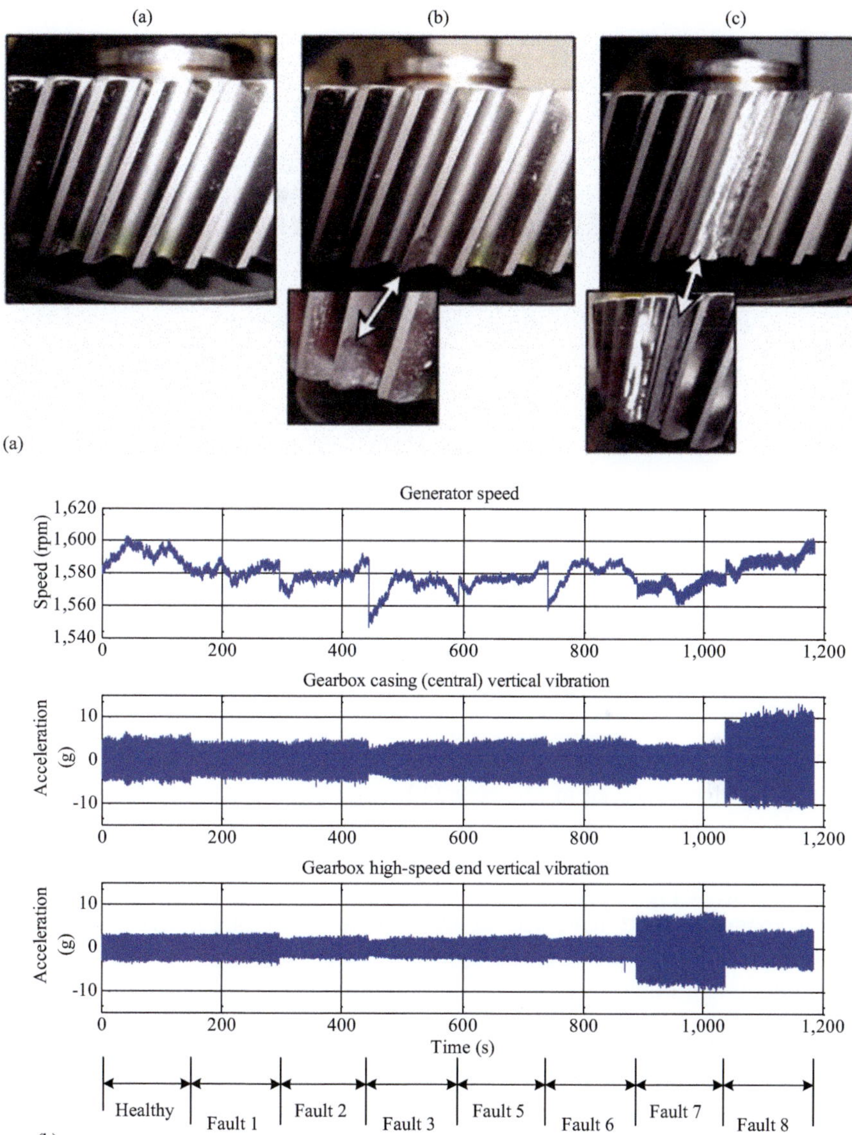


*Figure 9.20 From a variable speed wind turbine CMTR 2-stage gearbox, the detection of gear-tooth damage: (a) progressive gear tooth damage showing fault levels 1, 3 and 8; (b) raw electrical signals monitored; (c) narrow band analysis at machine rotational speed, $f_{rm}$, from a generator DE accelerometer*

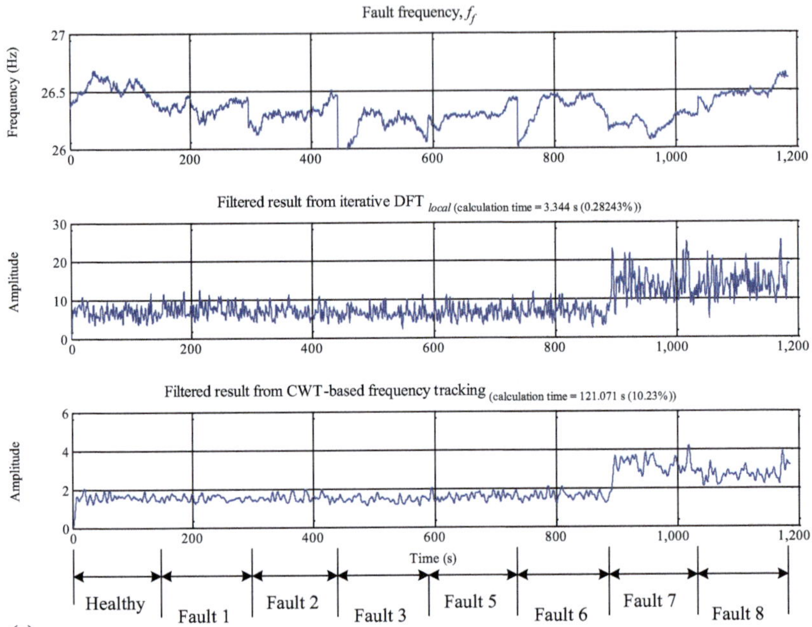


(c)

*Figure 9.20 (Continued)*

The volume of oil metallic debris shown in Figure 9.19(b) leads to gear pinion scuffing, like that observed in Figure 9.22(b).

Spectral vibration frequency analysis of the bearing in Figure 9.22(c) clearly shows the deterioration of this intermediate speed (IMS) pinion.

Note that in Figure 9.19(b) the debris in oil, not removed after that HSS bearing change, would have led to the kind of IMS pinion scuffing seen in Figure 9.22(b), so easily detectable by appropriate vibration sideband signal analysis, as shown in Figure 9.22(c).

#### 9.5.3.3 CMS summary

All these results show that CMS has strong potential to detect, diagnose and prognose faults in wind turbine drive trains, particularly in the generator, converter and gearbox and the off-shore wind industry should take note.

## 9.6 Data integration

### *9.6.1 Multi-parameter monitoring*

Despite the potential shown in the previous section, there is clear need for interpretation and the challenge is to achieve that detection, diagnosis and prognosis as automatically as possible to reduce manpower and access costs, Crabtree (2013) and Wilkinson (2008).

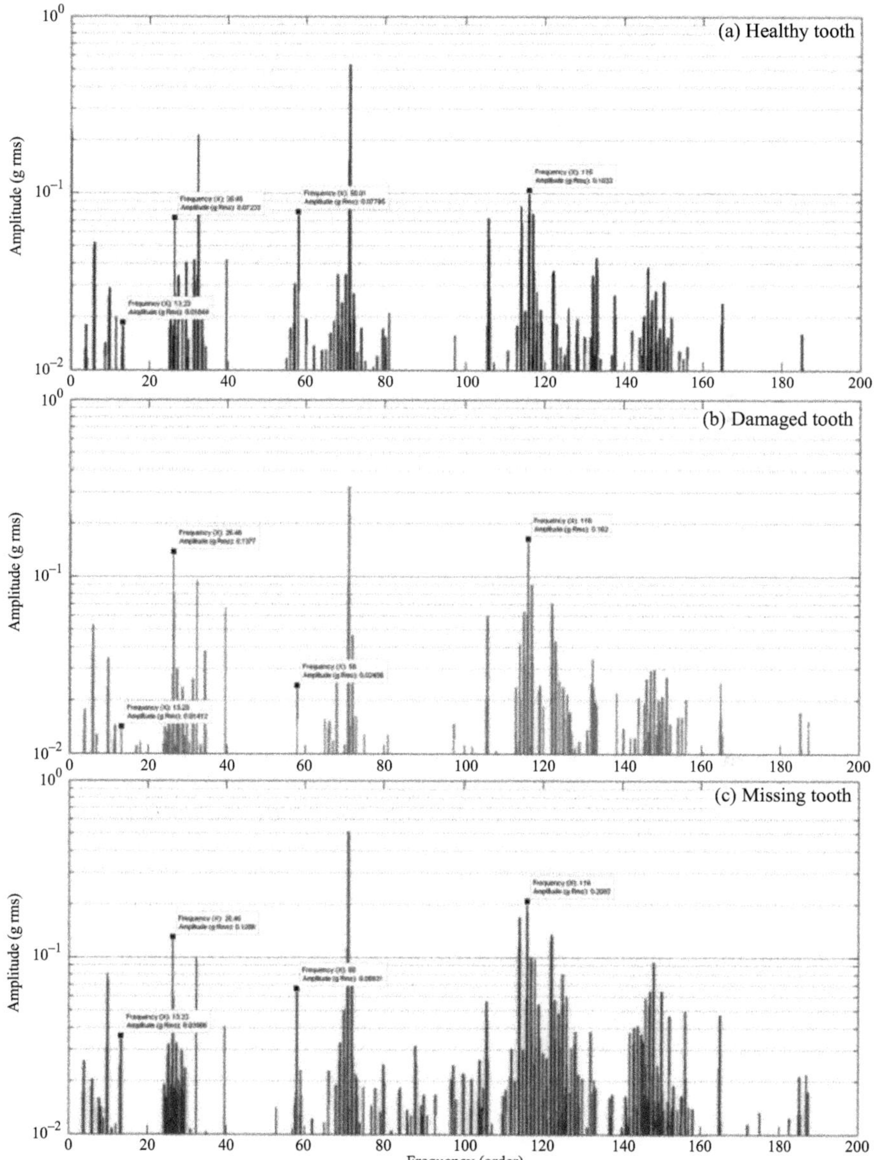


(a)

*Figure 9.21 From a variable speed wind turbine CMTR 2-stage gearbox, the detection of the same gear-tooth damage as Figure 9.20: (a) gearbox HSS vibration spectra of gear tooth damage; (b) gearbox HSS vibration spectral band amplitudes of gear tooth damage*

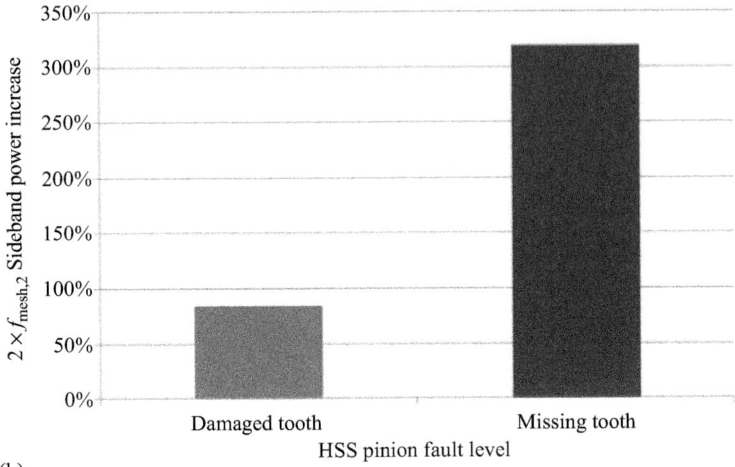

(b)

*Figure 9.21 (Continued)*

One aspect of monitoring which concerns operators is the reliability of the monitoring equipment and their detection reliability, exemplified by Sections 9.5.2 and 9.5.3.

The former is addressed by experience and the selection of systems from Appendices C and D. The latter depends upon the way the data are presented to the outside world and this can be seen from the figures in this chapter.

There is clear evidence that when a number of monitoring signals from disparate sources present confirmatory fault data, this is helpful and confidence-building to O&M Managers and Technicians alike. This is clear in the previous case of the gearbox bearing fault, as in Figure 9.15, where both vibration and debris counts were leading to the same conclusions.

This can be represented as follows, any CM signal sensor, e.g. for vibration or temperature, has a probability of detecting a fault in a sub-assembly, e.g. a gearbox.

The probability of accurate fault detection $P_{det}$ depends in part upon the sensor location $P_L$ and in part upon the sensor reliability $P_R$.

It has been reported in Tavner (2008), Zappalá *et al.* (2019) and Tavner *et al.* (2020) that relying on more than one CM sensor, e.g. $n$ sensors or multi-parameter monitoring, almost always increases the chances of successful incipient fault early detection, because if

$$P_{det\ n} = 1 - (1 - P_{Rn}P_{Ln})^n \tag{9.11}$$

It must be that, provided $P_{Rn}$ and $P_{Ln}$ have reasonable values, i.e. >50%:

$$P_{det\ n} > P_{det\ 1} = P_{R1}P_{L1} \tag{9.12}$$

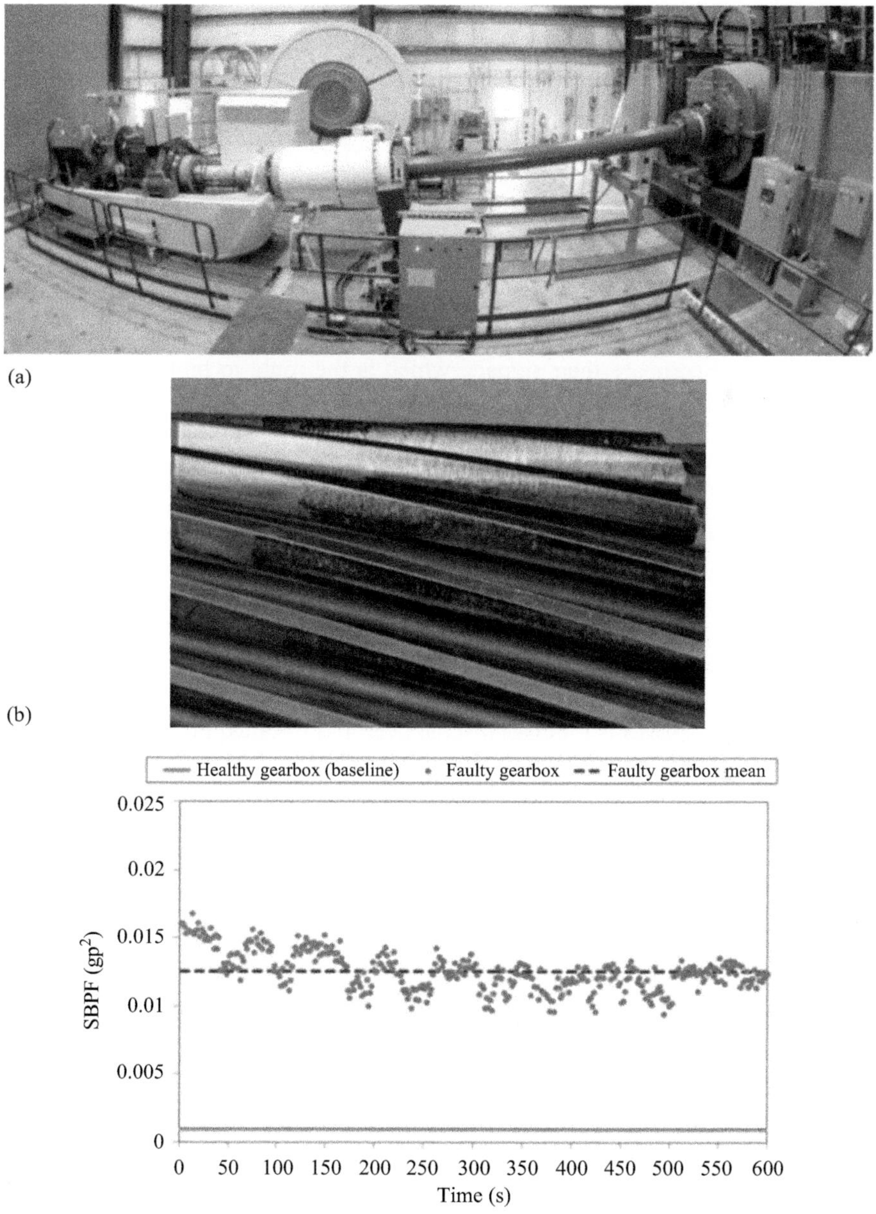


*Figure 9.22 Gearbox condition monitoring using NREL test rig: (a) NREL 750-kW wind turbine test rig; (b) gearbox high-speed pinion scuffing, due to oil contamination; (c) high-speed pinion accelerometer spectral analysis using sideband power factor (SBPF), demonstrating spectral change, before (lower line) and after the fault (upper line)*

That arises simply because of redundancy from sensor failure but more usually because sensors in different locations and different types raise detection probability and this must raise confidence in O&M Managers and Technicians that they are seeing a real effect.

The result of this is that there has been a tendency to swamp machines with sensors because data analysis and sensors are cheap and easy to monitor, particularly in SCADA, which can result in data overload as we are seeing in the wind industry.

However, there is a law of diminishing returns in (9.11), whilst two sensors may improve the $P_{det}$, five or six sensors produce much smaller improvement.

Therefore, operators are encouraged to reduce the number of sensors but raise their quality and compare their signals, which is the route to higher quality CM. This is the rationale for a greater integration of signal interpretation between SCADA and CMS with the objective of increasing the warning available from detection, which increases the prognostic horizon.

### *9.6.2 System architectures*

A barrier to truly integrated monitoring for the wind turbine is the current architecture, exemplified by Figure 9.1 and Appendices C and D, where monitoring systems are segregated, due to separations between wind turbines and their equipment OEMs.

SCADA data, both signals and alarms, are generated within the wind turbine controller, whereas CMS is purchased separately and installed on the wind turbine independently of its controller.

It is therefore physically difficult to integrate CMS and SCADA signals, notwithstanding their different bandwidths.

Some wind turbine controller OEMs, notably Mita Technik, Appendix C, are offering SCADA and CMS signal facilities within their controllers, where detection algorithms and alarm handlers can operate on both SCADA and CMS data, comparing trends and extending the prognostic horizon, Caselitz *et al.* (2005).

It is likely that the future of SCADA and CMS integration will lie in this direction, see Dao, Kazemtabrizi *et al.* (2019) and Dao, Kazemtabrizi *et al.* (2020).

## 9.7 Summary

This chapter has described the development of SCADA, SHM and CMS in wind turbines showing that the former is cheap and gives a breadth of coverage whereas the latter two are more costly but give great the depth of diagnosis, exemplified by the figures in this chapter.

The author has shown successful examples of SCADA, CMS and SHM monitoring giving reliable detection, diagnosis and prognosis of failure modes in the most important real wind turbine sub-assemblies in the field and on a wind turbine CMTR.

Good potential for fault detection with significant warning is possible with both SCADA and CMS, but there is clear evidence that integrating SCADA, CMS and SHM data would increase operators' and maintainers' field confidence by guiding them to the root cause of faults.

Finally, the chapter suggests how monitoring interpretations could be coordinated by integration between SCADA and CMS architectures to improve information for maintenance planning.

# Chapter 10
# Maintenance for off-shore wind farms

## 10.1 Introduction

The maintenance of wind farms has evolved organically from early days when many self-trained technicians maintained large numbers of small turbines on-shore in rural areas in Denmark, Germany and the USA, see Gipe (2016).

This has grown to a much larger industry, where wind turbine original equipment manufacturers (OEMs), like manufacturers Enercon and Siemens-Gamesa, trained their own staff, but also included engineers trained by large operators, like Ørsted (former DONG Energy).

The industry professionalised and was introduced to the more complex world of off-shore wind, dominated by the transport issues of mobilising staff to site and the need for data and maintenance methodologies.

This lead to an interaction between offshore wind investors and oil and gas energy concerns, who had off-shore experience working on very large assets, and the emergent wind industry, with knowledge of their own technologies but not of the off-shore logistics and their safety risks.

The costs involved in off-shore oil and gas maintenance regimes were initially unsupportable, for the lower value product throughput of inshore wind but gradually, experience and understanding between the oil and gas and the off-shore wind industry has led to improved maintenance methodologies with procedures based upon data and emphasis on accessibility and training becoming more important.

It is difficult to assess the true through-life maintenance costs of a new technology like off-shore wind but Crabtree *et al.* (2015) have provided a graphic from six different sources in Figure 10.1 comparing OPEX and CAPEX, suggesting that off-shore wind OPEX lies between 25% and 40% of the lifetime cost of energy (LCOE) for off-shore wind.

## 10.2 Maintenance methods

Maintenance methods can be categorised as shown in Figure 10.2. The maintenance strategy for off-shore wind is evolving. For on-shore wind power, the strategy is dominated by preventive maintenance, including planned maintenance scheduled by the wind turbine OEM maintenance manual instructions but affected by unplanned maintenance due to unscheduled stops of the wind turbine.

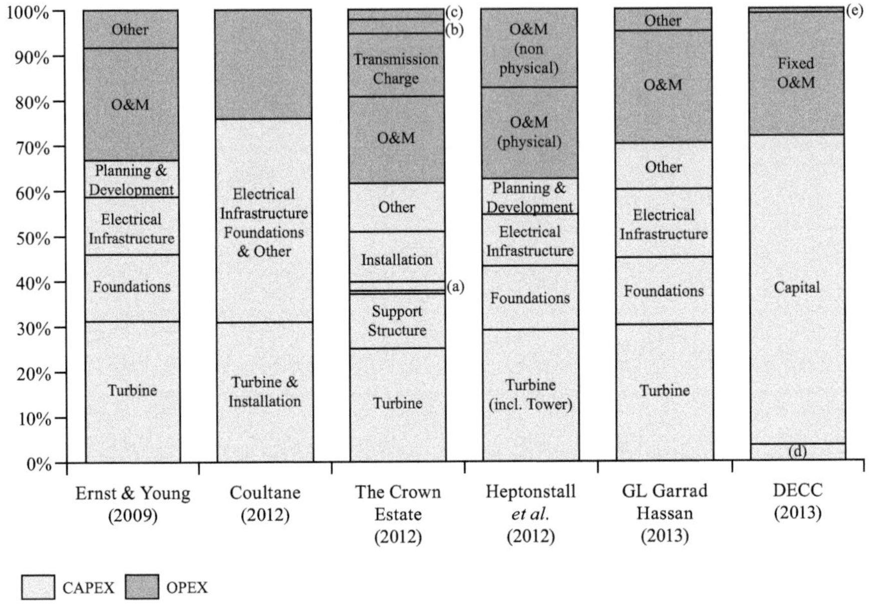


*Figure 10.1 Off-shore wind LCOE breakdown by CAPEX and OPEX from six studies, Crabtree et al. (2015): (a) array cables, (b) decommissioning, (c) seabed rent, (d) pre-development, (e) variable O&M*

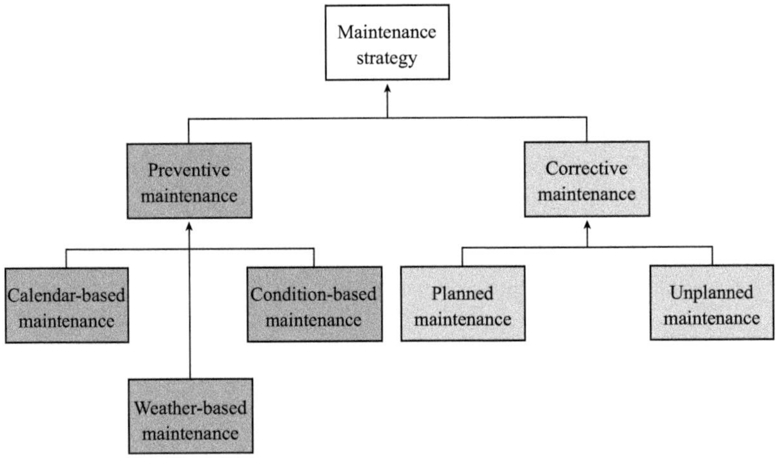


*Figure 10.2 Schematic overview of different wind farm maintenance strategies, Wiggelinkhuizen et al. (2008)*

Off-shore the exigencies of the weather and access mean that many preventive maintenance activities cannot be performed when scheduled and need to be done weather and calendar permitting and that induces a degree of planning and preventive action, which is creating a shift in off-shore wind O&M management towards a maintenance and asset management strategy.

## 10.3 Spares

Spares holdings have generally been the responsibility of wind turbine OEMs but as wind farm sizes have grown, the importance of having key spare sub-assemblies available for rapid change-out has become an important issue and this is of increasing concern off-shore where the window for repair, due to weather, logistic and other operational constraints, may be short. This issue has also expanded as maintainers have to deal not only with wind turbine spares but also those for the off-shore sub-stations and converter stations.

Spares holdings fall into two categories:

- Major spares with long manufacturing lead times, with a holding that relates to the maintenance and asset management strategy:
  - blades,
  - gearbox parts,
  - lubrication monitoring pack,
  - gearbox filters,
  - generator,
  - generator control modules,
  - hydraulic power pack,
  - inverter modules for wind turbine converter,
  - pitch motor mechanisms,
  - yaw motor mechanisms,
  - wind turbine transformer,
  - circuit breakers,
  - relays,
  - sub-station or converter station transformers,
  - inverter modules for HVDC converter station.
- Consumable spares in frequent and predictable demand, for which the holdings could be controlled as consignment stock. These spares could be summarised as follows:
  - lamps, buttons, switches and control relays;
  - wire and cable;
  - junction boxes;
  - pump motors;
  - filters;
  - heater units;
  - dehumidifiers;

- grease packs;
- lubricating oil packs.

## 10.4 Weather

The weather has major influence on off-shore wind farm maintenance, because of access and the issues presented in Chapter 3, primarily because of sea state as a result of wind speeds. Availability can reduce during the winter season, when wind speeds are generally higher and sea states worsen. This reduction in availability may be partly because of faults caused by worsening weather conditions but is primarily due to the fact that earlier wind turbine faults cannot be repaired because maintenance teams cannot get access to the asset.

However, wind turbine availability does not necessarily fall during worsening weather, which means that wind turbine OEMs and operators can avoid weather access effects by appropriate planning.

This means that wind turbine OEMs and operators must plan maintenance during periods of low wind speed when energy resource is less available, and access is easy.

Weather forecasting has therefore become an important aid to successful maintenance but that forecasting needs to be sufficiently reliable for the wind turbine OEM or operator to give at least 3 days' notice of significant weather changes. Early attempts at predicting weather using mobile phone weather resources proved rather unreliable.

Currently wind turbine OEMs and operators use local short-term weather window forecasts for this purpose, but national meteorological offices are developing tools to allow them to tailor their forecasts, accessing national data. This is a technique widely used by mariners, fishermen and oil and gas operators.

## 10.5 Access and logistics

### *10.5.1 Distance off-shore*

The ability to gain good access to the off-shore farms is pivotal to achieving the desired reliability and hence availability. This question of access has become more critical with the latest United Kingdom (UK) off-shore wind farm sites awarded in Rounds 3 and 4.

To give some idea of the increases involved, Table 10.1 gives the distances off-shore of several European wind farms to date after 2005.

Table 10.2 gives the distances off-shore for all of the UK Round 2 wind farms, coming into operation from 2012 to 2020,

Table 10.3 summarises the distance to shore of four UK East Coast harbours to two of the largest fields recently awarded by the Crown Estates in Round 3.

It is easy to see from Table 10.1 that physical access was not a major problem for those wind farm sites 3–23-km off-shore. A small vessel could take up to an

*Table 10.1 Off-shore wind farms >25× wind turbines, UK Round 1, Dutch, Danish and Swedish, 2005–12, distances from shore*

| Capacity (MW) | Turbine nos. | Wind farm | Min distance off-shore (km) | Max distance off-shore (km) | Country |
|---|---|---|---|---|---|
| 90 | 30 | Barrow | 7 | | UK |
| 90 | 25 | Burbo Bank | 5 | | UK |
| 90 | 30 | Kentish Flats | 8 | | UK |
| 60 | 30 | North Hoyle | 7 | | UK |
| 60 | 30 | Scroby Sands | 3 | | UK |
| 90 | 27 | Inner Dowsing | 5 | | UK |
| 97 | 30 | Lynn | 5 | | UK |
| 90 | 25 | Rhyl Flats | 8 | | UK |
| 90 | 30 | Robin Rigg A | 9 | | UK |
| 108 | 36 | Egmond aan Zee | 8 | 12 | NL |
| 120 | 60 | Prinses Amalia | 23 | | NL |
| 160 | 80 | Horns Rev | 14 | 20 | DK |
| 165.6 | 72 | Nysted | 6 | | DK |
| 110 | 48 | Lillgrund | 10 | | SE |
| **1,415** | **553** | **Sum/Average** | **9** | **16** | |

hour to reach the furthest wind turbine, and helicopter flying times would be measured in minutes, but then UK Round 1 and early 2 sites very rarely used expensive helicopter options for maintenance.

The UK Round 2 full progress is shown in Table 10.2.

However, with UK Rounds 3 and 4 sites at up to 210-km off-shore, Table 10.3, access will become more critical. An oil field support vessel (FSV) travelling from Blyth could take 10-h sailing time to the nearest edge of Dogger Bank and 17 h to the furthest edge of the site.

The advantages and disadvantages of various means of physically accessing increasingly large off-shore wind farms further from shore are set out next.

## *10.5.2 Installation access*

### 10.5.2.1 Mobile jack-up installations

Mobile jack-up installations are used during wind farm construction phases.

They give a fixed stable base for cranage to be able to precision lift larger components such as sub-station modules, wind turbine towers, nacelles and blades into position. They also have the advantage of being relatively unaffected by weather conditions once in place with the legs set down on the seabed and the main hull jacked out of the water.

They will be required during wind farm life for major refits, maintenance or repair jobs that will require heavy lifting capacity, such as turbine replacement or sub-station or converter station modification. The jack-up installation would be hired for the duration of the repair work.

*Table 10.2 Off-shore wind farms, UK Round 2, Progress 2012–20, distances from shore*

| Capacity (MW) | Turbine nos. | Wind farm | Turbine models | Mean distance off-shore (km) | Commission |
|---|---|---|---|---|---|
| 300 | 100 | Thanet | Vestas V90-2.0 MW | 11 | 2010 |
| 367 | 102 | Walney Phase 1 | Siemens-Gamesa SWT-3.6-107 | 14 | 2011 |
| | | Walney Phase 2 | | | 2012 |
| 504 | 140 | Greater Gabbard | Siemens-Gamesa SWT-3.6-107 | 23 | 2012 |
| 315 | 88 | Sheringham Shoal | Siemens-Gamesa SWT-3.6-107 | 17 | 2012 |
| 630 | 175 | London Array | Siemens-Gamesa SWT-3.6-120 | 20 | 2013 |
| 270 | 75 | Lincs | Siemens-Gamesa SWT-3.6-120 | 8 | 2013 |
| 389 | 108 | West of Duddon Sands | Siemens-Gamesa SWT-3.6-120 | 15 | 2014 |
| 219 | 73 | Humber Gateway | Vestas V112-3.0 MW | 15 | 2015 |
| 210 | 35 | Westermost Rough | Siemens=Gamesa SWT-6.0-154 | 10 | 2015 |
| 576 | 160 | Gwynt y Mor | Siemens-Gamesa SWT-3.6-107 | 20 | 2015 |
| 258 | 32 | Burbo Bank Extension | Vestas V164-8.0 MW | 6–13 | 2017 |
| 402 | 67 | Dudgeon | Siemens-Gamesa SWT-6.0-154 | 32 | 2017 |
| 659 | 40 | Walney Extension | MHI-Vestas 8.25 MW | 15 | 2018 |
| | 47 | | Siemens-Gamesa 7 MW | | |
| 573 | 91 | Race Bank | Siemens-Gamesa SWT-6.0-154 | 32 | 2018 |
| 400 | 116 | Rampion | MHI Vestas V112-3.45 MW | 13 | 2018 |
| 353 | 56 | Galloper | Siemens-Gamesa SWT-6.0-154 | 30 | 2018 |
| 588 | 84 | Beatrice | Siemens-Gamesa SWT-7.0-154 | 23 | 2019 |
| 1,218 | 174 | Hornsea 1 | Siemens-Gamesa SWT-7.0-154 | 120 | 2019 |
| 714 | 102 | East Anglia ONE | Siemens-Gamesa SWT-7.0-154 | 48 | 2020 |
| **8,945** | **1,865** | **Sum/Average** | | **24** | |

*Table 10.3 UK Rounds 3 and 4, distance from major UK East Coast ports to the largest proposed sites*

| Harbour – wind farm | Min distance off-shore (km) |
|---|---|
| Blyth – Z3 Dogger Bank | 201 |
| Blyth – Z4 Hornsea | 212 |
| Tyne – Z3 Dogger Bank | 197 |
| Tyne – Z4 Hornsea | 207 |
| Tees – Z3 Dogger Bank | 195 |
| Tees – Z4 Hornsea | 183 |
| Humber – Z3 Dogger Bank | 209 |
| Humber – Z4 Hornsea | 112 |
| **Average** | **190** |

*Figure 10.3 Example of a mobile jack-up installation*

For major or longer duration repairs, they provide a fixed platform to work from and can be connected directly to the wind turbine foundation by the means of adjustable gangways to allow for differential motion, due to sea-state or tide, giving continuous easy access between the workshop facilities on the jack-up rig and the wind farm asset. A jack-up rig is shown in Figure 10.3 during turbine erection but could also be used for large-scale in-service maintenance.

Advantages:

- Will achieve the required level of access needed year round.
- Oil and gas industry experience operating these types of vessel.

- Able to remain on location taking advantage of short weather windows.
- Capacity to carry a large range of spares and heavier components.
- Facilities onboard enable crews to achieve a longer more 'stable' shift patterns.
- Provides a stable platform for heavy lifts.

Disadvantages:

- Cost, likely to be greater than £2,000/h.
- Can only operate at one wind turbine or off-shore platform at a time.
- Requires good weather to jack-up/jack-down and move between locations.

#### 10.5.2.2 Fixed installations

Fixed installations are already in use on some off-shore wind farms. Their primary use is to house sub-stations and they were originally constructed using oil and gas platform techniques. To date they have not been continuously manned and may be used as refuges in the event of rapid change in weather conditions.

In far off-shore fields, it is likely that these installations may be manned all year round or at least for periods such as maintenance campaigns. The sub-station platform shown in Figure 10.4 is from the Horns Rev 2 wind farm off Denmark.

It is designed as a tubular steel foundation and building. It has a surface area of approx. $20 \times 28\ m^2$, placed some 14-m above mean sea level. The platform shown as an example accommodates the following technical installations:

- 36-kV switchgear;
- 36/150-kV transformer;
- 150-kV switchgear;

*Figure 10.4 Example of sub-station installation at Horns Rev 2.*

- SCADA, control and instrumentation system and communication unit;
- emergency diesel generator, including 2×50 t of fuel;
- sea-water-based fire-extinguishing equipment;
- staff and service facilities;
- helipad;
- crawler crane;
- man over-board boat (MOB).

For more remote fields, the staff and service facilities could easily be upgraded for permanent occupation. The MOB could also be upgraded to a transfer boat. The advantage of being on site is that short weather windows could be utilised. Minor wind turbine resets can be quickly achieved and more serious outages investigated and assessed quickly with the information passed back to shore for action.

## *10.5.3 Vessel access*

### 10.5.3.1 Without access systems

Vessels without access systems, Figure 10.5, have a cruising speed of 20 kn so take 1–5 h to get on-site to an inshore wind farm, where they then remain on standby till the maintenance crew need to return. They normally have a complement of around up to 12 personnel with 2 crew. Cooking and toilet facilities onboard make a pleasant working condition for the crews. Their catamaran hull design also makes them more stable. They have been successfully used in near shore wind farms up to 10–20-km off-shore in Rounds 1 and 2.

These vessels tend to be rated Marine and Coastguard Agency Class 2 allowing them to travel up to 111 km from a safe harbour. However, it is unlikely that they would be used for trips more than 74 km from shore because of the need for a sailing time of 4 h. It can be seen from Table 10.3 that these vessels would not be able to cover the Dogger Bank and Hornsea wind farms from the main North East coast ports likely to be used in UK Round 3.

Advantages of access transfer boats (Table 10.4):

- Simple marine engines, easily maintained.
- Low cost with fuel consumption at 100 L/h when cruising at a maximum 30 kn.
- Limited specialist training required for maintenance crews.
- Quick and responsive, already used on sites up to 10–20 km from shore.
- Could be used as an ‘infield’ vessel launching from a ‘mother ship’ or fixed platform.

Disadvantages of access transfer boats:

- Weather dependent, especially on sea state which must be <1.5-m $H_s$, making the achievement of >98% accessibility an impossibility.
- Transfer from the vessel to the tower is simple, the boat butts up against ladder and crew members jump onto the wind turbine ladder, see Richmond *et al.* (2018).
- Only limited amounts of equipment/tools can be transferred from the vessel to the wind turbine.

(a)

(b)

*Figure 10.5 Examples of access transfer boats*

#### 10.5.3.2 With access systems

To achieve the access levels needed to effectively operate a distant off-shore wind farm an oil FSV would be required, Figure 10.6(a). See Table 10.4 for calculated costs. The size of such a vessel, with dynamic positioning of a computer-controlled system to automatically maintain the position and heading of the vessel using its own propellers and thrusters, and a suitable access system, has been used successfully in the oil and gas industry to access unmanned off-shore platforms.

The vessel in Figure 10.6(a) has a dead weight of 4,577 t which is 90-m long with a deck length of 79 m, capable of taking 2,500-t deck cargo. The crane pictured is heave compensated and capable of lifting 200 t.

*Table 10.4 Calculation of hourly maintenance cost using transfer boats*

| | | |
|---|---|---|
| Hire and fuel costs | Spot market fuel cost £/tonne | £518 |
| | 12-h trip vessel rental and fuel costs average day rate | £1,727 |
| | Sail out and return journey, 2×74-km fuel cost | £207 |
| | *Based on 20-kn cruising speed giving 0.4 MT fuel used* | |
| | 8 h on location fuel cost | £207 |
| | *With no heavy seas and light sailing gives 0.4-MT fuel used* | |
| | **Total vessel hire and fuel costs/trip** | **£2,141** |
| Hours | Work hours/day estimated at 3×4 man crews×8 h | 24 h |
| | *This is based on 12-h shift less 4-h sailing out and return time* | |
| **Cost/h of transfer boat O&M work** | | **£89/h** |

Heave compensation is a hydro-pneumatic system that takes into account vessel heave to ensure that a crane hook remains stationary relative to the seabed or a fixed object external to the vessel. Maximum draft is 7.8 m. Maximum and cruising speeds are respectively 16.2/12.0 kn and respective fuel consumptions are 62/29 t/day. The vessel crew is 18 and up to 68 additional personnel can be accommodated if required. It has the ability to stay at sea for 5–7 weeks depending on sea state, conditions and fuel consumption.

The vessel in Figure 10.6(b) is purpose-designed for the off-shore wind industry, which has a dead weight of 4,367 t and 74 m long, as can be seen in the photo. It also has two access walkways to wind turbine platforms or sub-station, which are heave compensated, and the port walkway in Figure 10.6(b) is deployed onto a wind turbine transition piece.

There are several access systems being developed for use of these vessels including the following:

- Ampelmann, off-shore access system;
- personnel transfer system;
- sliding ladder (SLILAD);
- Momac Off-shore Transfer System.

Advantages of FSVs:

- Achieve required levels of wind farm access all year round.
- Experience in operating these vessels has already been in the off-shore oil and gas industry.
- Ability to remain on location in all weathers and therefore able to take advantage of short fair-weather windows.
- Capacity to carry a large range of spares and heavier components.

(a)

(b)

*Figure 10.6 Examples of field support vessels (FSV): (a) conventional off-shore oil and gas supply ship, FSV, BOURBON TOPAZ, with heave-adjustable cranes and derrick and simple access platforms to port and starboard, after the bridge; (b) purpose-built off-shore wind supply ship, FSV, SEAWAY MOXIE, with substantial maintenance crew accommodation with two more complex, heave-adjustable access platforms, to port and starboard*

- Facility to access off-shore wind turbine and sub-station platforms, see Figure 10.6(b).
- Facilities onboard for maintenance crew accommodation, allowing them to achieve a longer more 'stable' shift pattern, see Figure 10.6(b).

*Table 10.5 Calculation of estimated hourly maintenance of an oil and gas FSV*

| | | |
|---|---|---|
| Hire costs | Average day rate £10,000 | |
| | 14 days/trip vessel rental | £140,000 |
| Fuel costs | Spot market fuel £/tonne £300 | |
| | Sail out and return journey (2×222 km) fuel cost | £6,948 |
| | *Based on 12-kn cruising speed giving 29 t per 24 h fuel used* | |
| | 1 day in port fuel cost | £431 |
| | *1.5-t fuel for power generation* | |
| | 12 days on location fuel cost | £22,425 |
| | *With no heavy seas and light sailing gives 6.5 t per 24 h fuel used* | |
| | **Total vessel hire and fuel costs/trip** | **£169,804** |
| Hours | 12 days/trip | 432 h |
| | Work h/day estimated at 4 shifts×9 h 36 h | |
| | *This is based on 24-h working with 2×day and 2×night shifts* | |
| **Cost/h of FSV O&M work** | | **£393/h** |

Disadvantages of FSVs:

- Potential competition from oil and gas industry, see Figure 10.6(a), for the same vessels.
- Volatility in the day-rate based on demand due to oil and gas industry.
- Volatility in the FSV fuel bunker price as FSVs consume large amounts of fuel compared to small transfer boats, described earlier.

The calculation in Table 10.5 shows the hourly cost of maintenance using an oil and gas FSV. It assumes four crews, two crew by day/two crew by night, working two 12-h shifts, covering two wind turbines at each shift change. With pre-shift briefings and preparation, transfer times and rest periods during the shift, it is estimated that 9-h useful work can be achieved per crew per shift, based on oil and gas industry experience.

There are two volatile elements in this costing:

- First is the vessel day-rate which varies with daily demand and contract duration. The figure used in Table 10.5 is from the oil and gas industry vessel spot market for a 3-month contract.
- Second is the cost of fuel oil which varies with supply and demand.

Despite these variances, the previous calculation does give an indication that the hourly cost are much more favourable than helicopters, especially for more distant wind farms. But the main advantage of the FSV is its ability for 24-h working with two 12-h shifts giving an 8–10-h useful wind turbine work per shift. Such shift patterns are common in the oil and gas industry, so should not be problematic in the wind industry.

## 10.5.4 Aircraft access

### 10.5.4.1 Rotary wing

Although helicopters have been used as a means of transport to and from European and UK Rounds 1 and 2 wind farms, these were near-shore, $\leq 20$ km from land, for example Horns Rev 1, Figure 10.7.

The fact that because of visibility requirements, the dropping-off and recovery of maintenance crews would need to be done in daylight would also limit the time available for wind turbine work, especially in winter. Psychologically being left off-shore without cover, more than 2-h flying time away from base may prove difficult for maintenance crews to accept and result in important H&S issues if a casualty occurred. It should be noted that the oil and gas industry has tried to limit helicopters personnel movements, because historically and statistically this is the most dangerous aspect of an off-shore worker activity.

Another safety consideration is that the fields further off-shore cannot be covered by inshore lifeboats and there may, therefore, be a requirement to have a 'standby vessel' in field to provide safety cover when using helicopters adding to the cost. Such vessels are currently required in the oil and gas industry.

Examples of the cost per hour of maintenance for various types of helicopters are presented in Table 10.6. The figures show that smaller helicopters are cheaper to hire and run. However, it is likely that off-shore maintenance crews will not be less than three persons, for safety reasons. Also, the need for a helicopter winch operator indicates that the larger helicopters are likely to be necessary for maintenance at distant sites.

*Figure 10.7 Example of off-shore access to Vestas V80 wind turbines at Horns Rev by Eurocopter EC135, showing crew landing basket*

*Table 10.6 Calculation of hourly maintenance cost using helicopters compared to drones*

| Harbour – Wind farm | Min distance off-shore (km) | Max distance off-shore (km) |
|---|---|---|
| Blyth – Z3 Dogger Bank | 111 | 201 |
| Blyth – Z4 Hornsea | 105 | 212 |
| Tyne – Z3 Dogger Bank | 113 | 197 |
| Tyne – Z4 Hornsea | 98 | 206 |
| Tees – Z3 Dogger Bank | 103 | 195 |
| Tees – Z4 Hornsea | 77 | 183 |
| Humber – Z3 Dogger Bank | 107 | 209 |
| Humber – Z4 Hornsea | 29 | 112 |
| **Average** | **120** | **142** |
| Resources | S92 Helicopter | £30k/round trip off-shore |
| | Search & Rescue Helicopter | £9k/h |
| | Off-shore oil and gas helicopter personnel mobilisation | £50k/round trip to mobilise 2–3 persons |
| | Fixed wing drones | £3.5k/day |
| | Rotary wing drones | £2k/day |
| Costs | Flight time out (assume from inland +/−20 km) assuming Eurocopter EC135, 137-kn cruising speed | 1 h |
| | Take of landing and drop off/pick up | 0.5 h |
| | Flight time in | 1 h |
| | Total trip | 2.5 h |
| | Cost of seven-seater flight out £3,000 and return £6,000 | Cost of four-seater flight out £1,000 and return £2,000 |
| Time on site | Assuming shift 8 h and 3 h travel times = 5 h | |

However, it is noticeable that current off-shore wind turbines do not now seem to be fitted with the crew-landing baskets, previously visible in Figures 6.11 and 10.4, as fitted at Horns Rev 1.

Table 10.7 compares various helicopter rotor sizes and useful payload, which is proportional to the down-draft the aircraft can generate.

Large helicopters are expensive due to running/crew costs and are also in high demand from the oil and gas industry so the wind industry would be in direct competition for these machines. With the safety briefing, flying and winching time on-site the work period will be very limited, as will pay load for spares/tools. In

*Table 10.7 Comparison of various helicopter sizes*

| | Aircraft | No crew/ passengers | Rotor diameter (m) | Payload (kg) | Range (km) |
|---|---|---|---|---|---|
| Small | Bell 206B-3 | 1/4 | 10.2 | 674 | 693 |
| | Eurocopter EC135 | 1/7 | 10.2 | 1,455 | 635 |
| | MBB/Kawasaki BK 117 | 1/10 | 11.0 | 1,623 | 541 |
| Medium | Bell Twin Huey 212 | 2/13 | 14.6 | 2,119 | 439 |
| | Eurocopter EC155 B1 | 2/13 | 12.6 | 2,301 | 857 |
| | Sikorsky Spirit S-76 | 2/12 | 13.4 | 2,129 | 639 |
| Large | Bell 214-ST | 2/16 | 13.9 | 3,638 | 858 |
| | Sikorsky S-92 | 2/19 | 17.2 | 4,990 | 999 |
| | Eurocopter EC225 Super Puma MkII+ | 2/24 | 16.2 | 12,633 | 857 |
| Heavy lift | Boeing Chinook CH-47 | 3/55 | 18.3 (×2) | 12,495 | 2,252 |

Table 10.7, only two examples of small helicopters have been considered for the following for two reasons:

- First is rotor size, even for a small helicopter with rotor a diameter circa 10 m, an extended crew-landing basket is required to allow a safe stand-off for the helicopter rotor from the wind turbine blades.
- Second is the down-draft generated by the helicopter whilst hovering. In any larger helicopter than the sizes in Table 10.7, the strength of the down-draft may impose unacceptable stress on the wind turbine nacelle and crew landing basket.

The amount of equipment/spares that can be carried off-shore and lowered onto the wind turbine will possibly limit the maintenance to rudimentary servicing:

- They are still weather-dependent due to fog/wind/visibility.
- Can only drop off/pick up at the wind turbine in daylight.

Advantages of rotary wing aircraft:

- Experienced existing operators are available.
- Various aircraft sizes and capabilities are available.
- Quick access for assessing maintenance requirements or minor repairs.
- Suitable for close inshore wind farms where the helicopter can be quickly mobilised.
- Fast turn-around for emergency recovery of personnel direct to shore.
- They can operate independent of sea-state.
- Long operational experience with helicopters.

Disadvantages of rotary wing aircraft:

- Potential competition with oil and gas industry.

- Helicopter platforms on each wind turbine are expensive, even for large wind turbines.
- Costliness of operations.

#### 10.5.4.2 Fixed wing

Early experience in the United Kingdom is that fixed wing aircraft have not played a significant part in off-shore wind O&M work other than for surveillance. This may change with larger wind farms a long way off-shore, particularly in the case of off-shore floating wind farms, where distances to shore may be large.

Advantages of fixed wing aircraft:

- Experienced existing operators available.
- Various aircraft sizes and capabilities are available.

Disadvantages of fixed wing aircraft:

- Lack of knowledge of the potential usefulness of fixed wing aircraft for off-shore wind surveillance and sensing.
- Costliness of operations.

### *10.5.5 Drone access*

#### 10.5.5.1 Introduction

There has been an increasing interest by the off-shore wind industry in the use of drones, quite simply to lower costs by taking people out of the O&M loop. Drones could be used for the following:

- wind farm and turbine observation and surveillance to plan vessel movements;
- detailed structural survey photography;
- detailed sensing measurements, such as temperatures, deflections and gases.

The presence in large off-shore wind farms of substantial sub-station and converter station structures may also make it possible for drones, of limited range, to be deployed from those structures or even within structures such as sub-stations or even within large wind turbine nacelles, especially as we move to very large, 25 MW, floating wind turbines.

Information from drones could prove useful to maintainers, before mobilisation, by providing high-quality images and temperature sensing.

Figure 10.8 illustrates a fixed wing drone's surveillance capability off-shore.

#### 10.5.5.2 Fixed wing

Advantages of fixed wing drones:

- Various aircraft sizes and more capability becoming available.
- Long range capability >500 km.
- Low operational costs.

Disadvantages of fixed wing drones:

- Early state of fixed wing drone knowledge and usefulness for off-shore wind surveillance and sensing.

(a)

(b)

(c)

(d)

*Figure 10.8 Practical example of a fixed wing drone deployed for off-shore sensing and surveillance: (a) drone deployed from its container with local control room; (b) drone controlled from central control room; (c) drone operating around an off-shore asset; (d) drone's return to base*

*Table 10.8 Comparisons of O&M rates, range and payload for piloted and drone aircraft*

| Aircraft | Make | Type | Rate (£/h) | Range (km) | Payload (kg) |
|---|---|---|---|---|---|
| Fixed wing | Piloted Patana via or Diamond DA42 | Commercial, observation | 800 | 700 | Unknown |
| | Drone, Flylogix | Commercial, surveillance, sensing and load-carrying | 700 | 500 | 5–20 |
| Rotary wing | Piloted S92 | Commercial, load-carrying | 10,000 | 200 | 2,000 |
| | Piloted Eurocopter EC135 | Commercial, load-carrying | 3,000 | 150 | 272 |
| | Drone, Nordic, Schiebel Camcopter S-100 | Military, surveillance | 7,000 | 180 | 20 |
| | Drone | Commercial, surveillance, load-carrying | 1,000 | 5 | 5–100 |
| | Drone | Local access, surveillance, sensing and load-carrying | 100 | Unknown | 1–5 |

#### 10.5.5.3 Rotary wing

Early experience in off-shore wind with helicopters suggested that they are not a long-term cost-effective solution for access to turbine nacelles.

The use of FSVs is much more attractive and that has certainly been true during the construction phases and operations of Rounds 1 and 2 in the United Kingdom.

However, with the development of larger off-shore Rounds 3 and 4 off-shore wind farms, the need for faster more agile forms of access has switched from helicopters to fixed and/or rotary wing drones. The disadvantage of rotary wing drones is their short range, unless costly military versions are used. However, they may well find a place with off-shore wind farm sub-station, converter station fixed or floating installations and wind turbine nacelle structures, where they are of sufficient size and the drone is small enough.

A comparison of the use of fixed and rotary wing piloted aircraft and drones is given in Table 10.8.

Advantages of rotary wing drones:

- Various drone sizes becoming available.
- Small rotary wing drones could be beneficial for remote surveillance inside large off-shore structures, wind turbine nacelles, sub-stations and HVDC converter stations.
- Low operational costs.

Disadvantages of rotary wing drones:

- Short operational range, $<5$ km.

- Lack of knowledge of the usefulness of rotary wing drones for off-shore wind surveillance and sensing.

#### 10.5.5.4 Floating installations

Whilst helicopters have proven prohibitively expensive for routine O&M operations, they could prove useful during large maintenance campaigns to floating installations. Modern thinking for far floating off-shore wind farm routine maintenance is turning towards the following arrangements:

- Vessels to transport maintainers and provide a service hub.
- Fixed wing or rotary wing drones for visual and other sensing inspection.
- Small rotary wing drones for inspection inside enclosed spaces, for off-shore AC sub-stations or HVDC converter sub-stations or even large nacelles, where switchgear and inverter bridge components could be surveyed, inspected and monitored online, using thermographic, acoustic and partial discharge detection.

### *10.5.6 Access conclusions*

The previous analysis shows the following broad O&M hourly working costs for maintenance logistics:

- jack-up vessels for large campaign maintenance, >£2,000/h, dependent on sortie length;
- transfer vessels, £89/h, dependent on sortie length;
- FSVs, £393/h, dependent on sortie length;
- drone aircraft:
  - rotary wing:
    - small and simple: £100/h;
    - large and very intelligent: £6,000/h.
  - fixed wing:
    - small with surveillance and sensing capability: £700/h;
    - large with greater lift capacity: £800/h.
- full-size aircraft
  - small rotary wing: £3,000/h, dependent on sortie length;
  - large rotary wing: £10,000/h, dependent on sortie length.

The data show that whilst access transfer boats are a cost-effective solution for near-shore wind farms, they cannot be effective for the wind farms planned further off-shore, van Bussel *et al.* (2003).

Helicopters have been used successfully for some near-shore wind farms, Figure 10.4, but do not have sufficient range, lifting power and cost-effectiveness for further off-shore wind farms, where a change could be needed to larger heavy lift aircraft.

An alternative could be larger specialised FSVs, which could be cost-effective, jack-up vessels, or fixed installations combined with the wind farm sub-station infrastructure.

Each of these alternatives is being tried but it seems that fixed installations may prove to be the most cost-effective, supported by transfer vessels, rotary wing aircraft and/or drones.

An alternative future for distant off-shore wind farm accessibility will be purpose-built FSV vessels. With 20-year plus contracts and farms with wind turbine numbers potentially into three figures for wind farms to be financially worthwhile, building such vessels at the outset of a new development could be appropriate.

These vessels could be semi-submersible or catamaran-hull design for improved stability and high wave-height operability. They would also be dynamically positioned, with variable speed position control, to negate the need to anchor close to vulnerable seabed cables and utilities. Accommodation could be available for up to 100 marine crew technicians and specialists as required. A helideck would allow for crew changes by helicopter or medical evacuation if required. Such a vessel could be capable of staying on station for some months before returning to port for resupply. As these vessels are only in the early stages of deployment, the costs of purchase, hire or operation would be as yet unknown.

## 10.6 Data management for an integrated maintenance strategy

### *10.6.1 Sources and access to data*

Data to manage reliability, availability and maintenance come to the operator, maintenance staff and wind turbine OEM from a number of sources. These are set out in Table 10.9. Free access is not available to all these data for the operator because of the contractual arrangements in place between operator and OEM. However, it is clear that in order to meet items 7 and 8 in Table 10.9, data from items 4–6 should be integrated and measured against a baseline, which may be the performance of other off-shore wind farms but should also include measurement against item 1.

The challenge of Table 10.9 is how to integrate that data in an acceptable way so that operator, asset and maintenance management teams and maintenance technicians in the industry can work upon it, to meet the strategy and targets of items 7 and 8 and achieve a low cost of energy. This challenge is partly contractual but also technological and needs to feed the design and operation flow charts proposed in Figures 6.4 and 7.8, respectively. Sinha (2015), Doctoral Thesis, gives some guidance on the optimisation of off-shore wind farm maintenance.

The following sections develop such an off-shore wind farm knowledge management system.

*Table 10.9 Reliability, availability and maintenance data*

| Item | Data | Data owner | |
|---|---|---|---|
| | | **During warranty period** | **After warranty period** |
| 1 | Baseline reliability data about wind farm components from wind turbine OEMs and other wind farm component suppliers<br>This book is intended to provide some of these data | Wind turbine and wind farm component OEMs | Wind turbine and wind farm component OEMs |
| 2 | Wind turbine prototype test data | Wind turbine OEM | Wind turbine OEM |
| 3 | Wind farm component production test data | Operator | Operator |
| 4 | Wind farm commissioning data | Operator | Operator |
| 5 | SCADA and CMS from wind turbines and wind farm substation | Wind turbine OEM | Wind turbine OEM or Operator depending on maintenance contract |
| 6 | Wind farm maintenance records | Operator/wind turbine OEM | Operator |
| 7 | Asset management strategy | Operator | Operator |
| 8 | Contractual production targets | Operator/ Developer | Operator |

### *10.6.2 An off-shore wind farm knowledge management system*

#### 10.6.2.1 Structure and data flow in a wind farm

There are a number of closely interlinked industrial groups involved in the O&M of off-shore wind farms, which defines the structure of the data, aimed at the integration of O&M data as proposed in Chapter 8.

The work in the following sections are the product of many discussions within our Durham University Research Group and with other university and industry partners mentioned in the Acknowledgements, but most particularly the author is indebted to Christopher Crabtree in particular for leading this exercise.

This following structure may need adaptation to the organisation and conditions of individual operators and wind turbine OEMs. Parties can be grouped into six specific departments:

- Health monitoring (HM);
- Asset management (AM);
- Operations management (OM);
- Maintenance management (MM);
- Field maintenance (FM);
- Information management (IM).

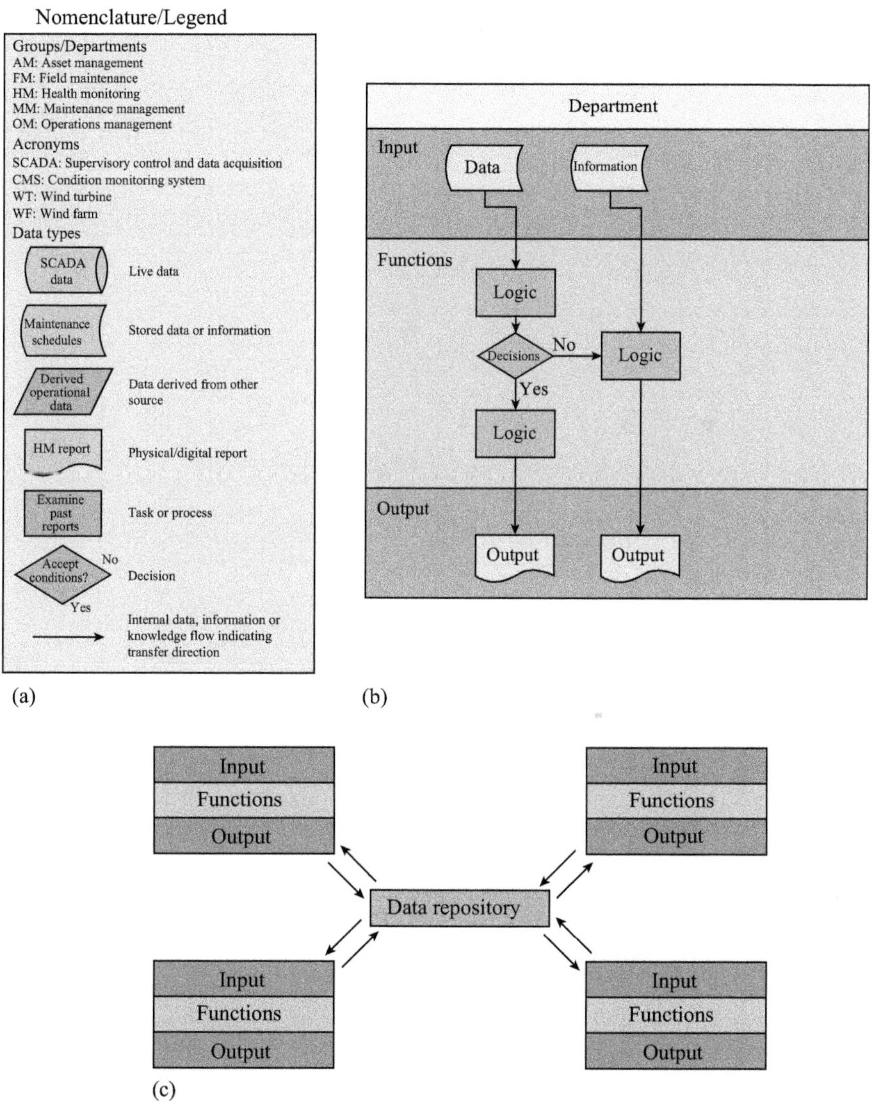


*Figure 10.9 Nomenclature, structure and organisation in a proposed off-shore wind farm knowledge management system; (a) nomenclature; (b) structure; (c) data flow*

The inputs, functions and outputs of each department are defined by means of a block diagram in the format shown in Figure 10.9.

The departments will be laid out in the following sections using the key/nomenclature and information flows shown in Figure 10.9. A wind farm will also produce various sets of live data shown in Figure 10.10.

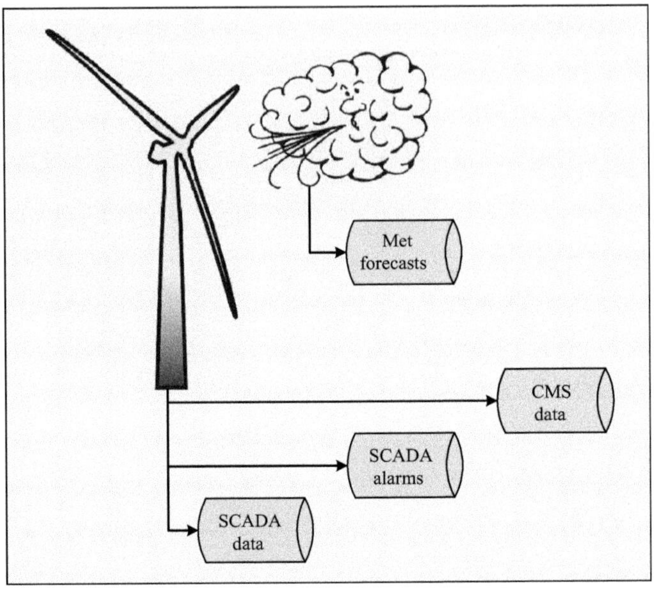


*Figure 10.10 Live data produced by an off-shore wind farms*

#### 10.6.2.2 Health monitoring

HM, Figure 10.11, is responsible for continuous monitoring of wind turbines to alert other groups to current and developing faults and advise on their severity (Table 10.10).

#### 10.6.2.3 Asset management

AM, Figure 10.12, is concerned with ensuring that operators and assets are operated in the most cost-efficient and valuable manner to secure the longest life cycle of profitable operation (Table 10.11).

#### 10.6.2.4 Operations management

OM, Figure 10.13, is concerned with achieving the required wind farm operation, meeting AM and grid requirements (Table 10.12).

#### 10.6.2.5 Maintenance management

MM, Figure 10.14, is concerned with implementing the requirements of AM, via OM, and responding to concerns raised by HM (Table 10.13).

#### 10.6.2.6 Field maintenance

FM staff, Figure 10.15, are responsible for the implementation of maintenance schedules and the confirmation of repair success (Table 10.14).

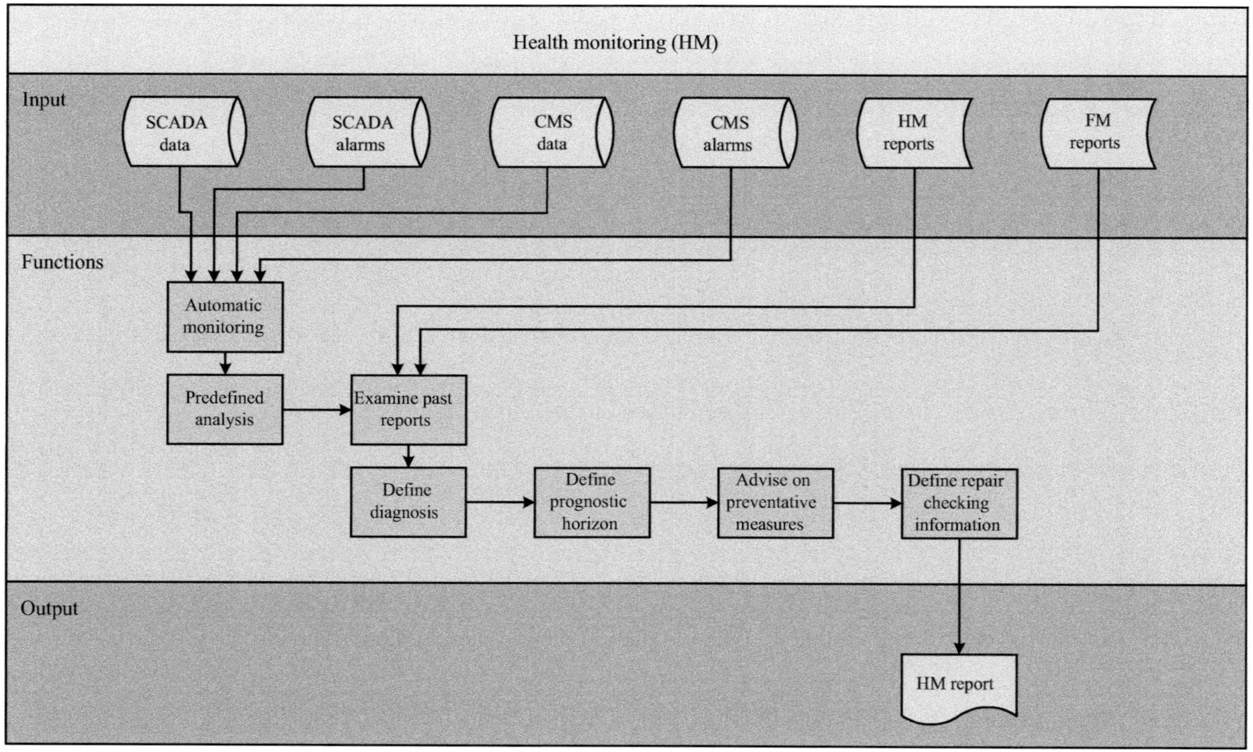


*Figure 10.11 Health monitoring structure*

*Table 10.10 Health monitoring department data*

| Inputs | Functions | Outputs |
|---|---|---|
| **Live data**<br>SCADA signals and alarms data<br>CMS signals and alarms data<br>**Stored information**<br>Field maintenance reports<br>Health management reports | 1. Apply expert knowledge to monitor wind turbine health via automated SCADA alarm and signal data, CMS alarm and signal data processing<br>2. Examine monitoring results and compare with historical FM reports and HM reports to identify completed repairs, known faults and further deterioration<br>3. Compare observed damage from FM reports with monitoring results to refine diagnostics<br>4. Generate HM reports, including fault development, expected time to failure, delaying measures and maintenance recommendations<br>5. Define information to allow FM to confirm repair success and include in HM reports | **Reports**<br>Health monitoring report |

#### 10.6.2.7 Information management

IM, Figure 10.16, handles the data produced by the wind farm (Table 10.15).

### *10.6.3 Complete system*

Finally the complete structure for these data is given in Figure 10.17 and represents a proposal for an off-shore wind farm knowledge management system.

For an example of integrated condition-based maintenance modelling, based on this IM structure, see Dao, Kazemtabrizi, Crabtree *et al.* (2021).

## 10.7 Summary

This chapter has shown how off-shore O&M needs to be managed and the key to that information and most importantly data from the operating plant.

It is clear that FSV vessels will continue to prove important for supporting off-shore wind, but experience with helicopters has proven to be costly and more attention should be paid to ensuring that FSVs have the right access equipment and that fixed wind drones are likely to be valuable far off-shore, whilst rotary wing drone may well prove valuable inside off-shore assets.

From the previous analysis, it is certain that we will continue to need to carry maintenance staff to wind turbines, sub-stations and converter stations but that improved methods for despatching those personnel to site, whether by vessel or helicopter, are needed. It also seems inevitable that there will be an increased use of drones, fixed or rotary wing, to collect the information necessary to deploy physical staff to remote off-shore locations in a safe and timely way and for the shortest period necessary to limit maintenance costs but that when those staff arrive, they have all the information at their fingertips to carry out an effective repair.

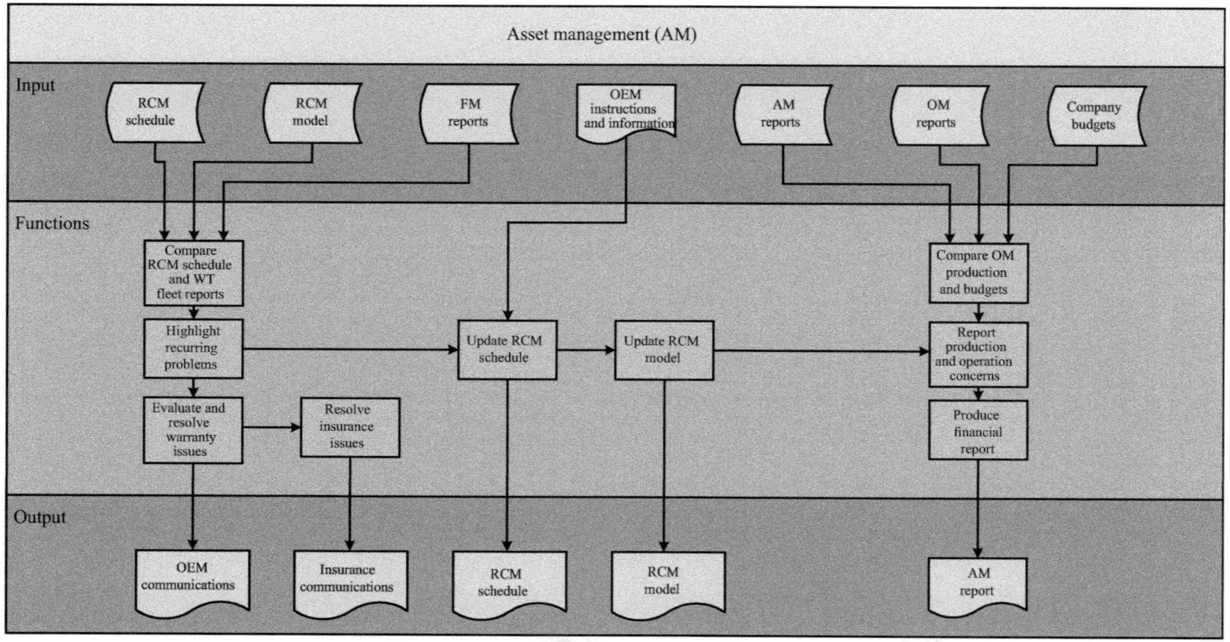


*Figure 10.12 Asset management structure*

*Table 10.11 Asset management data*

| Inputs | Function | Outputs |
|---|---|---|
| **Stored information**<br>Company budgets<br>RCM schedule<br>RCM model<br>FM reports<br>AM reports<br>OM reports<br>**External information**<br>OEM instructions and information | 1. Compare OM report outcomes with AM report plans and company budgets and query discrepancies<br>2. Examine reliability from FM reports and compare with RCM model<br>3. Ensure cost-effective use of assets using FM reports, OM reports and past AM reports<br>4. Produce finance reports for inclusion in AM reports<br>5. Communicate common and design/type failures with OEM and resolve warranty cases<br>6. Health and safety evaluations (HSE)<br>7. Deal with warranty issues<br>8. Deal with insurance issues | **Reports**<br>OEM communications<br>Insurance communications<br>RCM schedule<br>RCM model<br>AM report |

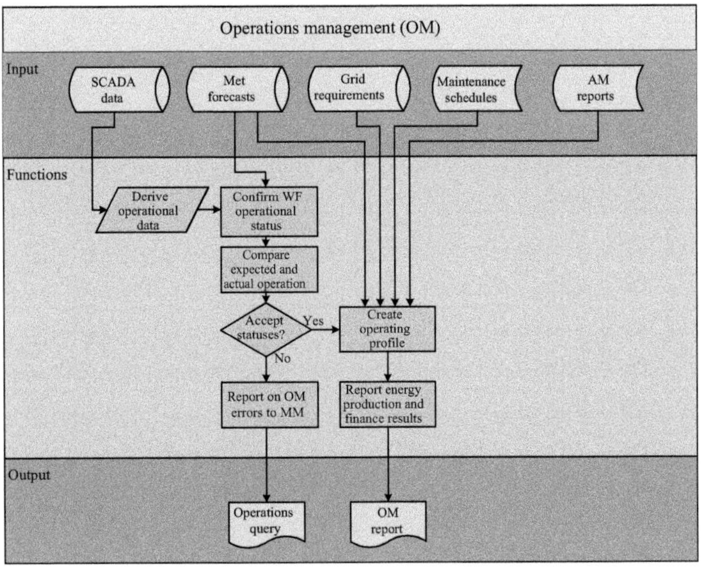


*Figure 10.13 Operations management structure*

*Table 10.12 Operations management data*

| Inputs | Functions | Outputs |
|---|---|---|
| **Live data**<br>SCADA data<br>Met forecasts<br>Grid requirements<br>**Stored information**<br>Maintenance schedules<br>Asset management reports | 1. Compare current WF operating conditions (derived from SCADA) with maintenance reports and query discrepancies with MM<br>2. Plan and implement WF operating schedules based on grid requirements, Met forecasts, maintenance schedules and AM report<br>3. Report financial results and energy generated compared to requirements in OM reports | **Reports**<br>O&M reports<br>**Direct reports**<br>Operations query |

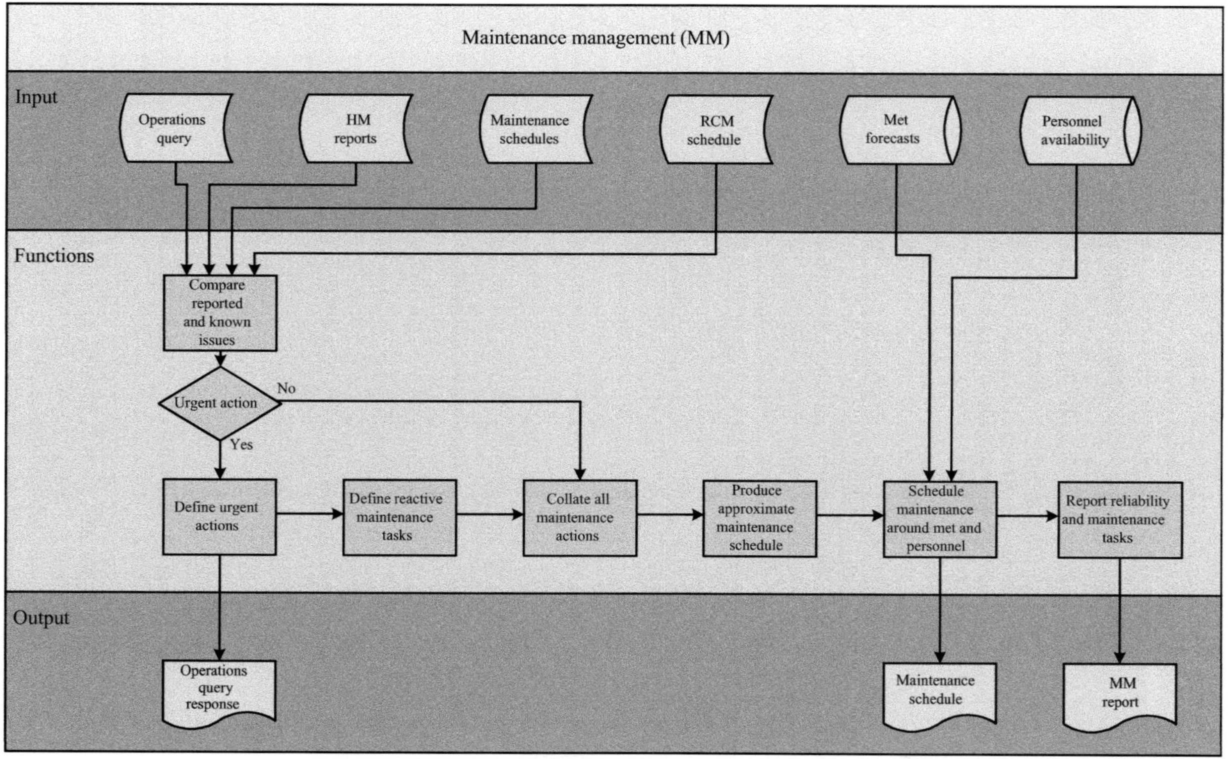


*Figure 10.14 Maintenance management structure*

*Table 10.13 Maintenance management data*

| Inputs | Functions | Outputs |
|---|---|---|
| **Live data**<br>Met forecasts | 1. Compare operations queries and HM reports with maintenance schedules (known issues) | **Reports**<br>Maintenance schedule |
| Personnel availability | 2. Respond to operations queries<br>3. Compare issues with RCM schedule | MM report |
| **Stored information** | 4. Detailed maintenance tasks, including preventative, reactive and RCM responses | Operations query responses |
| Maintenance schedules | 5. Produce approximate cost-effective maintenance schedule | |
| HM reports<br>RCM schedule | 6. Produce updated maintenance schedule based on met forecasts and personnel availability | |
| **Direct reports** | 7. Ensure RCM activities meet AM plans | |
| Operations query | 8. Report initial reliability figures in MM report | |

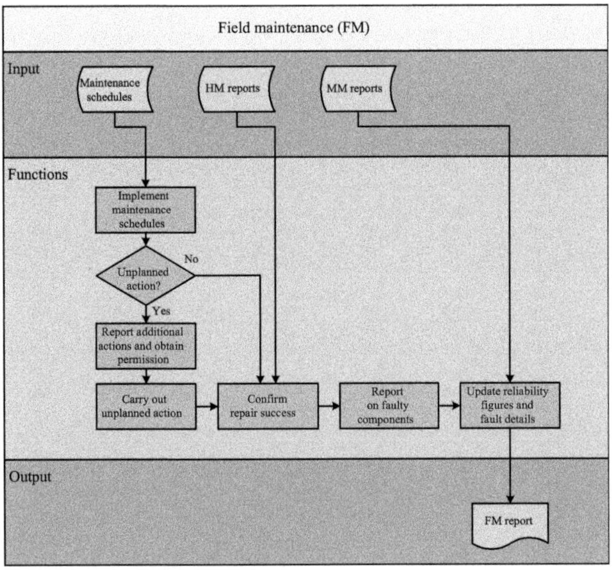


*Figure 10.15 Field maintenance management structure*

*Table 10.14 Field management data*

| Inputs | Functions | Outputs |
|---|---|---|
| **Stored information** | 1. Implement maintenance schedules | **Reports** |
| Maintenance management reports | 2. Report and resolve any faults or potential faults discovered during maintenance | Field management report |
| Maintenance schedules | 3. Confirm repair success against advice in HM reports | |
| Health management reports | 4. Update reliability figures and fault details from MM report and insert into FM report | |
| | 5. Report actions taken and results of faulty component examination in FM report | |

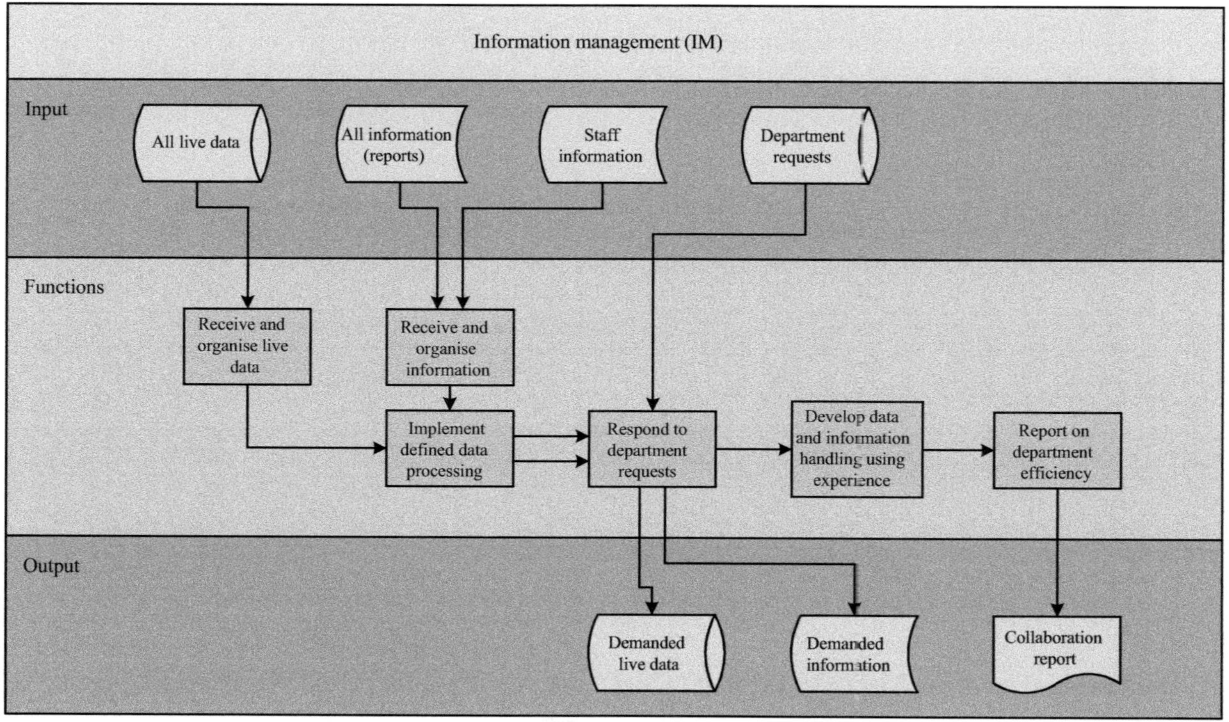


*Figure 10.16 Information management structure*

*Table 10.15 Information management data*

| Inputs | Functions | Outputs |
|---|---|---|
| All reports<br>All live data<br>Staff information<br>Department requests | 1. Receive live data and information from other departments<br>2. Manage and store reports in central repository<br>3. Alert departments to new reports<br>4. Provide on demand data and information based on department requests<br>5. Provide information analysis support to all departments<br>6. Perform database maintenance and updates<br>7. Control the removal of data for database maintenance<br>8. Realise effective communication between departments and report this efficiency and strategy in a collaboration report | Collaboration report<br>On-demand data and information |

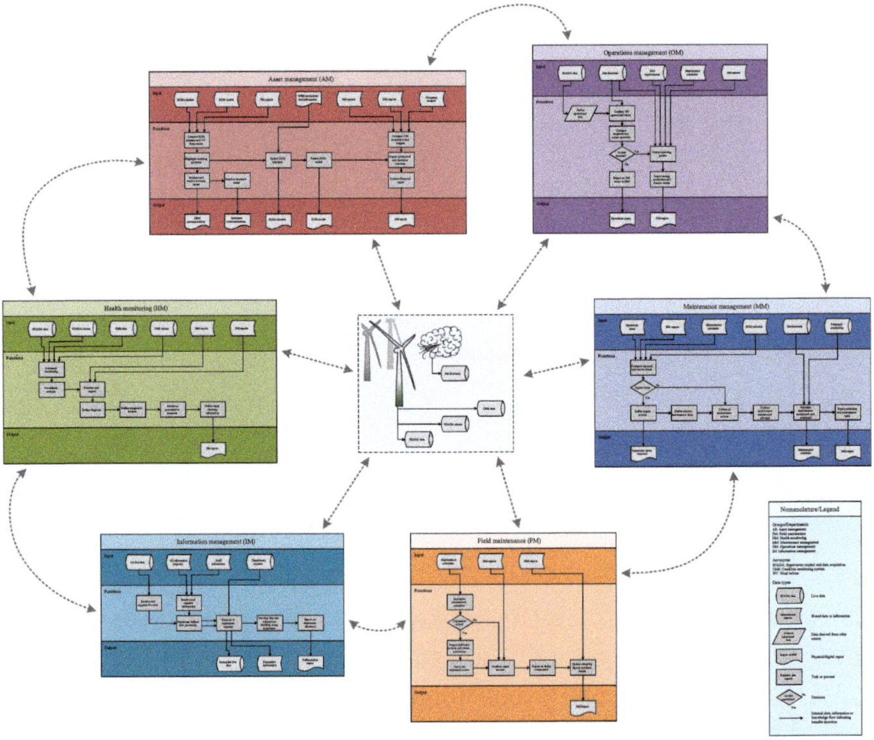


*Figure 10.17 Large off-shore wind farms: data management monitoring and maintenance*

The last section, especially, presents a proposal for an integrated maintenance strategy, showing the personnel, infrastructure and data issues associated with improving maintenance, availability and reliability for off-shore wind farms.

The key factors are shown to be mostly the presentation of appropriate data from the wind farms to staff to encourage their deployment in such a way that the costly infrastructure can be fully exploited.

A proposal has been presented for an off-shore wind farm knowledge management system to handle the SCADA and CMS monitoring data and wind farm accumulated reliability data, correlating it with the Maintenance Logs to provide an integrated system from which maintenance planning can proceed.

The training of staff and the availability of appropriate access infrastructure to carry this out will be dealt with in the next chapter.

# Chapter 11
# Production safety, training and qualification

## 11.1 Introduction

Chapter 1 emphasised the importance of trained staff for the off-shore wind industry. This is particularly important in the area of maintenance, most especially in off-shore facilities where access is restricted and support limited. Wind farm maintenance technicians and engineers require a special blend of skill and knowledge, including

- management, organisation and initiative;
- wind turbine and substation product knowledge;
- specialist expertise in mechanical, electrical, control and software technology knowledge;
- appropriate H&S working practices;
- survival training for the off-shore environment.

The formation of wind turbine technicians, engineers and graduates is of vital importance to instil appropriate wind turbine and substation product knowledge.

In the past, there has been a split in training provision among

- mainstream formation programmes by wind turbine original equipment manufacturers (OEMs) for their wind apprentice, engineers and graduates;
- specialised wind industry new employee induction programmes, for example BZEE and national wind energy associations, suitable for technicians retraining from other fields;
- specialised H&S and survival training to allow existing on-shore technicians to operate safely in the off-shore wind farm environment.

It is unlikely that purpose-designed apprentice formation schemes alone will be able to keep pace with the current rate of expansion in the industry, which also needs to attract technicians with relevant basic skills from other industries and retrain them for the wind industry.

Trained technicians may come into the wind industry from a variety of other relevant maintenance environments, such as power generation, automotive, oil and gas, aerospace and the armed forces. Their knowledge is making an important contribution to improving off-shore wind turbine maintenance quality as well as increasing the quantity of trained staff.

The off-shore environment is arduous and hazardous, and maintenance staff working off-shore must be well trained, not only to undertake the necessary maintenance tasks on our wind turbines, substations, converter stations and infrastructure but also to protect themselves and their colleagues from harm.

Those staff also need to have specific training to undertake this important and very interesting work and the qualifications to reflect their education, training and knowledge.

This has been a major part of our work on wind power in UK universities, and we aim to ensure that our students going into the off-shore wind industry are prepared for the work ahead.

## 11.2 Safety

When planning and undertaking maintenance activities on off-shore wind turbines substations and wind farms regardless of whether these activities are intrusive or non-intrusive, there are three key areas that must be considered, summarised in Figure 11.1, and their three overarching risks are as follows:

- risk of injury and/or long-term illness to person(s) undertaking maintenance,
- risk of damage to the plant or equipment subject to the maintenance,
- risk of damage to the equipment being used.

The first item earlier, risk of injury and/or long-term illness to person(s), is of paramount importance to all roles in the off-shore wind industry. Failure to address these hazards and associated risks could potentially result in harm to people, damage to plant or equipment, business interruption, organisational and reputational damage and unpredictable, costly financial penalties.

When planning and undertaking wind farm maintenance activities, regardless of whether these activities are intrusive or non-intrusive, there are three key areas that should be considered, summarised by the risks shown in Figure 11.1, and there are three overarching risks as follows:

- risk of injury and/or long-term illness to person(s) undertaking the CM activity;
- risk of damage to the plant or equipment subject to the CM activity;
- risk of damage to the CM equipment being used.

Failure to address these hazards and associated risks could potentially again lead to organisational and reputational damage and unpredictable, costly financial penalties.

The hazards presented to persons undertaking CM activities fall into two categories:

General safety hazards, associated with the working area and environment, e.g. slips, trips and falls, vehicle movements, working at height, working outside and lone working.

System-derived hazards, associated with the process, e.g. electricity, steam, gas, water, chemicals and stored energy. Most organisations use safe systems of work and working practices to manage risks to persons from both general

*Figure 11.1* *The main safety risks in carrying out wind turbine and wind farm maintenance: (a) electrical and mechanical hazards; (b) human trip, slip, falling object and immersion hazards; (c) hazards due to maintenance machinery*

safety and system-derived hazards. Safe systems of work can encompass training and competence requirements, requirements for risk assessments and method statements prior to work, setting to work procedures, application of safety rules for isolation and subsequent work or testing of plant items and returning equipment to service procedures. It is recommended that electrical CM tasks are planned and undertaken in accordance with a suitable safe system of work.

Through inappropriate application or incorrect set-up of electrical CM equipment, it is possible to damage plant or equipment subject to the activity.

This risk exists predominantly when the tool or technique applied is intrusive, for example the injection of excessive current or voltage into a piece of electrical equipment can lead to component, conductor or electrical insulation system damage.

Similarly, it may also be possible to damage the CM equipment through incorrect application or set-up, for example undertaking a voltage measurement with equipment that is not rated for that voltage.

It is, therefore, essential that prior to undertaking a CM activity, the specific application is assessed and deemed appropriate. Plant or equipment under test and the test equipment must be suitably rated for the activity. When setting up the monitoring equipment, test connections and test settings, for example injection voltages in insulation resistance tests, should be checked and confirmed as correct. The function of any in-built protection measures, such as overloads, should also be confirmed operational prior to commencing. For more complex CM activities, the preparation of a step-by-step procedure, detailing the specific equipment connections, settings and test durations may be appropriate.

## 11.3 Training and qualification

There are as yet no internationally agreed training and qualification standards for operatives of wind turbine and wind farm maintenance, although the industry has developed a very large body of product international standards, see Chapter 14, and there is a strong set of OEM and operator, national and international organisations.

However, there are well-developed training and qualification standards for vibration condition monitoring, with a categorisation, from ISO 18436 Part 2, of the training and qualifications required, which are instructive for any prospective standard for wind turbine and wind farm maintenance CM training and qualification, as follows:

'Individuals recognised or certified for vibration monitoring should be classified in one of four categories depending upon their qualifications. They shall have demonstrated competence in the concepts of machinery vibration CM and diagnostics of machines for their classification as described in the following sections of ascending competence'.

### *11.3.1 Category I – data collector*

Individuals meeting the requirements for Category I are recognised as being qualified to perform a range of simple single-channel machinery vibration CM and diagnostics of machines activities in accordance with ISO 17359 and ISO 13373-1. They should not be responsible, for example for the choice of sensor or for any analysis to be conducted, nor for the assessment of test results, except for identifying alert conditions against a pre-established alert setting or settings. They shall be qualified to

1. operate portable instrumentation on pre-assigned or pre-programmed routes;
2. acquire readings from permanently installed instrumentation;
3. input results into a database and download monitoring routes from a computer;
4. conduct testing under steady-state operating conditions following predefined procedures;
5. be able to recognise that no signal is present;
6. be able to compare overall or single-value vibration measurements against pre-established alert settings.

### *11.3.2 Category II – specialist*

Individuals certified to Category II are qualified personnel who are certified to perform industrial machinery vibration measurements and basic vibration analysis using single-channel measurements, with or without phase trigger signals, according to established and recognised procedures. Personnel certified to Category II require all the knowledge and skills expected of Category I and shall also be qualified to

1. select the appropriate machinery vibration measurement technique;
2. set up instruments for basic resolution of amplitude, frequency and time;
3. perform basic vibration analysis of machinery and components, such as shafts, bearings, gears, fans, pumps and motors, using spectrum analysis;
4. maintain a database of results and trends;
5. perform basic (single-channel) impact tests to determine natural frequencies;
6. classify, interpret and evaluate the test results (including acceptance tests) in accordance with applicable specifications and standards;
7. recommend minor corrective actions;
8. understand basic single-plane field balancing concepts;
9. be aware of some of the causes and effects of bad measurement data.

### *11.3.3 Category III – analyst*

Individuals certified to Category III are qualified to perform and/or direct and/or establish programmes for vibration CM and diagnostics of machines in accordance with ISO 17359 and ISO 13373-1. Personnel classified to Category III require all the knowledge and skills expected of personnel classified to Categories I and II and shall also be qualified to

1. select the appropriate machinery vibration analysis technique;
2. specify the appropriate vibration instrumentation hardware and software for both portable and permanently installed systems;
3. measure and perform diagnosis of single-channel frequency spectra, as well as time domain plots such as waveforms and orbits, under both steady-state and unsteady operating conditions, with or without a phase trigger;
4. establish vibration monitoring programmes, including the determination of machines for periodic/continuous monitoring, frequency of testing, route plans;
5. establish programmes for the specification of vibration levels and acceptance criteria for new machinery;
6. measure and analyse basic operating deflection shapes;
7. understand and be able to direct the use of alternative CM technologies, such as acoustic emission, thermography, motor current and oil analysis;
8. recommend field corrective actions, such as balancing, alignment and replacement of machine parts;
9. be able to use acceleration enveloping or demodulation;
10. perform basic single-plane field balancing;
11. report to management regarding programme objectives, budgets, cost justification and personnel development;
12. prepare reports for appropriate personnel on machine condition, recommend corrective action and report on the effectiveness of repairs;
13. provide instructions and technical direction to vibration trainees.

### *11.3.4 Category IV – expert*

Individuals certified to Category IV are qualified to perform and/or direct vibration CM and diagnostics of machines in accordance with ISO 17359 and ISO 13373-1 and all types of machinery vibration measurements and analysis. Personnel certified to Category IV require all the knowledge and skills expected of personnel certified to Categories I, II, and III and shall also be qualified to

1. apply vibration theory and techniques, including measurement and interpretation of multichannel spectral results such as frequency response functions, phase and coherence;
2. understand and perform signal analysis, including understanding of frequency and time domain processing, including orbits and their limitations;
3. determine the natural frequencies, mode shapes and damping of systems, components and assemblies;
4. determine the operating deflection shapes of machines and connected structures and recommend means for correction;
5. use generally recognised advanced techniques for vibration analysis, parameter identification and fault diagnosis;
6. apply the basic principles of rotor-bearing dynamics to vibration diagnosis;
7. conduct basic two-plane field balancing;

8. recommend advanced two-plane influence coefficient or static/couple balancing;
9. recommend corrective actions and/or design modifications, including component change or repair, isolation, damping, change of stiffness and change of mass;
10. provide technical guidance to vibration trainees;
11. interpret and evaluate published ISO codes of practice, International Standards and specifications;
12. recognise vibration caused by gas pulsation in machines such as reciprocating machines and screw compressors and to measure the necessary parameters and recommend means for correction.

Recommend corrective actions for resilient mounting and other holding-down and foundation problems.

# Chapter 12
# Overall summary conclusions

## 12.1 Reliability and availability in wind farm design

The author suggests that the task of improved reliability and availability for off-shore wind farms would be aided by the ready availability of more reliability data from original equipment manufacturers and operators to allow designs to be benchmarked against best practice. This is the kernel of the reliability proposals in Chapters 4 and 7 of this book.

In the early days, the wind industry manufacturers were secretive about performance to protect their intellectual property and champion individual improvement.

Operators have also been protective of wind farm performance data as those have contractual significance for their investors.

But the industry is now of a size and professionalism where it could find a way to share within the wind industry operational data, in a non-competitive way, to improve collective wind power performance, as the industry comes into direct competition with other renewable and non-renewable power sources.

The wind industry could share data to deal with the off-shore CAPEX and OPEX challenges and meet the competition head-on, Barberis Negra *et al.* (2007).

An important reliability and availability issue, in terms of cost, will be to determine maintenance cost-effectiveness.

Some operators are setting availability targets for off-shore wind farms. There may be dangers in this approach since higher availability can always be achieved with higher O&M investment. The better path would be to determine the optimal O&M costs to achieve an acceptable availability and that will vary from wind farm to wind farm, depending strongly on the location of the site, being affected by distance off-shore, local infrastructure and assets and their costs.

### *12.1.1 Importance of data*

To manage off-shore wind farms, data are essential; however, reliability data cannot achieve perfect accuracy, and there will always be a mismatch between the information available and that required to ensure perfect maintenance.

However, there have been various debates in the industry about the accuracy and effectiveness of data.

To quote the Introduction to reliability of this book 'the purpose of reliability data and analysis is not to define the way in which wind power plants should operate but to provide the numerical data to identify how their operational performance could be improved'.

The first debate about data came from mathematicians and statisticians, see Section 4.5, arguing first that wind turbine data did not consider total time on test, necessary for individual component reliability, but were censored data, measured every month or year for a complete machine. Then there were concerns about the validity of partial wind farm data, questioning whether wind turbine reliability could be inferred from incomplete data collected after installation. These detailed mathematical questions are important, as wind farm and wind turbine failure data are based almost entirely on turbine stoppages. However, a medical statistician has pointed out to the author that, in the absence of other information, all data are valuable, and each source must be combined with others to obtain an overall picture, a process of creating metadata.

The next debate came from the wind manufacturing industry, arguing that data must remain confidential and controlling its release. Wind farm operators take a more sanguine view, as reliability data can improve their profitability, but they still shrink from data sharing, except in controlled relationships.

However, researchers all over Europe have managed to collate valuable reliability data, such as Figure 12.1, which decisively demonstrate that turbine downtime is more important than individual component reliability or failure rate, an important research conclusion.

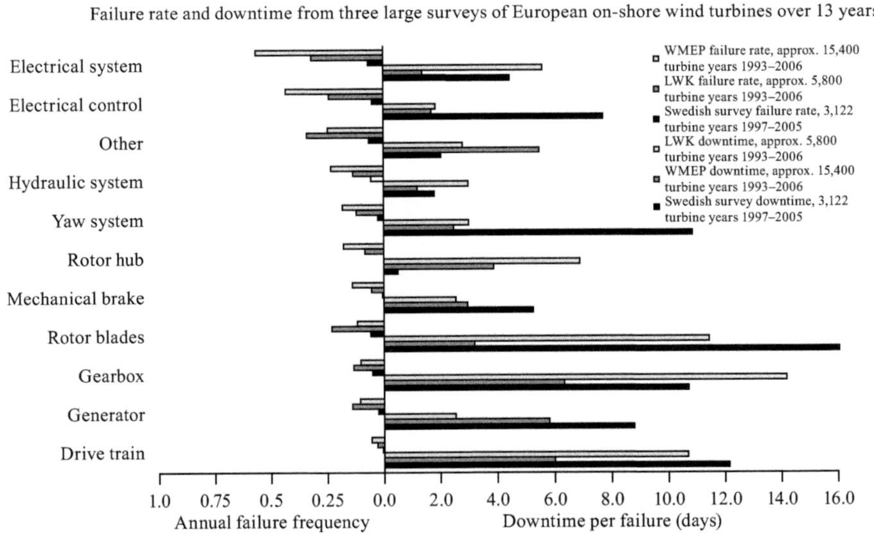


*Figure 12.1* *Wind turbine sub-assembly failure rate and downtime per failure for 20,000 turbine years of data from Swedish, LWK and WMEP surveys, Tavner (2017)*

However, an alternative debate from a wind turbine manufacturer, described in Section 4.10 as premature series failures (PSF), Figure 12.2, altered information to operators and should be resisted; no other industries, e.g. road, rail or air travel, countenance early failures as coming from PSF or such a cause.

A vital lesson from reliability data is that it can be used by aggregation to create larger metadata sets, from which reasonable wind farm predictions can be made, distinguishing wind turbine types and layouts. For example, which wind turbine configurations are more reliable and how to improve wind farm layouts to reduce capital expenditure and improve reliability and maintenance.

These are typical data benefits for improving outcomes, widely used by statisticians in the study of medical conditions and their treatments.

The author has worked on wind industry reliability for 20 years, but before that on power equipment reliability. At the time the UK power industry was subject, on the introduction of its largest power stations, to a high number of generator failures. It became a technical detective story and failure data were essential to resolve fundamental design, quality and operational issues. Analysis of data proved the way to advance. Within a few years, problems had been solved, generators were not changed but improved by repair and maintenance, going onto deliver 40–48 years of reliable service.

The off-shore wind industry is going through the same process, and there will be benefits from the collection and analysis of failure data but that cannot be achieved by operators, manufacturers or maintainers alone but by consortia.

## 12.2 Collating data

This book makes the proposition that the keys to higher availability and lower cost of energy for off-shore wind farms are as follows:

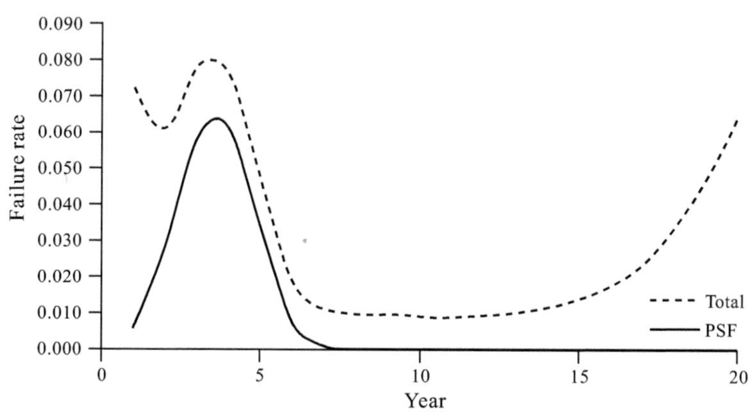


*Figure 12.2 A WT 'bathtub curve' incorporating premature serial failure (PSF) of some sub-assemblies, Stiesdahl et al. (2005)*

- Measurement of availability and reliability metrics against those expected for off-shore wind farms and their wind turbine components.
- A clear maintenance strategy to achieve those metrics.
- An asset management strategy to support that over the full asset life cycle.

Previous wind farm maintenance strategies were typified by Figure 12.3, without procedures.

A result without such an approach can be seen in Figure 12.4, taken from an early on-shore wind turbine survey showing measured wind turbine downtime on the left and maintenance on the right, the latter typified by corrective rather than predictive maintenance.

In Figure 12.4(b), maintenance is being equally time-distributed amongst sub-assemblies with no regard to the downtime consequences of sub-assembly failures, for example a gearbox, causing 22% of the downtime, is receiving only 8% of maintenance time, similar to hydraulics, which caused only 3% of downtime.

Whilst this approach may have been acceptable on-shore, where O&M time absorbed was facilitated solely by maintenance technicians accessing wind turbines from low-cost vans, it is clearly unacceptable off-shore where every maintenance visit incurs high boat, ship or even helicopter access time and costs, described in Chapter 10.

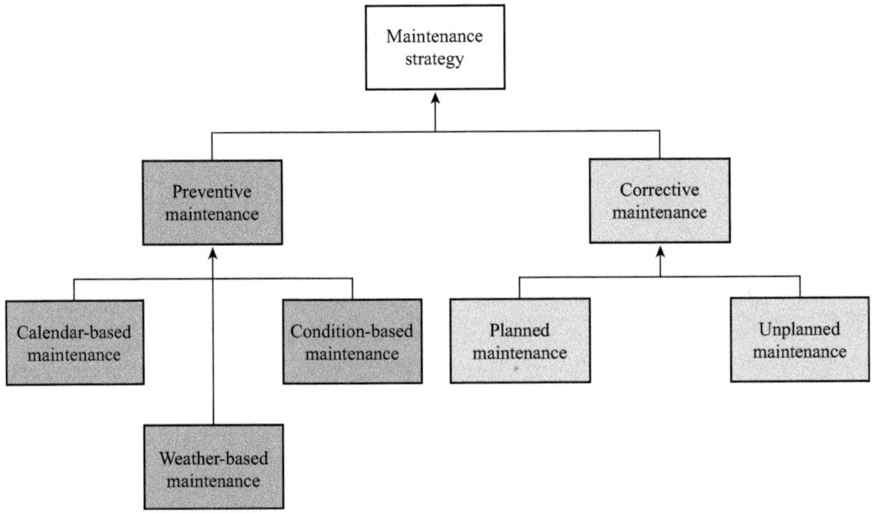


*Figure 12.3 Schematic overview of different wind farm maintenance strategies, highlighting on-shore on right, off-shore on left, Wiggelinkhuizen et al. (2008)*

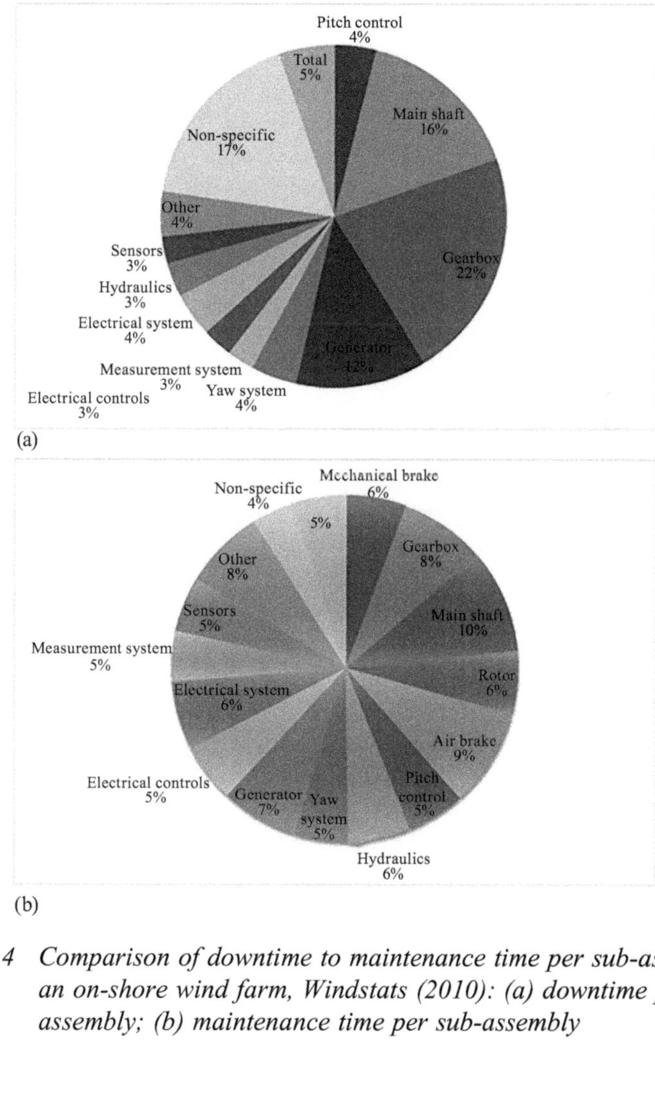


*Figure 12.4 Comparison of downtime to maintenance time per sub-assembly for an on-shore wind farm, Windstats (2010): (a) downtime per sub-assembly; (b) maintenance time per sub-assembly*

## 12.3 Operational planning for maintenance, RCM versus CBM

Figure 12.5, from Chapters 5 and 6, shows the sequence of design and manufacturing tasks needed to develop reliable off-shore wind turbines and highly available off-shore wind farms based on the reliability data collected.

Figure 12.5(a) concentrates upon off-shore wind turbine design for operation and Figure 12.5(b) upon complete wind farm maintenance strategy.

Both figures show the importance of data to provide the metrics to drive these processes to be deployed in off-shore wind as set out in Chapters 5 and 6.

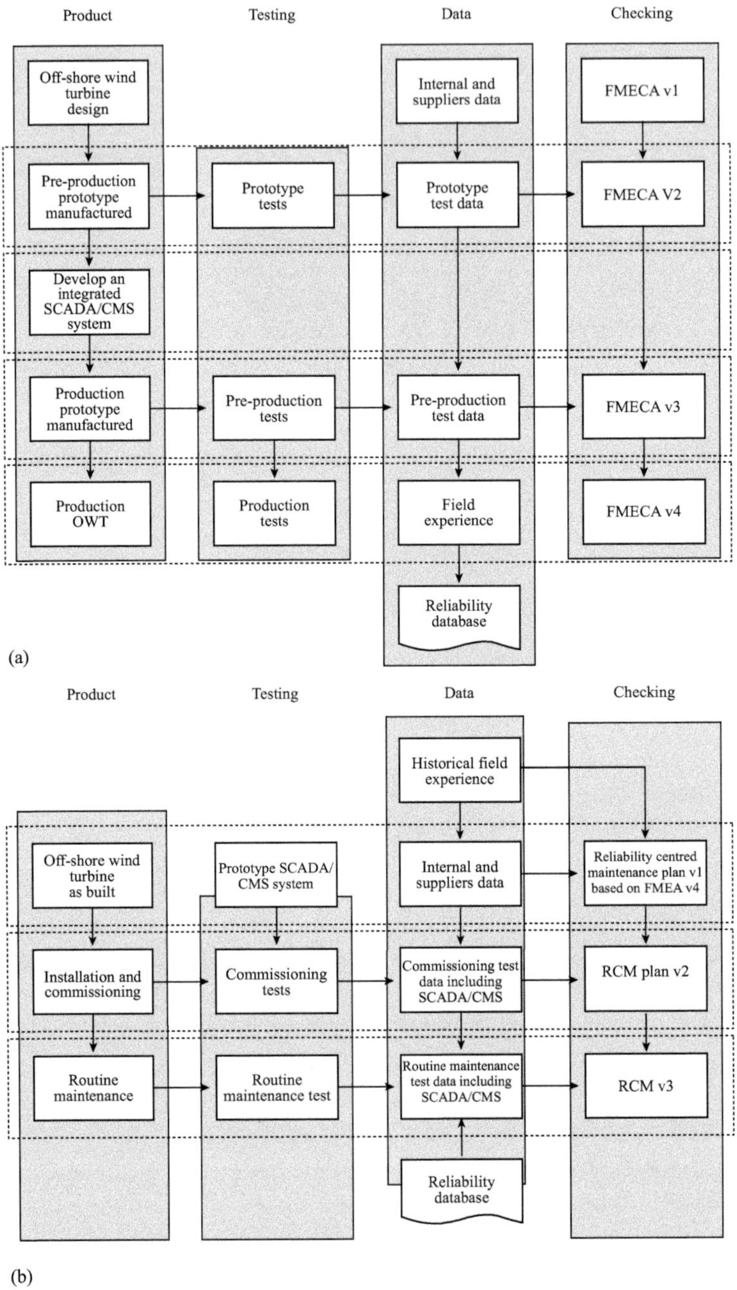


*Figure 12.5 Proposals to use FMEA and RCM as review tools during off-shore wind turbine design and manufacture for improved wind farm O&M: (a) FMEA as a design review tool; (b) FMEA and RCM together as an O&M review tool*

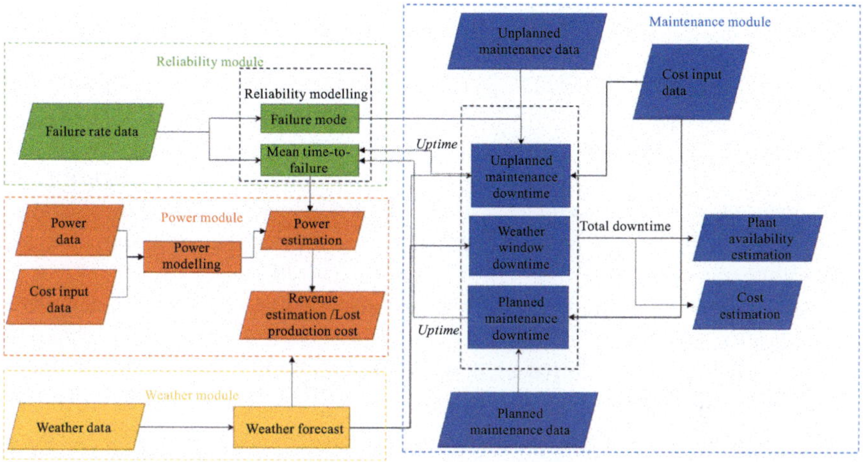


*Figure 12.6 Summary of a proposed workflow for an open operation and maintenance management tool, cf. Figure 4.7, Cervasco et al. (2021)*

Reliability-centred maintenance (RCM) is where wind turbine sub-assembly failure rate and downtime are used to drive maintenance activity, following Figure 12.5.

Therefore, from Figure 12.4, gearbox maintenance time should have been 22% of the total, bearing in mind that gearbox failures cause downtime. This distribution of maintenance must vary with time, depending on wind turbine and sub-assembly performances.

However, such an approach could be misguided if maintenance activity does not genuinely reduce failure sub-assembly rates and downtimes; how can this be determined?

It can only be determined by a clear understanding of sub-assembly history and performance. This can be achieved from the reports described in Chapter 10, i.e. RCM, Figure 12.5.

However, it could be achieved by monitoring wind turbine performance using methods like those described in Chapter 9, i.e. condition-based maintenance (CBM).

Wind turbines have exceptionally good monitoring cover, because of their unmanned remote robotic operation, but few operators make full use of monitoring information to manage maintenance, because of the volume and complexity of those data.

That approach must change off-shore. Data must be simplified and presented in a coordinated, comprehensible way, hence, the need for a data management system.

Data could then be used to drive RCM and CBM to raise availability and lower cost of energy, Figure 12.6. Both RCM and CBM drive the need for an off-shore wind farm knowledge management system.

## 12.4 Asset management

RCM and CBM address the ongoing operation of the wind farm but they cannot, on their own, secure the through-life wind farm reliable performance without longer term asset management, see Bertling *et al.* (2005).

The high capital cost of off-shore wind demands a rigorous operational regime to generate energy at an adequate price that not only recovers the investment cost but also, once payback is achieved, generates future revenue ensuring long-term profitability over the asset life.

These long-term benefits can only be secured by long-term asset management that is controlling the later part of the bathtub curve, Figure 12.7, where sub-assembly wear-out is controlled by planned change-out.

It seems clear that off-shore wind farms, with large numbers of identical or similar wind turbine assets, can benefit from planned change-out of the most vulnerable sub-assemblies: blades, gearboxes, generators, converters, transformers and even nacelles.

In fact, the change-out process could also embed sub-assemblies with improved operational and reliability performance.

## 12.5 Towards an integrated maintenance strategy, data management

An integrated approach to maintenance strategy has been proposed in the attached diagram, Figure 12.8.

## 12.6 Prospective costs of energy for off-shore wind

What has become clear from writing this book is that the higher capital and operational cost of off-shore wind means that much greater attention is being paid off-shore, than the wind industry was accustomed to on-shore, to making the asset work at a high availability, to achieve its projected payback targets.

This does not mean that payback for off-shore wind has been too difficult, since it was already being achieved at Baltic off-shore wind farms, with availabilities of 96%–98%, and at a growing number of North and Irish Sea off-shore wind farms moving towards higher availabilities, 90%–95%.

Experience with on-shore wind, Figure 12.9, has shown that, although initial capital costs were high, the distributed nature and repeatability of the technology were such that the capital cost learning curve for on-shore wind fell from 1990 to 2004, with rising installed capacity from 1 to 10 GW.

Therefore, many manufacturing, installation, operation and maintenance lessons were rapidly learnt.

From Figure 12.7, the learning curve for off-shore wind from 1990 to 2001 was clearly much slower than on-shore wind from 1990 to 2004.

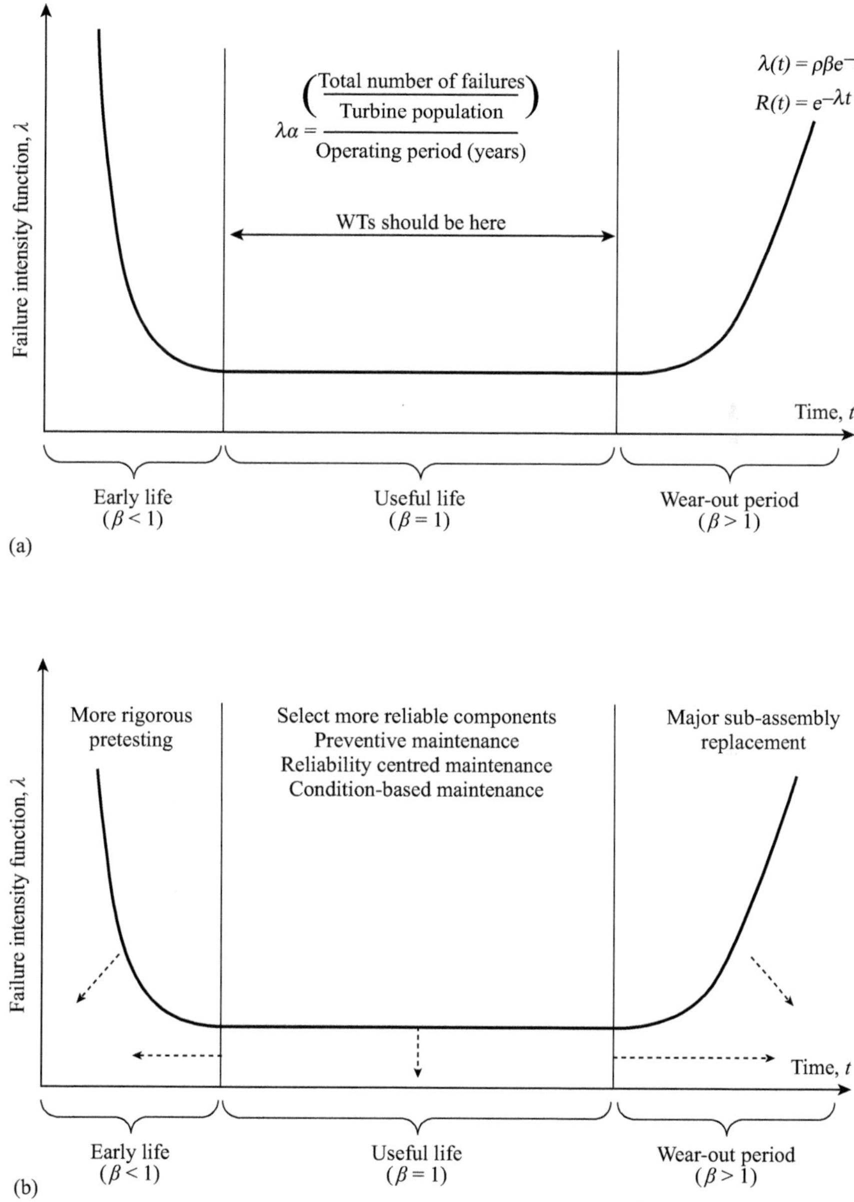


*Figure 12.7 (a) Bathtub curve of failure intensity showing and (b) effects on the curve of testing and maintenance*

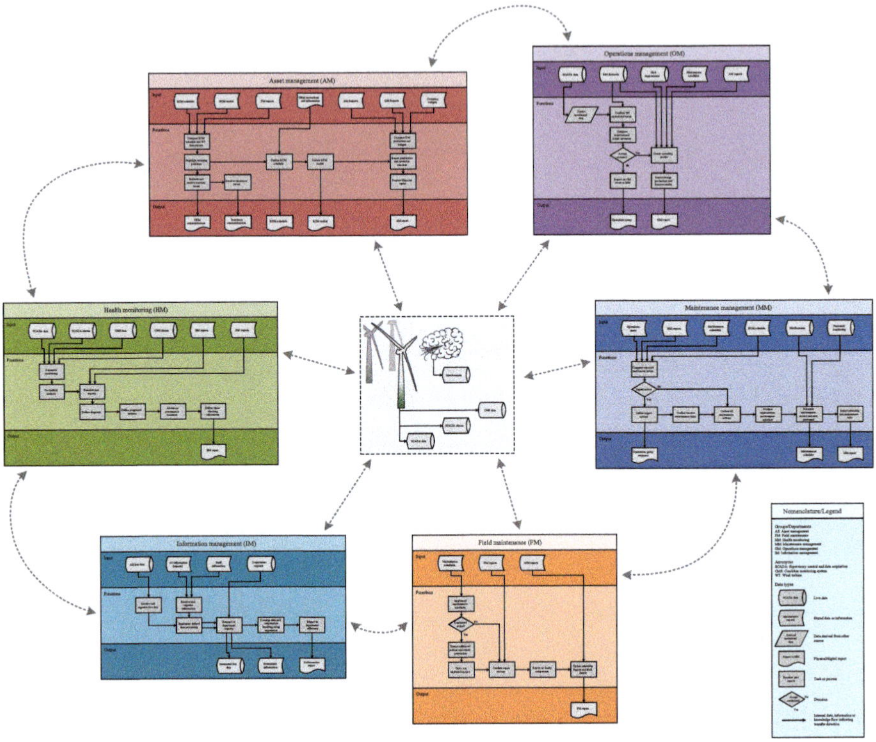


*Figure 12.8 Large off-shore wind farms: data management monitoring and maintenance*

But by 2017, off-shore wind has reached 15-GW installed capacity and its learning curve has progressed much further than Figure 12.7, approaching, paralleling and reducing on the costs of on-shore wind and probably also on those of natural gas combined cycle and pulverised coal plants.

This means that investors, developers and operators should now be looking more critically at the intrinsic reliability of off-shore wind farms than was ever considered on-shore.

The through-life maintenance costs of off-shore wind have been estimated by Crabtree *et al.* (2015) in a graphic Figure 12.10 from six different sources, suggesting that off-shore wind OPEX could lie between 25% and 40% of off-shore wind LCOE.

## 12.7 Production safety, training and qualification

The design of wind turbines is regulated by a certification process that ensures the strength and safe operation of the wind turbine designed. Furthermore, the wind turbine control systems are designed to ensure that safe operation. Stiesdal *et al.* (2005) said 'the classical principle of wind turbine control and monitoring is to

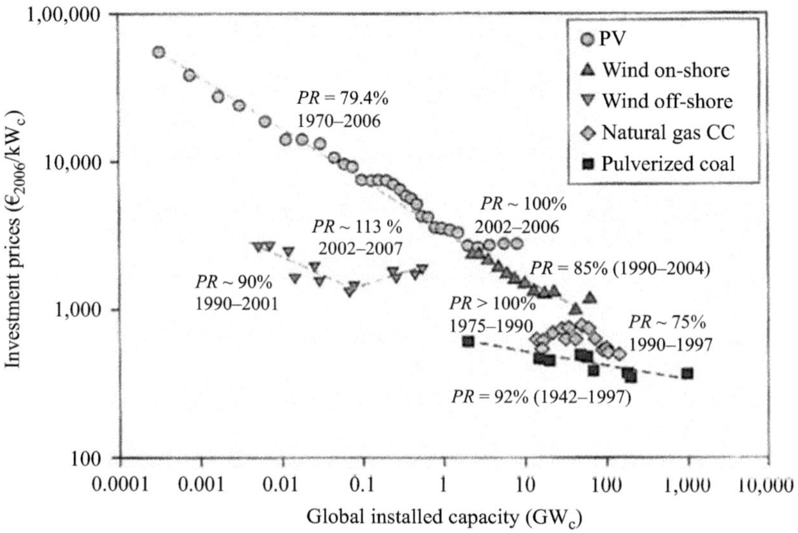


*Figure 12.9 Energy learning curve with on-shore and off-shore wind, Junjinger et al. (2010)*

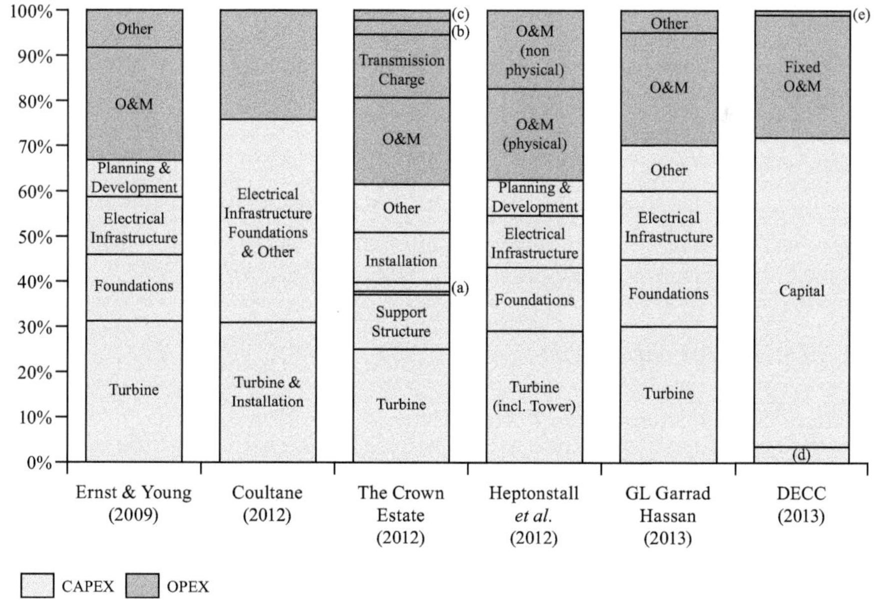


*Figure 12.10 Off-shore wind LCOE breakdown by CAPEX and OPE: (a) array cables; (b) decommissioning; (c) seabed rent; (d) pre-development; (e) variable O&M*

ensure that the wind turbine is always in a safe state – this is not automatically the same as ensuring that the operating time is maximised'.

Off-shore wind during installation and operation is also a potentially hazardous activity.

Developers and operators have, therefore, rightly adopted a strong certification- and H&S-oriented approach to the development of new wind turbines and to their installation and operation.

From 1997 to 2020, many H&S lessons have been learnt by staff in off-shore wind trained in this approach, teaching new staff drawn into the industry, as it moves to deeper and more distant waters.

As the wind industry matures, the current certification and H&S approaches are likely to change, as more stringent return on capital demands encourages a more vigorous, production-oriented approach, as happened in North Sea Oil & Gas production in the 1990s.

At this stage of industry development, the interaction among operators, asset managers, certifiers, insurers and investors will need to strengthen as attention shifts from capital investment issues to O&M and through-life costs.

For the expanding off-shore wind industry, a helpful innovation in this respect would be the development of a common structured training and qualification programme for maintenance operatives and managers.

This could be along the lines already developed by the vibration condition monitoring community, ISO 18436-2 (2014):

- Category I – data collector
- Category II – specialist
- Category III – analyst
- Category IV – expert

It could well be that in 10 years, we will see the off-shore wind industry developing this more structured operational environment, which will support the drive towards production, health and safety and certification.

## 12.8 Future prospects

The future prospects for off-shore wind power look favourable.

Early off-shore wind farms have demonstrated that high-energy resources are available off-shore at acceptable CAPEX costs, but with significant OPEX costs, emphasising the importance of the following:

- Designing high reliability wind farm assets that reduce initial CAPEX costs but also reduce ongoing prospective risk.
- Reducing the cost and risk of asset deployment and managing O&M to restrain OPEX costs.
- Within that framework, achieving low cost of energy figures with as high wind turbine availability as practicable from windy off-shore locations.

Experience with on-shore wind had shown that, though initial capital costs were high, the distributed nature and repeatability of turbine technology were such that capital cost learning curve fell with increasing installed capacity.

The same has now happened for fixed off-shore wind and from a considerably stronger energy resource. It is likely that the resource will be even stronger for floating off-shore wind installations and that means, with the impetus of scale, that CAPEX costs will fall further.

If the O&M lessons set out in this book are learned, it is certain that OPEX costs will also fall and that fixed and floating off-shore wind power will have a secure future.

This author knows that these lessons, which have been hard-learnt over the last 20 years and overcome by the wind industry's expansion and volume, could also be emulated by the tidal and wave renewable energy industries, which have had such a difficult embryonic stage, Hardisty (2009), Alcorn *et al.* (2014) and Tavner (2017).

# References


La Cour, P. (1900) *Forsogsmollen.* Det Nordiskke Forlag, Kopenhagen.

Betz, A. (1926) *Windenergie und ihre Ausnutzung durch Windmuhlen.* Vanderhoek & Ruprecht, Gottingen.

Putnam, P. C. (1947) *Power From the Wind.* Van Nostrand Reinhold, New York, NY, USA.

Hutter, U. (1954) *Die Entwicklung von Windkraftanlagen zur Stromerzeugung in Deutschland*, Vol 6(7) BWK.

van der Hoeven, I. (1957) Power spectrum of horizontal wind speed in the frequency range from 0.0007 to 900 cycles per hour. *Journal of Atmospheric Sciences*; 14(2): 160–164.

Goldberg, H. (1981) *Extending the Limits of Reliability Theory.* John Wiley & Sons, New York, NY, USA.

O'Donnell, P. and IEEE Reliability Working Group (1985) Report of large motor reliability survey of industrial and commercial installations. Parts I, II and III. *IEEE Transactions on Industry Applications*; 21: 853–872.

Albrecht, P. F., McCoy, R. M., Owen, E. L., and Sharma, D. K. (1986) Assessment of the reliability of motors in utility applications-updated. *IEEE Transactions on Energy Conversion*; 1: 39–46.

Schmid, J. and Klein, H. P. (1991) *Performance of European Wind Turbines.* Elsevier, Applied Science, London & New York, 167 pp.

Billinton, R. and Allan, R. N. (1992) *Reliability Evaluation of Engineering Systems: Concepts and Techniques.* 2nd Ed, Plenum Press, New York & London.

McIlveen, R. (1992) *Fundamentals of Weather and Climate.* 2nd Ed, Chapman & Hall, London.

Needham, J. (1994) *Science & Civilisation in China, Abridgement,* Vol 4, Cambridge University Press, Cambridge, UK, 333 pp.

Thorsen, O. V. and Dalva, M. (1995) A survey of faults on induction motors in offshore oil industry, petrochemical industry, gas terminals and oil refineries. *IEEE Transactions on Industry Applications*; 31(5): 1186–1196.

Billinton, R. and Allan, R. N. (1996) *Reliability Evaluation of Power Systems.* 2nd Ed, Plenum Press, New York, London.

Castro, S. F. and Allan, R. N. (1996) Generation availability assessment of wind farms. *IEE Proceedings. Part C. Generation, Transmission and Distribution*; 1043(5): 507–518.

IEEE Gold Book (1997) *Recommended Practice for Design of Reliable Industrial & Commercial Power Systems, IEEE Std 493-1997.* Piscataway, NJ, USA.

van Bussel, G. J. W. and Schöntag, C. (1997) Operation and maintenance aspects of large off-shore windfarms. *Proceedings of European Wind Energy Conference, EWEC1997.*

Ran, L., Gokani, S., Clare, J. C., Bradley, K. J., and Christopoulos, C. (1998) Conducted electromagnetic emissions in induction motor drive systems, Part 2: Frequency domain models. *IEEE Transactions on Power Electronics*; 13(4): 768–776.

WMEP – Wissenschaftliche Mess-und Evaluierungsprogramm (1996) *Federal German Ministry for Economics & Technology under the 250 MW Wind Test Programme.*

Karaki, S. H., Chedid, R. B., and Ramadan, R. (1999) Probabilistic performance assessment of wind energy conversion systems. *IEEE Transactions on Energy Conversion*; 4(2): 212–217.

Bellini, A., Filippetti, F., Franceschini, G., and Tassoni, C. (2000) Closed-loop control impact on the diagnosis of induction motors faults. *IEEE Transactions on Industry Applications IA*; 36(5): 1318–1329.

Mohammed, H. and Nwankpa, C. O. (2000) Stochastic analysis and simulation of grid-connected wind energy conversion system. *IEEE Transactions on Energy Conversion*; 15(1): 85–90.

Rigdon, S. E. and Basu, A. P. (2000) *Statistical Methods for the Reliability of Repairable Systems*. John Wiley & Sons, New York, NY, USA.

van Bussel, G. J. W. and Zaaijer, M. B. (2001) Reliability, availability and maintenance aspects of large-scale off-shore wind farms, a concepts study. *IMarE Conference*; 113(1): 119–126.

van Bussel, G. J. W., Henderson, A. R., Morgan, C. A., *et al.* (2001) *State of the Art and Technology Trends for Off-Shore Wind Energy: Operation and Maintenance Issues*. Delft University of Technology. Off-shore Wind. European Commission.

Bellini, A., Filippetti, F., Franceshini, G., *et al.* (2002) On-field experience with online diagnosis of large induction motors cage failures using MCSA. *IEEE Transactions on Industry Applications I*; 38(4): 1045–1053.

van Bussel, G. J. W. and Bierbooms, W. (2003) The DOWEC off-shore reference windfarm: Analysis of transportation for operation and maintenance. *Wind Engineering*; 27(5): 381–391.

Byrne, B. W. and Houlsby, G. T. (2003) Foundations for off-shore wind turbines. *Philosophical Transactions of the Royal Society of London. Series A*; 361(A): 2909–2930.

Karaki, R. and Billinton, R. (2004) Cost-effective wind energy utilization for reliable power supply. *IEEE Transactions on Energy Conversion*; 19(2): 435–440.

Musial, W., Butterfield, S., and Boone, A. (2004). *Feasibility of Floating Platform Systems for Wind Turbines*. National Renewable Energy Laboratory, NREL/CP-500-34874, Golden, CO, USA.

Ushiyama, I., Seki, K., and Miura, H. (2004) A feasibility study for floating off-shore windfarms in Japanese waters. *Wind Engineering*; 28(4): 383–397.

Bertling, L., Allan, R. N., and Eriksson, R. A. (2005) Reliability-centred asset maintenance method for assessing the impact of maintenance in power distribution systems. *IEEE Transactions on Power Systems*; 20(1): 75–82.

Caselitz, P. and Giebhardt, J. (2005) Rotor condition monitoring for improved operational safety of off-shore wind energy converters. *Journal of Solar Energy Engineering*; 127(2): 253–261.

IEA, NEA Report (2005) *Projected Costs of Generating Electricity*. OECD, France.

Stiesdal, H. and Madsen, P. H. (2005) Design for reliability. *European Off-shore Wind Conference*, Copenhagen.

Garcia, M. C., Sanz-Bobi, M. A., and del Pico, J. (2006) SIMAP: Intelligent system for predictive maintenance application to the health condition monitoring of a wind turbine gearbox. *Journal of Computers in Industry*; 57(6): 552–568.

Hau, E. (2006) *Wind Turbines-Fundamentals, Technologies, Application & Economics*. 2nd Ed., Springer-Verlag, Germany, 783 pp.

Lundberg, S. (2006) Wind Farm Configuration and Energy Efficiency Studies-Series DC versus AC Layouts. PhD Thesis, Chalmers University, Sweden.

Polinder, H., van der Pijl, F. F. A. de Vilder, G. J., and Tavner, P. J. (2006) Comparison of direct-drive and geared generator concepts for wind turbines. *IEEE Transactions on Energy Conversion*; 21(3): 725–733.

Tavner, P. J., Edwards, C., Brinkman, A., and Spinato, F. (2006) Influence of wind speed on wind turbine reliability. *Wind Engineering*; 30(1): 55–72.

Tavner, P. J., Xiang, J. P., and Spinato, F. (2006) Reliability analysis for wind turbines. *Wind Energy*; 10(1).

Walford, C. A. (2006) Wind turbine reliability understanding and minimizing wind turbine operation and maintenance costs. *Report SAND*, Sandia Labs, Albuquerque, NM.

Barberis, N. N., Holmstrøm, O., Bak-Jensen, B., and Sorensen, P. (2007) Aspects of relevance in off-shore wind farm reliability assessment. *IEEE Transactions on Energy Conversion*; 22(1): 159–166.

Birolini, A. (2007) *Reliability Engineering, Theory & Practice*. Springer, New York, NY, USA.

Hansen, A. D. and Hansen, L. H. (2007) Wind turbine concept market penetration over 10 years (1995–2004). *Wind Energy*; 10(1): 81–97.

Hochart, C., Fortin, G., Perron, J., and Ilinca, A. (2007) Wind turbine performance under icing conditions. *Wind Energy*; 11(4): 319–333.

Leithead, W. E. (2007) Wind energy. *Philosophical Transactions of the Royal Society of London. Series A*; 365: 957–970.

Ribrant, J. and Bertling, L. (2007) Survey of failures in wind power systems with focus on Swedish wind power plants during 1997–2005. *IEEE Transactions on Energy Conversion*; 22(1): 167–173.

Wangdee, W. and Billinton, R. (2007) Reliability assessment of bulk electric systems containing large wind farms. *International Journal of Electrical Power & Energy Systems*; 29(10): 759–66.

Faulstich, S., Durstewitz, M., Hahn, B., Knorr, K., and Rohrig, K. (2008) *Windenergie Report*. Institut für solare Energieversorgungstechnik, Kassel, Deutschland.

Harman, K., Walker, R., and Wilkinson, M. (2008) Availability trends observed at operational wind farms. *Proceedings of European Wind Energy Conference*, EWEC2008, Brussels.

Spinato, F. (2008) The Reliability of Wind Turbines. Doctoral Thesis, Durham University.

Tavner, P. J. (2008) Review of condition monitoring of rotating electrical machines. *IET Electric Power Applications*; 2(4): 215–247.

Wiggelinkhuizen, E., Verbruggen, T., Braam, H., Rademakers, L., Xiang, J., and Watson, S. (2008). Assessment of condition monitoring techniques for off-shore wind farms. *Journal of Solar Energy Engineering*; 130(3): 1004/1–1004/9.

Wilkinson, M. R. (2008) Condition Monitoring for Off-shore Wind Turbines. Doctoral Thesis, Newcastle University.

European Wind Energy Association (2009) *Wind Power the Facts*. EWEA, Brussels.

Hardisty, J. (2009) *The Analysis of Tidal Stream Power*. John Wiley & Sons, New York, NY, USA, 342 pp.

Guo, H., Watson, S., Tavner, P. J., and Xiang, J. (2009) Reliability analysis for wind turbines with incomplete failure data collected from after the date of initial installation. *Reliability Engineering and System Safety*; 94(2009): 1057–1063.

Manwell, J. F., McGowan, J. G., and Rogers, A. L. (2009) *Wind Energy Explained: Theory, Design and Application*. 2nd Ed., Wiley, Chichester, 680 pp.

Spinato, F., Tavner, P. J., van Bussel, G. J. W., and Koutoulakos, E. (2009) Reliability of wind turbine sub-assemblies. *IET Proceedings Renewable Power Generation*; 3(4): 1–15.

Twidell, J. W. and Gaudiosi, G. (Eds.) (2009) *Off-Shore Wind Power*, Multi-Science Publishing Co., UK, 300 pp.

Yang, W., Tavner, P. J., and Wilkinson, M. R. (2009) Condition monitoring and fault diagnosis of a wind turbine synchronous generator drive train. *IET Proceedings Renewable Power Generation*; 3(1): 1–11.

Zaher, A., McArthur, S. D. J., and Infield, D. G. (2009) On-line wind turbine fault detection through automated SCADA data analysis. *Wind Energy*; 12(6): 574–593.

Zhang, B. (2009) *Ancient Chinese Windmills*. Institute for the History of Natural Science, Chinese Academy of Sciences, Beijing, China.

Alewine, K. and Chen, W. (2010) Wind turbine generator failure modes analysis and occurrence. *Proceedings of Windpower*; May 24–26, Dallas, TX.

Arabian-Hoseynabadi, H., Oraee, H., and Tavner P. J. (2010) Failure modes and effects analysis (FMEA) for wind turbines. *International Journal of Electrical Power & Energy Systems*; 32(7): 817–824.

Arabian-Hoseynabadi, H., Oraee, H., and Tavner, P. J. (2010) Wind turbine productivity considering electrical sub-assembly reliability. *Renewable Energy*; 35(1): 190–197.

Arabian-Hoseynabadi, H., Tavner, P. J., and Oraee, H. (2010) Reliability comparison of direct-drive and geared drive wind turbine concepts. *Wind Energy*; 13(1): 62–73.

Coolen, F. P. A., Spinato, F., and Venkat, D. (2010) On modelling of grouped reliability data for wind turbines. *IMA Journal of Management Mathematics*; 21(4): 363–372.

Feng, Y., Tavner, P. J., and Long, H. (2010) Early experiences with UK round 1 off-shore wind farms. *Proceedings of Institution of Civil Engineers, Energy*; 163(EN4): 167–181.

Gray, C. S. and Watson, S. J. (2010) Physics of failure approach to wind turbine condition-based maintenance. *Wind Energy*; 13(5): 395–405.

Junjinger, M., van Sark, W., and Faaij, A. (Eds.) (2010) *Technological Learning in the Energy Sector – Lessons for Policy in the Energy Sector*. Edward Elgar Publishing, Cheltenham, UK; Northampton, MA, 332 pp.

Modarres, M., Kaminskiy, M. P., and Krivtsov, V. (2010) *Reliability Engineering and Risk Analysis: A Practical Guide*. 2nd Ed., CRC Press, Boca Raton, FL.

Richardson, P. (2010) Relating On-shore Wind Turbine Reliability to Off-shore Application. Master of Science Dissertation, Durham University.

Yang, W., Tavner, P. J., Crabtree, C. J., and Wilkinson, M. (2010) Cost-effective condition monitoring for wind turbines. *IEEE Transactions on Industrial Electronics*; 57(1): 263–271.

Watson, S. J., Xiang, J., Yang, W., Tavner, P. J., and Crabtree, C. J. (2010) Condition monitoring of the power output of wind turbine generators using wavelets. *IEEE Transactions on Energy Conversion*; 25(3): 715–721.

Alewine, K. and Chen, W. (2010) *Wind Turbine Generator Failure Modes Analysis and Occurrence*. Windpower, Dallas, TX.

Landwirtschaftskammer (LWK) (2010) Schleswig-Holstein, Germany.

Wilkinson, M. R., Hendriks, B., Spinato, F., Gomez, E., *et al.* (2010) Methodology and results of the ReliaWind reliability field study. *Proceedings of European Wind Energy Conference, EWEC2010*, Warsaw.

Tavner, P. J., Higgins, A., Arabian-Hoseynabadi, H., Long, H., and Feng, Y. (2010) Using an FMEA method to compare prospective wind turbine design reliabilities. *Proceedings of European Wind Energy Conference, EWEC2010*, Warsaw.

Yang, W., Tavner, P. J., Crabtree, C. J., and Wilkinson, M. (2010) Cost effective condition monitoring for wind turbines. *IEEE Transactions on Industrial Electronics*; 57(1): 263–271.

Beurksens, J. Ed. (2011) *Converting Off-shore Wind into Electricity – the Netherlands Contribution to Off-shore Wind Energy Knowledge*. Eburon Academic Publishers, the Netherlands.

Burton, T., Jenkins, N., Sharpe, D., and Bossanyi, E. (2011) *Wind Energy Handbook*. 2nd Ed., John Wiley & Sons, Ltd, Chichester, UK, 775 pp.

Crabtree, C. J. (2011) Condition Monitoring Techniques for Wind Turbines. Doctoral Thesis, Durham University.

Dent, C. J. and Gray, J. D. (2011) Component time-to-failure distribution estimation with limited statistical data: A critical survey. *IEEE Power & Energy Society General Meeting*, Detroit, MI, USA.

Faulstich, S., Lyding, P., and Tavner, P. J. (2011) Effects of wind speed on wind turbine availability. *Proceedings of the European Wind Energy Conference, EWEA2011*, Brussels, European Wind Energy Association.

Faulstich, S., Hahn, B., and Tavner, P. J. (2011) Wind turbine downtime and its importance for off-shore deployment. *Wind Energy*; 14(3): 327–337.

Jamieson, P. (2011) *Innovation in Wind Turbine Design*. John Wiley & Sons Ltd, Chichester, UK, 298 pp.

ReliaWind (2011) Deliverable D.2.0.4a-Report, Whole System Reliability Model, available from ReliaWind website.

Tavner, P. J., Faulstich, S., Hahn, B., and van Bussel, G. J. W. (2011) Reliability & availability of wind turbine electrical & electronic components. *European Journal of Power Electronics*; 20(4): 45–50.

Wilkinson, M. R., Hendriks, B., Spinato, F., *et al.* (2011) Measuring wind turbine reliability: Results of the ReliaWind Project. Scientific Track. *Proceedings of European Wind Energy Conference, EWEA2011*, Brussels.

Xie, K. and Billinton, R. (2011) Determination of the optimum capacity and type of wind turbine generators in a power system considering reliability and cost. *IEEE Transactions on Energy Conversion*; 26(1): 227–234.

Gasch, R. and Twele, J. (Eds.) (2012) *Wind Power Plants-Fundamentals, Design, Construction & Operation*. 2nd Ed., Springer Verlag, Berlin, Germany, 548 pp.

Qiu, Y., Feng, Y., Tavner, P. J., Richardson, P., Erdos, G., and Chen, B. (2012) Wind turbine SCADA alarm analysis for improving reliability. *Wind Energy*; 15(8): 951–966.

Tavner, P. J. (2012) *Off-shore Wind Turbines-Reliability, Availability & Maintenance*. 1st Ed., The Institution of Engineering and Technology, London, UK, 269 pp.

Wikipedia (2012) Beaufort Scale. http://en.wikipedia.org/wiki/Beaufort_scale.

Chen, B. D. (2013) Automated On-line Fault Prognosis for Wind Turbine Monitoring Using SCADA Data. Doctoral Thesis, Durham University.

Chen, B., Matthews, P. C., and Tavner, P. J. (2013) Wind turbine pitch faults prognosis using a-priori knowledge-based ANFIS. *Expert Systems with Applications*; 40(17) 6863–6876.

Delorm, T. M. (2013) Tidal Stream Devices: Reliability Prediction Models During Their Conceptual & Development Phases. Doctoral Thesis, Durham University.

Feng, Y., Qiu, Y., Crabtree, C. J., Long, H., and Tavner, P. J. (2013) Monitoring wind turbine gearboxes. *Wind Energy*; 16(5): 728–740.

Greaves, P. R. (2013) Fatigue Analysis and Testing of Wind Turbine Blades. Doctoral Thesis, Durham University.

Milan, P., Wächter, M., and Peinke, J. (2013) Turbulent character of wind energy. *Physical Review Letters*; 110(13): 138701.

Mukora, A. E. (2013) Learning Curves and Engineering Assessment of Emerging Energy Technologies: On-shore Wind. Doctoral Thesis, Edinburgh University.

Polinder, H., Ferreira, J. A., Jensen, B. B., Asger B., Atallah, K., and McMahon, R. A. (2013) Trends in wind turbine generator systems. *IEEE Journal of Emerging and Selected Topics in Power Electronics*; 1(3): 174–185.

Tavner, P. J., Greenwood, D. M., Whittle, M. W. G., Gindele, R., Faulstich, S., and Hahn, B. (2013) Study of weather & location effects on wind turbine failure rates. *Wind Energy*; 16(2): 175–187.

Whittle, M. (2013) Wind Turbine Generator Reliability: An Exploration of the Root Causes of Generator Bearing Failures. Doctoral Thesis, Durham University.

Zaggout, M. N. (2013) Wind Turbine Generator Condition Monitoring via the Generator Control Loop. Doctoral Thesis, Durham University.

Alcorn, R. and O'Sullivan, D. (Eds.) (2014) *Electrical Design for Ocean Wave and Tidal Energy Systems, IET Renewables Series 17*. The Institution of Engineering and Technology, London, 254 pp.

Carroll, J., McDonald, D., and McMillan, D. (2014) Reliability comparison of wind turbines with DFIG and PMG Drive Trains. *IEEE Transactions on Energy Conversion*; 30(2): 663–670.

Sorensen, J., Steenbergen, R. D. J. M., van Gelder, P., Miraglia, S., and Ton Vrouwenvelder, A. C. W. M. (2014) *Safety, Reliability and Risk Analysis: Beyond the Horizon*. CRC Press, Taylor & Francis, Boca Raton, FL, USA.

Wyllie, P. B. (2014) Electrothermal Modelling for Doubly Fed Induction Generator Converter Reliability in WindPower. Doctoral Thesis, Durham University.

Zaggout, M., Tavner, P. J., Crabtree, C. J., and Ran, L. (2014) Detection of rotor electrical asymmetry in wind turbine doubly-fed induction generators. *IET Renewable Power Generation*; 8(1): 1–9.

Zappalá, D., Tavner, P. J., Crabtree, C. J., and Sheng, S. (2014) Side-band algorithm for automatic wind turbine gearbox fault detection and diagnosis. *IET Renewable Power Generation*; 8(4): 380–389.

Zappalá, D. (2014) Advanced Algorithms for Automatic Wind Turbine Condition Monitoring. Doctoral Thesis, Durham University.

Chung, H. S., Wang, H., and Blaabjerg, F., (Eds.), Tavner, P. J., Contributor (2015) *Reliability of Power Electronic Converter Systems*. The Institution of Engineering and Technology, London, UK, 489 pp.

Chen, B. D., Matthews, P., and Tavner, P. J. (2015) Automated on-line fault prognosis for wind turbine pitch systems using supervisory control and data acquisition. *IET Proceedings of Renewable Power Generation*; 9(5), 503–513.

Crabtree, C. J., Zappala, D., and Hogg, S. I. (2015) Wind energy: UK experiences and off-shore operational challenges. *Proceedings of the Institution of Mechanical Engineers, Part A: Journal of Power and Energy*; 229(7): 727–746.

Godwin, J. L. (2015) Exploiting Robust Multivariate Statistics and Data Driven Techniques for Prognosis and Health Management. Doctoral Thesis, Durham University.

OREDA (2015) *OREDA Handbook*, Vol I & II, 6th Ed., OREDA, Norway.

Sinha, Y. (2015) Optimisation of Off-shore Wind Farm Maintenance. Doctoral Thesis, Robert Gordon University.

Serrano-González, J. and Lacal-Arántegui, R. (2016) Technological evolution of onshore wind turbines—A market-based analysis. *Wind Energy*; 19: 2171–2187.

Delorm, T. M., Lu, Y., Christou, A., and McCluskey, P. (2016) Comparisons of off-shore wind turbine reliability. *Proceedings of the Institution of Mechanical Engineers, Part O: Journal of Risk and Reliability*: 1–14.

Gipe, P. (2016) *Wind Energy for the Rest of Us: A Comprehensive Guide to Wind Power and How to Use It*. Wind Works, Bakersfield, CA USA, 491 pp.

Ng, C. and Ran, L. (Eds.) (2016) *Off-Shore Wind Farms-Technologies, Design & Operation*. Woodhead, Elsevier, Oxford, 634 pp.

Dahmani, O., Bourguet, S., Machmoum, M., Guerin, P., Rhein, P., and Josse. L. (2017) Optimization and reliability evaluation of an off-shore wind farm architecture. *IEEE Transactions on Sustainable Energy*; 8(2): 542–550.

Hogg, S. and Crabtree, C. J. (Eds.) (2017) *UK Wind Energy Technologies*. Routledge, Earthscan, Abingdon, UK, 290 pp.

Lavely, A. W. (2017) Effects of Daytime Atmospheric Boundary Layer Turbulence on the Generation of Non-steady Wind Turbine Loadings and Predictive Accuracy of Lower Order Models. Doctoral Thesis, Pennsylvania State University.

Nandi, T. N., Herrig, A., and Brasseur, J. G. (2017) Non-steady wind turbine response to daytime atmospheric turbulence. *Philosophical Transactions of the Royal Society of London. Series A*; 375(2091).

Tavner, P. J. (2017) *Wave & Tidal Generation Devices-Reliability & Availability*. IET Energy Engineering, London, UK, 304 pp.

Leimeister, M., Kolios, A., and Collu, M. (2018) Critical review of floating support structures for off-shore wind farm deployment. *Journal of Physics: Conf. Series*; 1104: 1–11.

Leimeister, M. and Kolios, A. (2018) A review of reliability-based methods for risk analysis and their application in the off-shore wind industry. *Renewable and Sustainable Energy Reviews*; 91: 1065–1076.

Musial, W., Beiter, P., Spitsen, P., Nunemaker, J., and Gevorgian, V. (2018) *Off-shore Wind Technologies Market Report*. US Department of Energy, USA.

Richmond, M., Balaam, T., Causon, P., *et al.* (2018) Multi-criteria decision analysis for benchmarking human-free lifting solutions in the off-shore wind energy environment. *Energies*; 11(5): 1–21.

Young, D. G., Ng, C., Oterkus, S., Li, Q., and Johanning, L. (2018) Assessing the mechanical stresses of dynamic cables for floating off-shore wind applications. *Journal of Physics: Conference Series*; 1102(012016).

BVG Associates, Crown Estate, ORE Catapult (2019) *Guide to an Off-shore Wind Farm*. BVG Associates, distributed by ORE Catapult, Glasgow, UK, 128 pp.

Dao, C. D., Kazemtabrizi, B., and Crabtree, C. J. (2019) Wind turbine reliability data review and impacts on levelized cost of energy. *Wind Energy*; 22(12): 1848–1871.

Fischer, K., Pelka, K., Puls, S., *et al.* (2019) Exploring the causes of power-converter failure in wind turbines based on comprehensive field-data and damage analysis. *Energies*; 12(593): 1–27.

Hou, P., Zhu, J., Ma, K., Tang, G., Hu, W., and Chen, Z. (2019) A review of off-shore wind farm layout optimization and electrical system design methods. *Journal of Modern Power System and Clean Energy*; 7: 975–986.

Huang, Q., Wang, X., Fan, J., Zhang, X., and Wang, Y. (2019) Reliability and economy assessment of off-shore wind farms. *Journal of Engineering*; 2019 (16): 1554–1559.

Huang, Q., Wang, X., Qian, T., Yao, L., and Wang, Y. (2019) Reliability assessment of DC series–parallel off-shore wind farms. *Journal of Engineering*; 2019(16): 1457–1461.

Musial, W., Beiter, P., Nunemaker, J., Ahmann, L., and Busch, J. (2019) *Oregon Off-shore Wind Site Feasibility and Cost Study*. National Renewable Energy Lab, NREL/TP-5000-74597, Golden, CO, USA.

Shah, R., Preece, R., Barnes, M., and Carmona Sánchez, J. (2019) Techno-economic evaluation of power electronics-assisted transmission system frequency regulation. *Journal of Engineering*; 2019(17): 4138–4142.

Stetco, A., Dinmohammadi, F., Zhao, X., *et al.* (2019) Machine learning methods for wind turbine condition monitoring: A Review. *Renewable Energy*; 133: 620–635.

Papatzimos, A. K., Thies, P. R., Lonchampt, J., Joly, A., and Dawood, T. (2019) Data-informed lifetime reliability prediction for off-shore wind farms. *IEEE International Conference on Prognostics and Health Management* (ICPHM).

van Hertem, D., Leterme, W., Chaffey, G., *et al.* (2019) Substations for future HVDC grids: Equipment and configurations for connection of HVDC network elements. *IEEE Power and Energy Magazine*; 17(4): 56–66.

Zappalá, D., Sarma, N., Djurović, S., Crabtree, C. J., Mohammad, A., and Tavner, P. J. (2019) Electrical & mechanical diagnostic indicators of wind turbine induction generator rotor faults. *Renewable Energy*; 131: 14–24.

Brigham, K., Zappalá, D., Crabtree, C. J., and Donaghy-Spargo, C. (2020) Simplified automatic fault detection in wind turbine induction generators. *Wind Energy*; 23(4): 1135–1144.

Dao, C. D., Kazemtabrizi, B., and Crabtree, C. J. (2020) Offshore wind turbine reliability and operational simulation under uncertainties. *Wind Energy*; 23(10): 1919–1938.

Papatzimos, A. K. (2020) Data-driven Operations & Maintenance for Off-shore Wind Farms: Tools and Methodologies. Doctoral Thesis, Exeter University.

Lin, Z., Cevasco, D., and Collu, M. (2020) A methodology to develop reduced-order models to support the operation and maintenance of off-shore wind turbines. *Applied Energy*; 259(114228).

Mohammed, A., Hu, B., Hu, Z., *et al.* (2020) Distributed thermal monitoring of wind turbine power electronic modules using FBG sensing technology. *IEEE Sensors Journal*; 20(17): 9886–9894.

Musial, W., Beiter, P., and Nunemaler, J. (2020) *Cost of Floating Off-shore Wind Energy Using New England Aqua Ventus Concrete Semi-submersible Technology*. National Renewable Energy Lab, NREL/TP-5000-75618, Golden, CO, USA.

Sanchez, J. C., Marjanovic, O., Barnes, M., and Green, P. R. (2020) Secondary model predictive control architecture for VSC-HVDC networks interfacing wind power. *IEEE Transactions on Power Delivery*; 35(5): 2239–2341.

Tavner, P. J., Ran, L., and Crabtree, C.J. (2020) *Condition Monitoring or Rotating Electrical Machines*. 3rd Ed., IET Energy Engineering, London, UK, 400 pp.

Gideon, R. A. and Bou-Zeid, E. (2021) Collocating off-shore wind and wave generators to reduce power output variability: A multi-site analysis. *Renewable Energy*; 163: 1548–1559.

Cevasco, D., Koukoura, S., and Kolios, A. J. (2021) Reliability, availability, maintainability data review for the identification of trends in off-shore wind energy applications. *Renewable and Sustainable Energy Reviews*; 136: 110414.

Dao, C. D., Kazemtabrizi, B., Crabtree, C. J., and Tavner, P. J. (2021) Integrated condition-based maintenance modelling and optimisation for offshore wind turbines. *Wind Energy*; 24(1): 1–19.

Tavner, P. J. and Zappalá, D. (2021) Weather, uncertainty & reliability. *Proceedings of ITI Conference on Turbulence.*

# Standards

IEC 60812 (2006) Analysis techniques for system reliability – Procedure for failure mode and effects analysis (FMEA).
IEC 61400-1 (2010) Wind turbines – Part 1: Design Requirements.
IEC 61400-2 (2013) Small wind turbines.
IEC 61400-3 (2010) Wind energy generation systems – Part 3: Design requirements for off-shore wind turbines.
IEC 61400-4 (2012) Wind energy generation systems – Part 4: Design requirements for wind turbine gearboxes.
IEC 61400-6 (2020) Wind energy generation systems: Tower and foundation design requirements.
IEC 61400-11 (2012) Wind energy generation systems: Acoustic noise measurement techniques.
IEC 61400-12-1 (2017) Wind energy generation systems: Power performance measurements of electricity producing wind turbines.
IEC 61400-12-2 (2016) Wind energy generation systems: Power performance of electricity-producing wind turbines based on nacelle anemometry.
IEC 61400-13 (2015) Wind energy generation systems: Measurement of mechanical loads.
IEC 61400-14 (2005) Wind energy generation systems: Declaration of apparent sound power level and tonality values.
IEC 61400-15-1 (Draft) Wind energy generation systems: Site suitability input conditions for wind power plants.
IEC 61400-21 (2008) Wind energy generation systems: Measurement and assessment of power quality characteristics of grid connected wind turbines.
IEC 61400-21-2 (2019) Wind energy generation systems: Measurement and assessment of electrical characteristics – Wind power plants.
IEC 61400-22 (2010) Wind energy generation systems: Conformity testing and certification.
IEC 61400-23 (2014) Wind energy generation systems: Full-scale structural testing of rotor blades.
IEC 61400-24 (2010) Wind energy generation systems: Lightning protection.
IEC 61400-25-1 (2006) Wind energy generation systems: Communications for monitoring and control of wind power plant – Overall description of principles and models.
IEC 61400-25-2 (2015) Wind energy generation systems: Communications for monitoring and control of wind power plants – Information models.

IEC 61400-25-3 (2015) Wind energy generation systems: Communications for monitoring and control of wind power plants – Information exchange models.

IEC 61400-25-4 (2008) Wind energy generation systems: Communications for monitoring and control of wind power plants – Mapping to communication profile.

IEC 61400-25-5 (2006) Wind energy generation systems: Communications for monitoring and control of wind power plants – Conformance testing.

IEC 61400-25-6 (2010) Wind energy generation systems: Communications for monitoring and control of wind power plants – Logical node classes and data classes for condition monitoring.

IEC 61400-26-1(2019) Wind energy generation systems: Availability for wind energy generation systems.

IEC 61400-26-2 (2014) Wind energy generation systems: Production-based availability for wind turbines.

IEC 61400-27-1 (2015) Wind energy generation systems: Electrical simulation models – Part 27-1 – Wind turbines

IEC 61400-29 Wind energy generation systems: Marking and lighting of wind turbines.

IEC 61400-40 Wind energy generation systems: Electromagnetic compatibility (EMC) – Requirements and test methods.

IEC 61400-50-3 Wind energy generation systems: Use of nacelle-mounted lidars for wind measurements.

IEC 61400-101 Wind energy generation systems: General requirements for wind turbine plants.

DNV GL (2010) Guideline for the certification of wind turbines.

DNV GL (2012) Guideline for the certification of off-shore wind turbines.

DNV GL (2013) Guideline for the certification of condition monitoring systems for wind turbines.

IEEE Std 493 (1997) Gold Book – Recommended practice for design of reliable industrial and commercial power systems. IEEE Press, Piscataway.

ISO 13373-1 (2002) Condition monitoring and diagnostics of machines – Vibration condition monitoring: General procedures.

ISO 14224 (2006) Petroleum, petrochemical and natural gas industries – Collection and exchange of reliability and maintenance data for equipment.

ISO 16079-1(2017) Condition monitoring and diagnostics of wind turbines – Part 1: General guidelines.

ISO 16079-2 (2020) Condition monitoring and diagnostics of wind turbines – Part 2: Monitoring the drivetrain.

ISO 17359 (2018) Condition monitoring and diagnostics of machines, General guidelines.

ISO 18436-2 (2014) Condition monitoring and diagnostics of machines – Requirements for qualification and assessment of personnel: Vibration condition monitoring and diagnostics.

MIL-STD-1629 (1980) Military standard procedures for performing a failure mode, effects and criticality analysis. US Department of Defense.

MIL-HDBK-189 (1981) Reliability growth management. US Department of Defense, Washington.

SAE J 1739 (2009) Potential failure mode and effects analysis in design, potential failure mode and effects analysis in manufacturing and assembly processes.

SMC Regulation 800 Acquisition management failure modes, effects, and criticality analysis (FMECA), US Department of the Air Force.

OREDA, Off-shore reliability data:

- OREDA (1984) Off-shore reliability data handbook. VERITEC – Marine Technology Consultants, PennWell Books, 1st edition.
- OREDA (1997) Off-shore reliability data handbook. SINTEF Industrial Management. Det Norske Veritas, Norway, 3rd edition.
- OREDA (2002) Off-shore reliability data handbook. SINTEF Industrial Management. Det Norske Veritas, Norway, 4th edition.

VGB PowerTech, Guideline (2014) Reference designation system for power plants, RDS-PP, Part 32, Wind Power Plants.

# *Appendix A*
# WMEP operators' report form

This report form was prepared by Faulstich *et al.* (2008).

********* (Company name)**
**Maintenance and repair report**

Work done on: ___ Date Month Year

Report No: ___

Post code

Plant ID number

Operator

Wind Turbine manufacturer and model

**Cause of malfunction**

- ☐ High wind
- ☐ Grid failure
- ☐ Lightning
- ☐ Icing
- ☐ Malfunction of control system
- ☐ Component wear or failure
- ☐ Component loosening
- ☐ Other cause
- ☐ Unknown cause

**Cause work**

- ☐ Scheduled maintenance (only examination and functional check)
- ☐ Scheduled maintenance with replacement of worn components or repair of faults
- ☐ Unscheduled maintenance or repair after malfunction

**Effect of malfunction**

- ☐ Overspeed
- ☐ Overload
- ☐ Noise
- ☐ Vibrations
- ☐ Reduced power output
- ☐ Causing follow up damages
- ☐ Plant stoppage
- ☐ Other consequences

**Down times**

☐ Not stopped ☐ Stopped

From

To

Date Month Year

Reading of hour counter

**Removal of malfunction**

Faultless operation without later repair:

☐ Control reset ☐ Changing control parameters

Repaired or replaced components:

☐ **Rotor hub** ☐ **Gearbox**

**Down times**

☐Not stopped ☐Stopped

From | | |

| | |

Date Month Year

Reading of hour counter

**Cost according to calculation**

Material £

Labour £

Journey £

Total cost (incl.tax) £

**Comments**

**The Operator**

Place/Date

Signature

**Removal of malfunction**

Faultless operation without later repair:

☐ Control reset ☐ Changing control parameters

Repaired or replaced components:

☐ **Rotor hub**
- ☐Hub body
- ☐Pitch mechanism
- ☐Pitch bearing

☐ **Rotor blades**
- ☐Blade bolts
- ☐Blade shell
- ☐Aerodynamic brakes

☐ **Generator**
- ☐Windings
- ☐Brushes
- ☐Bearings

☐ **Electrical system**
- ☐Inverter
- ☐Fuses
- ☐Switches
- ☐Cables/connections

☐ **Sensors**
- ☐Anemometer/Wind vane
- ☐Vibration switch
- ☐Temperature switch
- ☐Oil pressure switch
- ☐Power sensor
- ☐Rev counter

☐ **Control system**
- ☐Electronic control unit
- ☐Relay
- ☐Measurement cables and connections

☐ **Gearbox**
- ☐Bearings
- ☐Gearwheels
- ☐Gear shaft
- ☐Sealings

☐ **Mechanical brakes**
- ☐Brake disc
- ☐Brake pads
- ☐Brake shoe

☐ **Drive train**
- ☐Rotor bearings
- ☐Drive shafts
- ☐Couplings

☐ **Hydraulic system**
- ☐Hydraulic pump
- ☐Pump motor
- ☐Valves
- ☐Hydraulic pipes/hoses

☐ **Yaw system**
- ☐Yaw bearings
- ☐Yaw motor
- ☐Wheels and pinions

☐ **Structural components/ Housing**
- ☐Foundation
- ☐Tower/Tower bolts
- ☐Nacelle frame
- ☐Nacelle cover
- ☐Ladder/lift

**Main component exchanged**

Please check if complete component is exchanged

☐ Nacelle
☐ Rotor blades
☐ Rotor hub
☐ Gearbox
☐ Generator
☐ Yaw system
☐ Tower
☐ Control system cabinet
☐ Grid transformer

# *Appendix B*
# Reliability data collection

## B.1 Introduction

### *B.1.1 Background*

Wind turbine manufacturers, operators, maintainers and investors agree that it is essential for wind turbines to have a high reliability to achieve a high-capacity factor and availability and thereby deliver electricity at a low cost of energy.

An important factor in achieving those objectives is that, when designed, wind turbines should have the highest possible reliability. The European wind industry achieved wind turbine availabilities in 2015 on-shore of 96%–97% and off-shore of 90%–95%, and at the time of this Edition those figures are probably on-shore 98% and off-shore 97%. It would be desirable to raise these availabilities and design for reliability would contribute to that aim.

An important requirement of design for reliability is to be able to measure, predict and analyse wind turbine reliability using accurately defined mean time to failure, mean time to repair and mean time between failures data for wind turbines. These standard terms are defined by International Standards and are listed in Section 1.7.1.

The definition of the terminology and taxonomy of wind turbines and the collection of reliability data and its interrelationship with wind turbine design, defined by IEC 61400, needs to be standardised. It is also clear that in order to increase wind turbine reliability more and higher, quality reliability data are needed from the wind industry, within limits of commercial confidentiality.

Appendix B was a proposal from the EU FP7 ReliaWind project for the standardisation of

- taxonomy of the wind turbine,
- English terminology for the naming of components,
- methods for collecting reliability data from wind turbines in the field,
- method for reporting failures from wind turbines in the field.

The purpose of these standardisations is for the improvement wind turbine reliability in the field, to raise wind turbine availability and lower the consequent cost of energy.

These issues also affect other industries, including off-shore oil and gas, power generation, transportation, military and aerospace.

An example of reliability data collected from the first of these industries, oil and gas, is shown by OREDA.

A standard for the collection of reliability data from that industry also exists, ISO 14224 (2006).

### *B.1.2 Previously developed wind industry methods*

The most detailed previous public domain wind turbine data collection campaign was funded from 1996 to 2006 by the German Federal Ministry for Economics & Technology under the 250-MW Wind Test Programme, which included the Wissenschaftliche Mess-und Evaluierungsprogramm (WMEP), Scientific Measurement and Evaluation Programme, administered now by Fraunhofer IWES Institute, WMEP (1996). This built-on earlier work by Schmid *et al.* (1991), a standard failure report form, was used by wind turbine operators for return to IWES and that is shown in Appendix A. Schmid *et al.* (1991) gave valuable examples of data collection forms, the proposals later have drawn from Appendix A and this experience.

## B.2 Standardising wind turbine taxonomy

### *B.2.1 Introduction*

This section summarises the general principles and guidelines on which the taxonomy will be based and is derived from a deliverable prepared for ReliaWind by the author and other consortium members.

The taxonomy should be adaptable for application to the common reliability analyses needed for wind turbines, such as failure mode and effects criticality analysis, Arabian-Hoseynabadi *et al.* (2010), failure rate Pareto analysis, reliability growth analysis and Weibull analysis.

The intention of adopting such a taxonomy would be to overcome current deficiencies of the data collection which can be summarised as follows:

- inconsistency of naming of the systems, sub-systems, assemblies, sub-assemblies and components of wind turbines;
- intractability of the systems being monitored;
- unspecified wind turbine technology or concept;
- problems of confidentiality between parties exchanging data.

### *B.2.2 Taxonomy guidelines*

A wind turbine taxonomy is a structure which names the main features of a wind turbine in a standard terminology exemplified in Figure B.1.

- The taxonomy must be reliability-oriented, particularly in respect of analysis. It is agreed that such an approach is the best compromise between the various

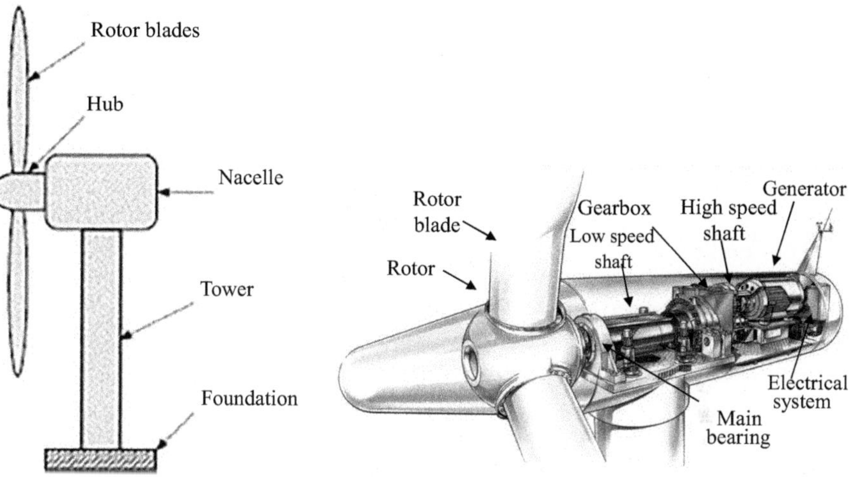


*Figure B.1 Example of a wind turbine and its nacelle layout showing the terminology*

industry needs, which leads to a different system breakdown, grouping and terminology that would be achieved by simple components description.

- The taxonomy will include all the wind turbine concept components in five levels. Data will be retrieved on the basis of a concept code, which allows wind turbine model mapping within the taxonomy for any given data set.
- The taxonomy is based on a Danish concept wind turbine. That is an upwind, three-bladed, horizontal axis, un-ducted wind turbine. Other concepts could be included upon the achievement of significant industrial uptake by this taxonomy.
- At the highest level, outside the taxonomy, the wind turbine concept should be identified by a code – for example indicating stall-, active stall- or pitch-regulated, fixed or variable speed, geared- or direct-drive, doubly fed induction, induction or squirrel cage induction or wound or permanent-magnet synchronous generator. Therefore, each item in the taxonomy will be clearly linked to a code associated with each unique wind turbine concept.
- The taxonomy should also inform the structure of the monitoring input/output (I/O) applied to the wind turbine, whether that is for signal condition and data acquisition (SCADA) or condition monitoring system signals and alarms because the taxonomy will be used to focus on SCADA and service log data available from operational wind farms. Therefore, the terminology of components in the I/O list of the SCADA, IEC 61400-25-6, should agree with the component names used in the taxonomy earlier.
- The taxonomy shall be organised in five indented levels. Each level should be justified with a brief description that shall include the rationale for the level grouping and intended use.

- The first five indented levels of the taxonomy must comply with the table given in Table B.2. The taxonomy may not reach the lower level components, for example to individual electronic capacitors, but an analyst could add additional lower levels, if needed, but they must be compatible with the upper five levels of the taxonomy. Analysts could also add additional elements in levels 1–5, if absolutely necessary for their purpose, denoted by the prefix CUSTOM, although this customisation is strongly to be discouraged.
- The taxonomy will have a short code alpha-numeric designation for each item at each level. It is anticipated that the construction of the designation could follow the guidance of VGB PowerTech, Guideline (2014) which adopts an alpha-numeric code although some in the wind industry prefer a word code.
- The lowest level components will be grouped according to the following two concepts:
  - Functional grouping for the signalling, supervisory and control components, examples: pitch encoder grouped with control and communication system, LV electrical systems grouped together.
  - Positional grouping for mechanical components, examples: gearbox, pitch system, blade, frequency converter, generator, blade.
- For example: generator temperature sensor and pitch encoder are both components of the monitoring system. This segregation is necessary due to the nature of wind turbine systems that signalling, supervisory and control components tend to spread throughout the wind turbine, whereas mechanical devices are located in a specific position within the wind turbine. This is exemplified in Table B.1.
- In the case of ambiguity the designation will follow the order above: first the functional groupings, second the positional grouping.
- At the lowest indented level the component name should have no ambiguity with similar components of different assemblies. For example: the pitch pinions and the yaw pinions.

*Table B.1 Examples of parts groupings*

| Functional grouping | Positional grouping |
|---|---|
| Control and communication system | Generator |
| Lightning protection system | Pitch system |
| 110-V electrical auxiliary system | Gearbox |
| 220-V electrical auxiliary system | Yaw system |
| 400-V electrical auxiliary system | Blade |
| Wind turbine power system | Hub |
| SCADA system | Main shaft set |
| Collection system | Foundation |
| Grid connection | Tower |
| Hydraulics system | |

### *B.2.3 Taxonomy structure*

The structure of system, sub-system, assembly, sub-assembly and component that should be adopted is shown in Figure B.2. The wind turbine itself is considered the system.

Examples of this terminology are shown in Table B.2.

A full taxonomy, listing sub-systems, assemblies and sub-assemblies is provided in Section B.6.

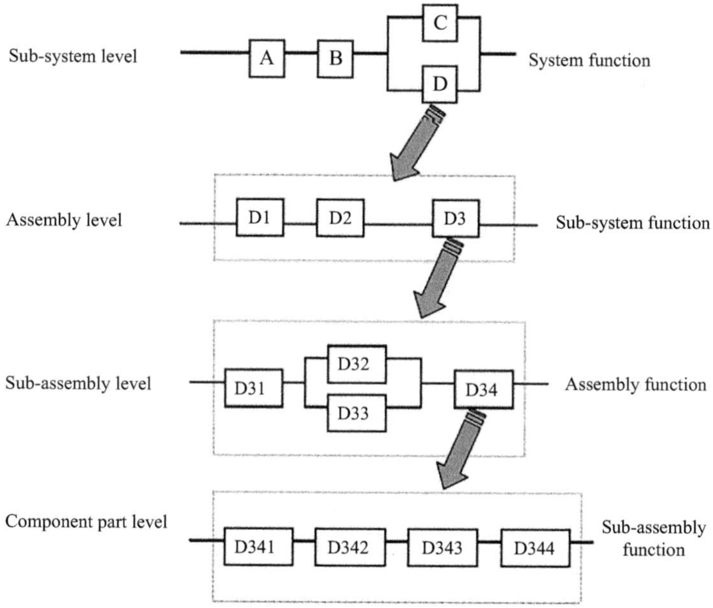


*Figure B.2 Example of system, sub-system, assembly, sub-assembly and component structure*

*Table B.2 Examples of application of terminology*

| System | Sub-system | Assembly | Sub-assembly | Component |
|---|---|---|---|---|
| Wind turbine | Rotor | Electrical pitch system | Pitch motor | Brush |
| Wind turbine | Drive train module | Gearbox | Gearbox | Stage 1 planetary wheel |
| Wind turbine | Electrical module | Frequency converter | Power electronics | IGBT |

## B.3 Standardising methods for collecting wind turbine reliability data

In ReliaWind, it was proposed that reliability data from wind turbines should be collected in the five organised tables (Tables B.3–B.7).

This list of events in Table B.3 will be exhaustive within the following criteria:

- The event required manual intervention to restart the machine.
- The event resulted in downtime $\geq 1$ h.
- There will be no missing events or missing time periods; or if there are, the missing time periods will be noted and the reasons stated.
- Every cell in the table should have either a data value or be filled with N/A, not available.

Table B.4 would be derived entirely from Table B.3 and no new information is added.

It is thought unlikely that enough detail will be available in Table B.3 to permit the calculation of failure rate on a per component basis.

Failure rates should be reported per year as standard but information from Table B.7 will allow calculation per operational period in a year, per GWh in a year, per revolution, or some other metric, depending on what information is available for the particular wind farm.

Confidentiality may require that this information be aggregated on a wind farm, rather than wind turbine basis.

This information could be presented graphically for example as shown in Table B.7.

Table B.5 is also derived entirely from Table B.4 and no new information is added.

The downtime will be given in units of hours.

Confidentiality may require that this information be aggregated on a wind farm, rather than wind turbine basis.

This information could be presented graphically for example as shown in Figure 2.1(b).

There would be two versions of Table B.6:

- Table B.6 (A) will have all values stated exactly and to maintain end user confidentiality will remain private.
- Table B.6 (B) will be available to a ReliaWind but will be less specific about machine characteristics, with identifiable parameters categorised into appropriate ranges to make anonymous the data as shown in the example earlier.

The control-type column will be populated from a standard list.

Further columns may be added to this table depending on what information is available for each wind farm.

Confidentiality requirements may mean that the information in Table B.7 could not be publicly available.

*Table B.3 Events*

| Wind farm | Turbine ID | Date and time of event | Time to repair TTR (h) | Actual repair time ART (h) | System | Sub-system | Assembly | Sub-assembly | Component | Failure mode | Root cause | Maintenance category | Severity category | Additional information |
|---|---|---|---|---|---|---|---|---|---|---|---|---|---|---|
| A | 1 | 2008-04-01 11:28:01 | 54.2 | N/A | Wind turbine | Drive train | Gearbox assembly | Gearbox | N/A | N/A | N/A | 4 | 3 | N/A |
| A | 23 | 2008-04-24 01:56:11 | 168.4 | 3.5 | Wind turbine | Rotor | Pitch system | N/A | N/A | N/A | N/A | 3 | 2 | N/A |
| B | 2 | 2008-04-25 08:43:24 | 2.5 | 1 | Wind turbine | Power | Generator assembly | Generator | Stator phase b winding | Open | Over current | 1 | 1 | Series failure |

| Field | Definition |
|---|---|
| Wind farm | Confidentiality will require that this be an anonymous identifier |
| Turbine ID | Turbine identifier within the wind farm |
| Time of event | Time stamp in ISO form yyyy-mm-dd hh:mm:ss |
| Mean downtime (MDT) | Total number of hours during which the turbine was not operational, i.e. includes all the time needed to restore the wind turbine to an operating condition |
| Time to repair (TTR) | Actual number of hours completing the repair, i.e. excludes logistics associated with the repair action such as having the component delivered to site, arranging technicians' time |
| System structure | The system structure used is that set out in Section B.2 |
| Sub-system | Select from an approved list, see Section B.6. It should usually be possible to ascribe a failure to a particular sub-system |
| Assembly | Select from an approved list Section B.6. It should usually be possible to ascribe a failure to a particular assembly |
| Sub-assembly | Select from an approved list Section B.6. It may not always be possible to ascribe a failure to a particular component |
| Part | Select from an approved list. It may not always be possible to ascribe a failure to a particular component |
| Failure mode | The particular way in the failure occurred, independent of the reason for failure. This may be subjective, but very useful if available |
| Root cause | The cause of failure. This may be subjective, but very useful if available |

*(Continues)*

*Table B.3* (*Continued*)

| **Field** | **Definition** |
|---|---|
| Maintenance category | A description of the maintenance impact of the failure:<br>1. Manual restart<br>2. Minor repair<br>3. Major repair<br>4. Major replacement |
| Severity category | A description of the severity of the failure based on MIL-STD-1629 (1980) which relates to the ability of the system to carry out the function for which it was defined safely and efficiently:<br>1. Minor<br>2. Marginal<br>3. Critical<br>4. Catastrophic |
| Additional information | Pertinent comments where available |

*Table B.4 Failure rates*

| Wind farm | Turbine | Sub-system | Assembly | Year | | | |
|---|---|---|---|---|---|---|---|
| | | | | 1 | 2 | 3 | ... |
| A | 1 | Drive train | Gearbox | 0 | 2 | 1 | ... |
| A | 1 | Power | Generator | 2 | 1 | 2 | ... |
| A | 1 | Rotor | Pitch | 1 | 2 | 1 | ... |
| ... | ... | | ... | ... | ... | ... | ... |

*Table B.5 Downtime*

| Wind farm | Turbine | Sub-system | Assembly | Year | | | |
|---|---|---|---|---|---|---|---|
| | | | | 1 | 2 | 3 | ... |
| A | 1 | Drive train | Gearbox | 24 | 5 | 1 | ... |
| A | 1 | Power | Generator | 65 | 4 | 2 | ... |
| A | 1 | Rotor | Pitch | 21 | 5 | 5 | ... |
| ... | ... | | ... | ... | ... | ... | ... |

For a wind farm to be included in the survey it is desirable that the site contains at least 15 turbines that have been running for at least 2 years since commissioning.

Data for the tables earlier should be provided by wind turbine operators.

## B.4 Standardising downtime event recording

The approach recommended is to describe and classify downtime events as stoppages of duration $\geq 1$ h and requiring at least a manual restart, categorising downtime events as follows:

- Category 1: manual restart
- Category 2: minor repair
- Category 3: major repair
- Category 4: major replacement

## B.5 Standardising failure event recording

### *B.5.1 Failure terminology*

When a failure has occurred, it is important to record the details of that failure. In the WMEP Failure Report Form, shown in Appendix A, a simple tick box approach was adopted.

This provides insufficient detail for maintenance and root cause analysis purposes and the following approach is suggested for recording failures in detailed

*Table B.6 Wind farm configuration*

| Wind farm | Turbines | Rated power (MW) | Mean wind speed (m/s) | Mean turbulence intensity | Hub height (m) | Rotor diameter (m) | Terrain type | Control type | ... |
|---|---|---|---|---|---|---|---|---|---|
| A | 20–40 | 1–2 | 6–8 | 0.25–0.50 | 60 | 40 | Off-shore | A | ... |
| B | 0–20 | 2–3 | 8–10 | 0.50–0.75 | 55 | 30 | On-shore exposed | B | ... |
| ... | ... | ... | ... | ... | ... | ... | ... | ... | ... |

*Table B.7 Additional turbine information*

| Wind farm | Month | Energy generated (GWh) | Revolutions | ... |
|---|---|---|---|---|
| A | 2008-01 | 50 | $1.544 \times 10^5$ | ... |
| A | 2008-02 | 70 | $2.422 \times 10^5$ | ... |
| ... | ... | ... | ... | ... |

fault or maintenance logs, taken from the recommendations in Faulstich *et al.* (2008). The terminology to be used should be that shown in Section B.6, which is consistent with the proposed structure Section B.2 and overall taxonomy Section B.6. The failure modes suggested there include those identified by work package partners of ReliaWind.

## *B.5.2 Failure recording*

The intention of this section is to provide a broad method of failure recording, rather than trying to capture every different possible failure mode. For example a bearing failure could encompass

- inner race failures;
- outer race failures;
- cage failures;
- element failures.

The recommended failure recording terminology is in part recursive, referring successively to the component, sub-assembly, assembly, sub-system, system, defined in Section B.2, in turn.

For example using Section B.7 for recording a gearbox epicyclic bearing failure, the failure description should take the following format:

- Bearing failure: Planet bearing: Epicyclic part: Gearbox: Drive train: Wind turbine.

## *B.5.3 Failure location*

Location indicators are needed for components, such as bearings, where several may be found in a single assembly or sub-assembly; in that case, the following rules could be followed, using the previous failure example:

- If more than one epicyclic stage exists in a gearbox, the first stage is that closest to the wind turbine rotor and so on.
- In a parallel shaft gear train, the pinion drives and the gear is driven.
- The two ends of a gearbox are the rotor end or generator end.
- Where there are two bearings on a gearbox shaft that closer to the gear should be referred to as the inner bearing, and that further from the gear as the outer bearing.
- Generator bearings should be defined as drive end and non-drive end.

## B.6 A detailed wind farm taxonomy

The following taxonomy is not intended to be exhaustive but rather indicative of the detail that O&M engineers must apply to analyse the reliability and maintainability issues of a complex wind farm.

| System | Sub-system | Assembly | Sub-assembly | Component |
|---|---|---|---|---|
| Wind farm | Operational infrastructure | | | |
| | Maintenance infrastructure | | | |
| | On-shore sub-station | Grid connection | Sub-station transformers | |
| | | | Sub-station switchgear | |
| | | | Sub-station and cable communication and control | |
| | | | EHV cable link shore-side | |
| | HV export shore connection | HV subsea cables | | |
| | | Shore connection redundancy | | |
| | Off-shore sub-station | Shore connection | EHV cable link off-shore-side | |
| | | | Sub-station transformers | |
| | | | Sub-station switchgear | |
| | | | Sub-station and cable communication and control | |
| | | Meteorological station | | |
| | Support structure | Foundation | Gravity-based foundation | Concrete |
| | | | | Steel reinforcement |
| | | | Monopile | Corrosion protection |
| | MV collection system | MV subsea cables | | |
| | | Collection cable redundancy | | |
| Wind turbine | Support structure and foundation | | Gravity-based foundation | Concrete |
| | | | | Steel reinforcement |
| | | | Monopile | Corrosion protection |
| | | | | Pile |
| | | | | Transition piece |
| | | | | Subsea pile |
| | | | | Corrosion protection |

*(Continues)*

(*Continued*)

| System | Sub-system | Assembly | Sub-assembly | Component |
|---|---|---|---|---|
| | | | Jacket or twisted jacket | Steel reinforcement |
| | | | | Piles |
| | | | | Steel structure |
| | | | | Corrosion protection |
| | | | On-shore | Concrete |
| | | | | Nuts and bolts |
| | | | | Piles |
| | | | | Steel reinforcement |
| | | | Space frame/tripod | Corrosion protection |
| | | | | Piles |
| | | | | Structures |
| | Tower | | Access equipment | Ladder |
| | | | | Landing pad |
| | | | | Lightning protection |
| | | | Access equipment | Ladders |
| | | | | Landing pad |
| | | | | Climb assist |
| | | | Tower details | Maintenance crane |
| | | | | Nuts and bolts |
| | | | | Paint/coating |
| | | | | Tower section |
| | | | | Lightning protection |
| Wind turbine | Rotor module | Nacelle | Nacelle structure | |
| | | | Nacelle sealing | |
| | | | Nacelle dehumidifier | |
| | | Blades | Blade lightning protection termination | Blade lightning protection termination |
| | | | Blade lightning down-conductor | Blade lightning down-conductor |
| | | | De-icing system | De-icing system |
| | | | Leading edge bond | Leading edge bond |
| | | | Nuts and bolts | Nuts and bolts |
| | | | Paint and coating | Paint and coating |
| | | | Root structure | Root structure |
| | | | Sandwich shell | Sandwich shell |
| | | | Spar box | Spar box |
| | | | Spar cap | Spar cap |
| | | | Spar web | Spar web |
| | | | Trailing edge bond | Trailing edge bond |
| | | Hub | Exit hatch | Exit hatch |
| | | | Nose cone | Nose cone |
| | | Pitch system | Pitch cabinet | Battery |
| | | | | Battery charger |

(*Continues*)

(*Continued*)

| System | Sub-system | Assembly | Sub-assembly | Component |
|---|---|---|---|---|
| | | | | Heater |
| | | | | Local controller |
| | | | | Switchboard |
| | | | Pitch drive | Motor |
| | | | | Motor cooling |
| | | | | Motor cooling system |
| | | | | Motor drive |
| | | | | Pinion |
| | | | | Pitch bearing |
| | | | | Pitch gearbox |
| | | | Pitch sensors | Position encoder |
| | | | | Temperature sensor |
| | | | | Voltmeter |
| Wind turbine | Drive train module | Gearbox | Cooling system | Radiator |
| | | | Gears | Hollow shaft |
| | | | | Planet carrier |
| | | | | Planet gear |
| | | | | Ring gear |
| | | | | Spur gear |
| | | | | Sun gear |
| | | | Housing | Bushing |
| | | | | Case |
| | | | | Mounting |
| | | | | Torque arm system |
| | | | Lubrication system | Hose |
| | | | | Motor |
| | | | | Motor |
| | | | | Primary filter |
| | | | | Pump |
| | | | | Reservoir |
| | | | | Seal |
| | | | | Secondary filter |
| | | | Sensors | Debris |
| | | | | Oil level |
| | | | | Pressure 1 |
| | | | | Pressure 2 |
| | | | | Temperature |
| Wind turbine | Drive train module | Generator | Cooling system | Cooling fan |
| | | | | Filter |
| | | | | Hose |
| | | | | Radiator |
| | | | Lubrication system | Pump |
| | | | | Reservoir |
| | | | Rotor | Commutator |
| | | | | Exciter |
| | | | | Resistance controller |
| | | | | Rotor lamination |
| | | | | Rotor winding |
| | | | | Slip ring |

(*Continues*)

(*Continued*)

| System | Sub-system | Assembly | Sub-assembly | Component |
|---|---|---|---|---|
| | | | Sensors | Core temperature sensor |
| | | | | Encoder |
| | | | | Wattmeter |
| | | | Stator | Magnet |
| | | | | Stator lamination |
| | | | | Stator winding |
| | | | Structural and mechanical | Front bearing |
| | | | | Housing |
| | | | | Rear bearing |
| | | | | Shaft |
| | | | | Silent block |
| Wind turbine | Drive train module | Gearbox | Low speed side | Axial bearing |
| | | | | Compression coupling |
| | | | | Connector plate |
| | | | | Main bearing seal |
| | | | | Main bearing temperature sensor |
| | | | | Main shaft |
| | | | | Radial bearing |
| | | | | Rotor lock |
| | | | | Slip ring |
| | | | High speed side | Coupling |
| | | | | Rotor lock |
| | | | | Shaft |
| | | | | Transmission shaft |
| | | | Mechanical brake | Caliper |
| | | | | Disc |
| | | | | Pad |
| | | | | Transmission lock |
| | | | Sensors | High-speed sensor |
| | | | | Low-speed sensor |
| | | | | Position sensor |
| Wind turbine | Electrical module | Auxiliary electrical system | Electrical services | 24 DC feeder |
| | | | | Auxiliary transformer |
| | | | | Breaker |
| | | | | Cabinet |
| | | | | Fan |
| | | | | Fuse |
| | | | | Grid protection relay |
| | | | | Light |
| | | | | Mechanical switch |
| | | | | Power point |
| | | | | Protection cabinet |
| | | | | Pushbutton |
| | | | | Relay |
| | | | | Space heater |

(*Continues*)

(*Continued*)

| System | Sub-system | Assembly | Sub-assembly | Component |
|---|---|---|---|---|
| | | | | Surge arrester |
| | | | | Thermal protection |
| | | | | UPS |
| | | | Lightning protection system | Air termination |
| | | | | Bonding element |
| | | | | Earth connector |
| | | | | Earth termination |
| | | | | Sliding contact |
| | | | | Spark gap system |
| | | | | Surge arrester |
| Wind turbine | Electrical module | Control and communication | Ancillary equipment | Breaker |
| | | | | Cabinet temperature sensor |
| | | | | Cable |
| | | | | Contactor |
| | | | Communication system | Analogue I/O unit |
| | | | | Digital I/O unit |
| | | | | Ethernet module |
| | | | | Field bus master |
| | | | | Field bus slave |
| | | | | Frequency unit |
| | | | Condition monitoring system | Condition cables |
| | | | | Data logger |
| | | | | Protocol adapter card for data logger |
| | | | | Sensors |
| | | | Controller hardware | Controller power supply |
| | | | | CPU |
| | | | | Internal communication system |
| | | | | Main I/O unit |
| | | | | Watch dog unit |
| | | | | Closed loop control software |
| | | | | Supervisory control software |
| | | | Safety chain | Emergency button |
| | | | | Max speed switch |
| | | | | Power switch |
| | | | | Short circuit switch |
| | | | | Vibration switch |
| | | | | Watch dog switch |
| | | | | Wind-up switch |
| Wind turbine | Electrical module | Frequency converter | Converter auxiliaries | Auxiliary power supply |

*(Continues)*

(*Continued*)

| System | Sub-system | Assembly | Sub-assembly | Component |
|---|---|---|---|---|
| | | | | Cabinet |
| | | | | Cabinet sensors |
| | | | | Communication and interface unit |
| | | | | Control boards |
| | | | | Generator side fan |
| | | | | Grid side fan |
| | | | | Measurement unit |
| | | | | Power supply |
| | | | | Power supply 24 V |
| | | | | Tachometer adapter |
| | | | | Cabinet heating system |
| | | | | Thermostat |
| | | | Converter power bus | Branching unit |
| | | | | Capacitor |
| | | | | Contactor |
| | | | | Generator side converter |
| | | | | Generator side power module |
| | | | | Grid side converter |
| | | | | Grid side power module |
| | | | | Inductor |
| | | | | Load switch |
| | | | | Pre-charge unit |
| | | | Power conditioning | Common mode filter |
| | | | | Crowbar |
| | | | | DC chopper |
| | | | | Generator side filter |
| | | | | Line filter assembly and voltage limiter unit |
| Wind turbine | Electrical module | Power electrical system | Measurements | Voltage |
| | | | | Current |
| | | | Power circuit | Cables |
| | | | | Machine contactor |
| | | | | Machine transformer |
| | | | | MV busbar/ isolator |
| | | | | MV switchgear |
| | | | | Soft start electronics |
| | Nacelle module | Hydraulics system | Hydraulic power pack | Motor |

(*Continues*)

(*Continued*)

| System | Sub-system | Assembly | Sub-assembly | Component |
|---|---|---|---|---|
| | | | | Pump |
| | | | | Pressure valve |
| | | | | Filter |
| | | | Actuator | Bushing |
| | | | | Cylinder |
| | | | | Hose/fitting |
| | | | | Hydraulic linear drive |
| | | | | Limit switch |
| | | | | Linkage |
| | | | | Miscellaneous hydraulics system |
| | | | | Position controller |
| | | | | Proportional valve |
| | | | | Pump |
| | | | Torque converter | |
| | | | Differential | |
| | | | Viscous coupling | |
| Wind turbine | Nacelle module | Nacelle auxiliaries | Meteorological sensors | Anemometer |
| | | | | Wind vane |
| | | | Nacelle sensors | Emergency vibration sensor |
| | | | | Yaw encoder |
| | | | Dehumidifier | |
| | | | Safety system | Beacon |
| | | | | Down conductor |
| | | | | Fall arrester |
| | | | | Firefighting system |
| | | | | Nacelle cover metallic mesh |
| | | | | Lightning protection termination |
| | | | | Service crane |
| | | Nacelle structure | Bedplate | Bolts |
| | | | | Cast or welded structure |
| | | | Cover | Fibreglass |
| | | | | Hatch |
| | | | Generator frame | Bolts |
| | | | | Cast or welded structure |
| | | Yaw system | Yaw brake | Yaw brake calipers |
| | | | | Yaw brake disc |
| | | | | Yaw brake hoses |
| | | | | Yaw brake paths |
| | | | | Damper |
| | | | | Yaw bearing |
| | | | | Yaw gearbox |
| | | | | Yaw motor |
| | | | | Yaw pinion |

*(Continues)*

(*Continued*)

| System | Sub-system | Assembly | Sub-assembly | Component |
|---|---|---|---|---|
| | | | Yaw sensors | Wind-up counter |
| | | | | Yaw encoder |
| Wind turbine | Rotor module | Blade | Blade lightning protection termination | |
| | | | Blade lightning down-conductor | |
| | | | De-icing system | De-icing system |
| | | | Leading edge bond | Leading edge bond |
| | | | Nuts and bolts | Nuts and bolts |
| | | | Paint and coating | Paint and coating |
| | | | Root structure | Root structure |
| | | | Sandwich shell | Sandwich shell |
| | | | Spar box | Spar box |
| | | | Spar cap | Spar cap |
| | | | Spar web | Spar web |
| | | | Trailing edge bond | Trailing edge bond |
| | | Hub | Exit hatch | Exit hatch |
| | | | Nose cone | Nose cone |
| | | Pitch system | Pitch cabinet | Battery |
| | | | | Battery charger |
| | | | | Heater |
| | | | | Local controller |
| | | | | Switchboard |
| | | | Pitch drive | Motor |
| | | | | Motor cooling |
| | | | | Motor cooling system |
| | | | | Motor drive |
| | | | | Pinion |
| | | | | Pitch bearing |
| | | | | Pitch gearbox |
| | | | Pitch sensors | Position encoder |
| | | | | Temperature sensor |
| | | | | Voltmeter |
| Wind turbine | Support structure | Foundation | Gravity-based foundation | Concrete |
| | | | | Steel reinforcement |
| | | | Monopile | Corrosion protection |
| | | | | Pile |
| | | | | Transition piece |
| | | | On-shore | Concrete |
| | | | | Nuts and bolts |
| | | | | Piles |
| | | | | Steel reinforcement |
| | | | Space frame/tripod | Corrosion protection |
| | | | | Piles |

*(Continues)*

(*Continued*)

| System | Sub-system | Assembly | Sub-assembly | Component |
|---|---|---|---|---|
| | | | | Structures |
| | | Tower | Access equipment | Ladder |
| | | | | Landing pad |
| | | | | Lightning protection |
| | | | Tower | Climb assist |
| | | | | Maintenance crane |
| | | | | Nuts and bolts |
| | | | | Paint/coating |
| | | | | Tower section |

## B.7 Detailed wind turbine failure terminology

| Sub-system | Assembly | Sub-assembly or component | Failure or failure mode from ReliaWind |
|---|---|---|---|
| Foundation | Monopile | | Scour, erosion, corrosion |
| | Tripod | | Erosion, corrosion |
| | Gravity base | | Scour, erosion |
| | Transition piece | | Grout slippage, grout loss |
| | Jacking brackets | | Fatigue |
| | Bolts | | Fatigue, corrosion, erosion |
| Tower | Structure | | Fatigue, corrosion |
| | Bolts | | Corrosion, overload, fatigue |
| | Climbing system | | Corrosion, overload |
| | Lift | | Motor failure, interlock failure |
| Rotor module | Rotor | Rotor hub | Fracture, corrosion |
| | | Rotor blade | Mechanical imbalance, aero-dynamic imbalance |
| | | Spars | Cracking, debonding from skin |
| | | Coating | Roughening, impact damage |
| | | Lamination | Debonding, lightning damage, impact damage |
| | | Leading edge | Erosion, ice build up |
| | | Trailing edge | Debonding, ice build up |
| | | Tip brake | Loss of tip |
| | | Tip brake wire | Snagging, broken |
| | Pitch system | | Pitch bearing failure, seizure, over-load, motor failure |

*(Continues)*

(*Continued*)

| Sub-system | Assembly | Sub-assembly or component | Failure or failure mode from ReliaWind |
|---|---|---|---|
| | | | Hydraulic oil contamination, hydraulic oil leakage, hydraulic pump failure |
| | | | Slip ring wear |
| | | | Blade mismatch, aerodynamic imbalance |
| Nacelle module | Yaw system | | Yaw bearing failure, yaw ring wear, yaw ring distortion or damage, yaw motor failure, yaw brake failure, yaw brake seizure, yaw alignment error |
| | Hydraulics system | Hydraulic power pack, motor | Winding failure, over-temperature |
| | | Hydraulic power pack, pump | Over-temperature, seal failure |
| | | Hydraulic power pack, pressure valve | Seal failure |
| | | Hydraulic power pack, filter | Blockage |
| | Anemometer | | Ice build-up, seizure, calibration drift, impact damage |
| | Wind vane | | Ice build-up, seizure, calibration drift, impact damage |
| | Electrical system | | See sub-systems |
| | Access system | | Wear, looseness, breakage |
| | Generator supports | | Cracking, bending, looseness |
| Drive-train | Main bearing | | Bearing failure, misalignment, lubrication |
| | Main shaft | | Cracking, permanent bend |
| | Mechanical brake | | Pad wear, overheating, disc wear, hydraulic failure |
| | Gearbox | Gearcase | Fracture |
| | | Suspension | Wear, looseness |
| | | Torque arm | Wear, looseness |
| | | Lubrication system | Loss of lubricant, contaminated lubricant, aged lubricant, lubricant system failure, lubrication pump failure, blocked lubrication filters, blocked jets |
| | | Epicyclical part, planet carrier | Lubrication |
| | | Epicyclical part, planet bearing | Bearing failure, lubrication |

(*Continues*)

(*Continued*)

| Sub-system | Assembly | Sub-assembly or component | Failure or failure mode from ReliaWind |
|---|---|---|---|
| | | Epicyclical part, planet gear | Tooth failure, lubrication |
| | | Epicyclical part, internal gear | Tooth failure, lubrication, fracture |
| | | Epicyclical part, sun gear | Tooth failure, lubrication |
| | | Epicyclical part, shaft | Cracking, journal damage |
| | | Parallel shaft part, gear | Tooth failure, lubrication |
| | | Parallel shaft part, bearing | Bearing failure, lubrication |
| | | Parallel shaft part, pinion | Tooth failure, lubrication |
| | | Parallel shaft part, shaft | Cracking, journal damage |
| | | High speed shaft | Cracking, permanent bend |
| Drive-train | Generator | Coupling | Misalignment, perishing, wear |
| | | Rotor | Fracture |
| | | Rotor windings | Shorted turn, earth failure, broken bar |
| | | Stator | Shorted turn, earth failure |
| | | Stator windings | Shorted turn, earth failure |
| | | Bearings | Bearing failure, lubrication failure |
| | | Stator cooling system | Blockage, over-temperature |
| | | Slip rings | Brush wear, over-temperature |
| | | Encoder | Encoder failure |
| Electrical module | Frequency converter | Power electronics | Component failure, joint failure |
| | | Grid-side filter | Component failure |
| | | Grid-side inverter | IGBT failure, over-temperature |
| | | DC link | Capacitor failure |
| | | Generator-side inverter | IGBT failure, over-temperature |
| | | Generator-side filter | Component failure |
| | | Crowbar | Thyristor failure |
| | | Crowbar resistor | Over-temperature, component failure, fuse failure |
| | | DC chopper | Component failure |
| | | DC chopper resistor | Over-temperature, component failure, fuse failure |
| | Transformer | Windings | Winding failure, over-temperature |
| | | Core | Over-temperature |
| | | Oil system | Oil deterioration, over-temperature |
| | Switchgear | | Circuit breaker failure |

# *Appendix C*
# Commercial SCADA systems

This survey was prepared by Drs B D Chen, D Zappalá, C J Crabtree and the Author.

## C.1 Introduction

A wind farm's existing SCADA data stream is a valuable resource, which can be monitored by wind turbine OEMs, operators and other experts to observe and, hence, optimise the performance of the wind turbine. In order to conduct an efficient SCADA data analysis, data analysis tools are required.

This survey discusses commercially available SCADA systems which are currently being applied in the wind turbine industry.

## C.2 SCADA data

SCADA systems are a standard installation in large wind turbines and wind farms, their data being collected from individual wind turbine controllers. A SCADA system assesses the status of the wind turbine and its sub-systems using sensors fitted to the wind turbine, such as anemometers, thermocouples and switches. The signals from these instruments are monitored and recorded at a low data rate, usually at 10-min intervals. The SCADA data show the operating condition of a wind turbine. Many large wind turbines are now fitted with condition monitoring systems (CMS), which monitor sensors associated with the rotating drive train, such as accelerometers, proximeters and oil particle counters. The CMS is normally separate from the SCADA and collects data at much higher data rates.

By analysing SCADA data, we are able to observe the relationship between different signals and, hence, deduce the health of wind turbine sub-assemblies. It would prove beneficial, from the perspective of utility companies, if the data could be analysed and interpreted automatically to support the operators in identifying defects.

## C.3 Commercially available SCADA data analysis tools

Table C.1 provides a summary of the available SCADA systems based on information collected from Internet. It should be noted that the table is accurate up to

*Table C.1 Table of commercially available SCADA systems*

| Ref | Product and company information | | | Product details | |
|---|---|---|---|---|---|
| | Product name | Company | Country of origin | Description | Main function |
| 1 | BaxEnergy WindPower Dashboard | BaxEnergy GmbH | Germany | Offers an extensive and comprehensive customisation for full integration of SCADA system applications and SCADA software | Real-time data acquisition and visualisation, alarm analytics and data reporting |
| 2 | CitectSCADA | Schneider Electric Pty. Ltd. | Australia | CitectSCADA is a reliable, flexible and high performance system for any industrial automation monitoring and control application | Graphical process visualisation, superior alarm management, built-in reporting and powerful analysis tools |
| 3 | CONCERTO | AVL | Austria | A commercially available analysis and post-processing system capable of handling large quantities of data | One tool for manifold applications, all data post-processing tasks within one tool, advanced data management |
| 4 | ENERCON SCADA system | ENERCON | USA | The Enercon SCADA System is used for data acquisition, remote monitoring, open-loop and closed-loop control for both individual wind turbine and wind farms. It enables the customer and Enercon Service to monitor the operating state and to analyse saved operating data | Requesting status data, storing operating data, wind farm communication and loop control of the wind farm |
| 5 | Gamesa WindNet | Gamesa | Spain | The WindNet SCADA system in a wind farm is configured with a basic hardware and software platform based on Windows technology. The user interface is an easy-to-use SCADA application with specific options for optimal supervision and control of a wind farm including devices like wind turbines, meteorological masts and a sub-station | Supervision and control of wind turbines, meteorological masts. Alarm and warning management, report generation and user management |
| 6 | GE – HMI/ SCADA – iFIX 5.1 | GE (General Electric Co.) | USA | iFIX is a superior proven real-time information management and SCADA solution. It is open, flexible and scalable which includes impressive next-generation visualisations, a reliable control engine and a powerful built-in historian | Real-time data management and control, information analysis |

| | | | | | |
|---|---|---|---|---|---|
| 7 | GH SCADA | GL Garrad Hassan | Germany | GH SCADA has been designed by Garrad Hassan in collaboration with wind turbine OEMs, wind farm operators, developers and financiers to meet the needs of all those involved in wind farm operation, analysis and reporting | Remote control of individual wind turbine. On-line data viewing. reports and analysis |
| 8 | ICONICS for Renewable Energy | ICONICS Inc. | USA | Provide portals for complete operations, including energy analytics, data histories and reports, geo SCADA with meteorological updates | Portals for complete operations, data histories and reports, geo SCADA with meteorological updates |
| 9 | InduSoft Wind Power solutions | InduSoft | USA | InduSoft Web Studio software brings you a powerful HMI/SCADA package that can monitor and adjust any operating set point in the controller or PLC | Wind turbine monitoring, maintenance assistance and control room |
| 10 | INGESYS Wind IT | IngeTeam | Spain | INGESYS Wind IT makes it possible to completely integrate all the wind power plants into a single system. It provides advanced reporting services | Advanced reporting, client/server architecture, standard protocols and formats |
| 11 | reSCADA | Kinetic Automation Pty. Ltd. | USA | Targeting and specialising in renewable energy industries. Will save time, effort and cost in developing HMI/SCADA | Office 2007 GUI style, data visualisation, diary and mapping tools |
| 12 | SgurrTREND | SgurrEnergy | UK | SgurrEnergy provide a variety of wind monitoring solutions to evaluate the wind resource potential at your prospective wind farm site, offering a one-stop shop for all mast services from planning application, data collection and mast decommissioning to wind analysis services for energy yield prediction, project layout and design services | Wind monitoring, process and archive the data and reporting services |
| 13 | SIMAP | Molinos del Ebro, S.A. | Spain | SIMAP is based on artificial intelligence techniques. It is able to create and dynamically adapt a maintenance calendar for the wind turbine that it is monitoring. The new and positive aspects of this predictive maintenance methodology have been tested in wind turbines | Continuous collection of data, continuous processing information, failure risk forecasting and dynamical maintenance scheduling |

(Continues)

*Table C.1 (Continued)*

| Ref | Product and company information | | | Product details | |
|---|---|---|---|---|---|
| | **Product name** | **Company** | **Country of origin** | **Description** | **Main function** |
| 14 | WindCapture | SCADA Solutions | Canada | WindCapture is a SCADA software package used for monitoring, controlling and data collection and reporting for wind turbine generators. It was designed and tailored to the demands of OEMs, operators, developers and maintenance managers of wind energy project and facilities | Real-time data reporting with the highest degree of accuracy. Advanced GUI |
| 15 | Wind Systems | SmartSignal | USA | SmartSignal analyses real-time data and detects and notifies wind farms of impending problems, allowing owners to focus on fixing problems early and efficiently | Model maintenance, monitoring services and predictive diagnostics |
| 16 | Wind Asset Monitoring Solution | MATRIKON | Canada | The solution bridges the gap between instrumentation and management systems to enable and sustain operational excellence by retrieving and better managing data that are not often readily accessible | Monitor and manage all remote assets, leverage and integrate with SCADA and CMMS |

2011 but may not be definitive. The products are arranged alphabetically by product name and each has a number; quick summary of Table C.1 shows the following:

- Three products are developed in association with wind turbine OEMs (4, 5, and 6).
- Two products are developed by renewable energy consultancies (7 and 12).
- Nine products are developed by industrial software companies, including manufacturers of the wind turbine controllers (1, 2, 3, 8, 9, 11, 14, 15 and 16).
- One product is developed by wind turbine operator (13).
- One product is developed by an electrical equipment provider (10).

Among these 15 products, Gamesa WindNet (5) and Enercon SCADA System (4) are wind farm cluster management systems. Both provide a framework for data acquisition, remote monitoring, open/closed loop control and data analysis for both individual wind turbines and wind farms. The Enercon SCADA System was launched in 1998 and is now used in conjunction with more than 11,000 wind turbines. Gamesa WindNet consists of a wide area network system for wind farms connected to an operational centre.

GE – HMI/SCADA – iFIX 5.1 (6) was developed by General Electric Co. (GE), also a wind turbine OEM. It is ideally suited for complex SCADA applications. The software also enables faster, better intelligent control and visibility of wind farm operations.

GH SCADA (7) and SgurrTREND (12) were developed by renewable energy consultancies in collaboration with wind turbine OEMs, wind farm operators, developers and financiers to meet the needs of all those involved in wind farm operation, analysis and reporting.

CONCERTO (2) is not specialised for SCADA data analysis. It is a generic data post-processing tool focusing on quick and intuitive signal analysis, validation, correlation and reporting for any kind of acquired data. Gray *et al.* used it to perform an analysis of wind turbine SCADA data.

SIMAP (13) is based on artificial intelligence techniques. The new and positive aspects of this predictive maintenance methodology have been tested on wind turbines. SIMAP has been applied to a wind farm owned by a Spanish wind energy company called Molinos del Ebro, S.A., Garcia *et al.* (2006).

INGESYS Wind IT (10) was developed by IngeTeam, an electrical equipment provider. The system aims to integrate wind power plants into a single system and then optimise wind farm management. INGENSYS Wind IT also provides an advanced reporting service for power curve analysis, faults, alarms and customer reports.

The other products – BaxEnergy WindPower Dashboard (1), CitectSCADA (3), ICONICS for Renewable Energy (8), InduSoft Wind Power (9), reSCADA (11), WindCapture (14), Wind Systems (15) and MATRIKON's Wind Asset Monitoring Solution (16) were developed by industrial software companies, which integrated SCADA systems, and they aim to provide reliable, flexible and high performance applications for wind turbine automation, monitoring and control.

## C.4 Summary

From this survey we conclude that:

- There are a wide variety of commercial SCADA systems available to the Wind Industry.
- Most of the commercially available SCADA systems are able to analyse real-time data.
- The performance analysis techniques used in SCADA systems vary from tailored statistical method to artificial intelligence.
- Successful SCADA systems could provide wind farm cluster management. They provide a framework for data acquisition, alarm management, reporting and analysis, production forecasting and meteorological updates.
- Some built-in diagnostics techniques are able to diagnose the sub-assembly failure of wind turbine.

Finally, it should be noted that the development of SCADA systems is aimed to provide a reliable, flexible and high performance for wind turbine automation monitoring and control. The industry is already noting the importance of operational parameters such as load and speed and so techniques may begin to adapt further to the wind turbine environment leading to more reliable wind turbine diagnostics solution.

*Appendix D*

# Commercial condition monitoring systems

This survey was prepared by Drs C J Crabtree, D Zappalá and the Author.

## D.1 Introduction

As wind energy assumes greater importance in remote and off-shore locations, effective and reliable condition monitoring (CM) techniques are required. Conventional CM methods used in the power generation industry have been adapted by a number of industrial companies and have been applied to wind turbines commercially.

This survey considers commercially available condition monitoring systems (CMS) currently applied in the wind industry. Information has been gathered over several years from conferences and websites and includes information available from product brochures, technical documents and discussion with company representatives. The research was carried out as part of the Supergen Wind Energy Technologies Consortium whose objective is to devise a comprehensive CMS for practical application on wind turbines. The survey also identifies some of the advantages and disadvantages of existing commercial CMS alongside discussion of access, cost, connectivity and commercial issues surrounding the application of wind turbine CMS.

## D.2 Reliability of wind turbines

Quantitative studies of wind turbine reliability have recently been carried out based on publically available data referred to in Chapter 4. These studies have shown wind turbine gearboxes to be a mature technology with constant of slightly deteriorating reliability with time. This would suggest that wind turbine gearboxes are not an issue; however, surveys by WMEP and LWK have shown that gearboxes exhibit the highest downtime per failure among on-shore sub-assemblies. This is shown graphically in Figure 3.5 where we clearly see a consistently low gearbox failure rate between two surveys with high downtime per failure. Similar results have also been shown for the Egmond aan Zee wind farm, from NoordZee data, where gearbox failure rate was not high but the downtime and resulting costs are. The poor early reliabilities for gearbox and drive train reliability components have led to an emphasis in wind turbine CMS on drive train components and, therefore, on vibration analysis.

The high downtime for gearboxes derives from complex repair procedures. Off-shore wind turbine maintenance can be a particular problem as this involves specialist equipment such as support vessels and cranes but has the additional issue of potentially unfavourable weather and wave conditions. ReliaWind developed a systematic and consistent process to deal with detailed commercial data collected from operational wind farms. This includes the analysis of 10-min average supervisory control and data acquisition (SCADA) data as discussed earlier, automated fault logs and operation and maintenance reports. However, more recent information on wind turbine reliability and downtime, especially when considering off-shore operation, suggests that the target for wind turbine CMS should be widened from the drive train towards wind turbine electrical and control systems Tavner, Faulstich *et al.* (2011).

As a result of low early wind turbine reliability, particularly in larger wind turbines and as a result of the move off-shore, interest in CMS has increased. This is being driven forward by the insurer Germanischer Lloyd who published guidelines for the certification of CMS DNV GL (2013) and certification of wind turbines both on-shore DNV GL (2010) and off-shore DNV GL (2012).

## D.3 Monitoring of wind turbines

Wind turbines are monitored for a variety of reasons. There are a number of different classes into which monitoring systems could be placed and these are shown in Figure D.1, showing the general layout and interaction of the various classes.

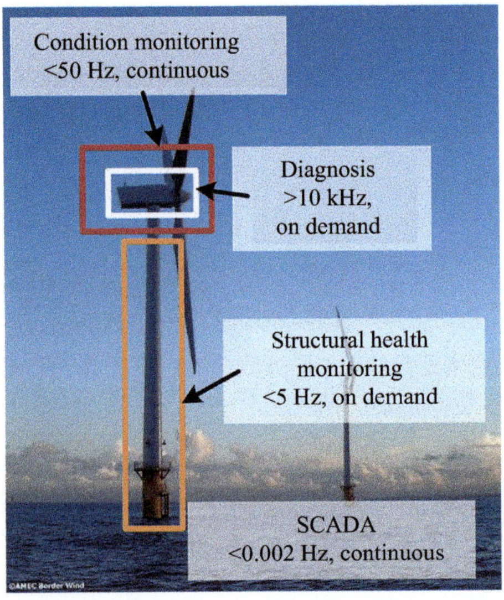


*Figure D.1 Structural health and condition monitoring of a wind turbine*

First, we have SCADA systems. Initially these systems provided measurements for wind turbine energy production and to confirm that the wind turbine was operational through 5- to 10-min averaged values transmitted to a central database. However, SCADA systems can also provide warning of impending malfunctions in the wind turbine drive train. The 10-min averaged signals usually monitored in modern SCADA systems include

- active power output (and standard deviation over 10-min interval);
- anemometer-measured wind speed (and standard deviation over 10-min interval);
- gearbox bearing temperature;
- gearbox lubrication oil temperature;
- generator winding temperature;
- power factor;
- reactive power;
- phase currents;
- nacelle temperature (1-h average).

This SCADA configuration is designed to show the operating condition of a wind turbine but not necessarily give an indication of the health and a wind turbine. However, more up-to-date SCADA systems include additional alarm settings based not only on temperature but also on vibration transducers. Often we find several vibration transducers fitted to the wind turbine gearbox, generator bearings and the turbine main bearing. The resultant alarms are based on the level of vibration being observed over the 10-min average period. Research is being carried out into the CM of wind turbines through SCADA analysis, Yang, Tavner, Crabtree, Wilkinson (2010) and Crabtree (2011).

Second, there is the area of structural health monitoring (SHM). These systems aim to determine the integrity of the wind turbine tower and foundations. SHM is generally carried out using low sampling frequencies below 5 Hz.

While SCADA and SHM monitoring are key areas for wind turbine monitoring, this survey will concentrate on the remaining two classes of CM and diagnosis systems.

Monitoring of the drive train is often considered to be most effective through the interaction of these two areas. CM itself may be considered a method for determining whether a wind turbine is operating correctly or whether a fault is present or developing. A wind turbine operator's main interest is likely to be in obtaining reliable alarms based on CM information which can enable them to take confident action with regard to shutting down for maintenance. The operator need not know the exact nature of the fault but would be alerted to the severity of the issue by the alarm signal. Reliable CM alarms will be essential for any operator with a large number of wind turbines under its ownership. On this basis, CM signals should not need to be collected on a high frequency basis as this will reduce bandwidth for transmission and space required for storage of data.

Once a fault has been detected through a reliable alarm signal from the CMS, a diagnosis system could be activated either automatically or by a monitoring engineer to determine the exact nature and location of the fault. For diagnosis systems,

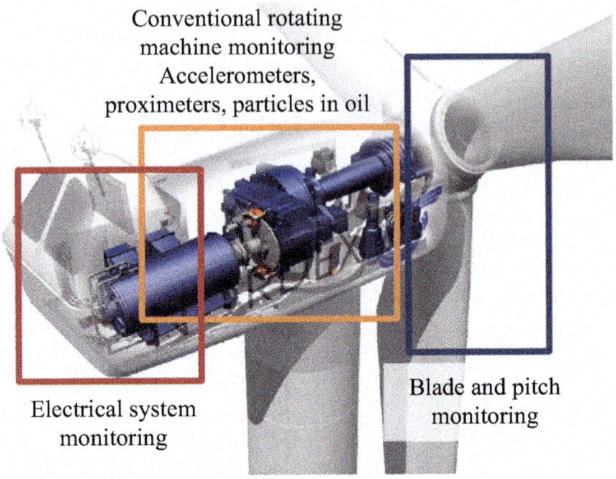


*Figure D.2 Layout of three areas for condition monitoring within the nacelle of a wind turbine*

data recorded at a high sampling frequency are required for analysis; however, this need only be collected on an intermittent basis. The operational time of the system should be configured to provide enough data for detailed analysis but not to flood the monitoring system or data transmission network with excess information.

Finally, Figure D.2 gives an indication of three sections of a wind turbine which may require monitoring based on reliability data such as that in Crabtree *et al.* While each of the three areas is shown as separate entities, CM must blur the boundaries between them in order to provide clear alarms and, subsequently, diagnostic information.

Many of the CMS included in this survey are a combination of CMS and diagnostic systems due to the high level of interaction that can exist between the two types of system.

## D.4 Commercially available condition monitoring systems

Provides a summary of a number of widely available and popular CMS for wind turbines. The information has been collected from interaction with CMS and wind turbine OEMs and product brochures over a long period of time and is up-to-date up to 2011. However, since some information has been acquired through discussion with sales and product representatives and not from published brochures, it should be noted that the table is only as accurate as the given information. The systems in Table D.1 are arranged alphabetically by product name.

*Table D.1 Table of commercially available condition monitoring systems*

| | Product and company information | | | Product details (based on available literature and contact with industry including EWEC 2008, 2009, 2010, 2011) | | | | |
|---|---|---|---|---|---|---|---|---|
| Ref. | Product | Supplier or manufacturer (known users) | Country of origin | Description | Main components monitored | Monitoring technology | Analysis method(s) | Data rate or sampling frequency |
| 1 | ADAPT.wind | GE Energy | USA | Up to 150 static variables monitored and trended per WT. Planetary cumulative impulse detection algorithm to detect debris particles through the gearbox planetary stage. Dynamic energy index algorithm to spread the variation over five bands of operation for spectral energy calculations and earlier fault detection. Alarm, diagnostic, analytic and reporting capabilities facilitate maintenance with actionable recommendations. Possible integration with SCADA system. | Main bearing, gearbox, generator | Vibration (accelerometer)<br>Oil debris particle counter | FFT frequency domain analysis<br>Time domain analysis | – |
| 2 | APPA System | OrtoSense | Denmark | Oscillation technology based on interference analysis that replicates the human ear's ability to perceive sound. | Main bearing, gearbox and generator | Vibration | Auditory perceptual pulse analysis (APPA) | – |
| 3 | Ascent | Commtest | New Zealand | System available in three complexity levels. Level 3 includes frequency band alarms, machine template creation, statistical alarming. | Main shaft, gearbox, generator | Vibration (accelerometer) | FFT frequency domain analysis<br>Envelope analysis<br>Time domain analysis | – |
| 4 | Brüel & Kjaer Vibro | Brüel & Kjaer (Vestas) | Denmark | Vibration and process data automatically monitored at fixed intervals and remotely sent to the diagnostic server. User-requested time waveforms for frequency and time series analysis. Time waveform | Main bearing, gearbox, generator, nacelle.<br>Nacelle temperature. | Vibration<br>Temperature sensor<br>Acoustic | Time domain<br>FFT frequency analysis | Variable up to 40 kHz.<br>25.6 kHz. |

*(Continues)*

*Table D.1 (Continued)*

| | Product and company information | | | Product details (based on available literature and contact with industry including EWEC 2008, 2009, 2010, 2011) | | | | |
|---|---|---|---|---|---|---|---|---|
| Ref. | Product | Supplier or manufacturer (known users) | Country of origin | Description | Main components monitored | Monitoring technology | Analysis method(s) | Data rate or sampling frequency |
| | | | | automatically stored before and after user-defined event allowing advanced vibration post-analysis to identify developing faults. | Noise in the nacelle | | | |
| 5 | CMS | Nordex | Germany | Start-up period acquires vibration 'fingerprint' components. Actual values automatically compared by frequency, envelope and order analysis, with the reference values stored in the system. Some Nordex turbines also use the Moog Insensys fibre optic measurement system. | Main bearing, gearbox, generator | Vibration (accelerometer) | Time domain based on initial 'fingerprint' | – |
| 6 | Condition based maintenance system (CBM) | GE (Bently Nevada) | USA | This is built upon the Bently Nevada ADAPT.wind technology and System 1. Basis on System 1 gives monitoring and diagnostics of drive train parameters such as vibration and temperature. Correlate machine information with operational information such as machine speed, electrical load and wind speed. Alarms are sent via the SCADA network. | Main bearing, gearbox, generator, nacelle<br>Optional bearing and oil temperature | Vibration (accelerometer) | FFT frequency domain analysis<br>Acceleration enveloping | – |
| 7 | Condition based monitoring system | Bachmann Electronic GmbH | Austria | Up to nine piezoelectric acceleration sensors per module. Basic vibration analysis with seven sensors. PRÜFTECHNIK solid | Main drive train components<br>Generator | Vibration (accelerometer) | Time domain<br>FFT frequency analysis | 24-bit res<br>190 kHz sample rate per channel |

| | | | | | | | | |
|---|---|---|---|---|---|---|---|---|
| | | | | borne sound sensors used for low frequency diagnostics of slowly rotating bearings on the WT LSS. Three channels for the ±10 V standard signal per module. Fully integration in Bachmann's M1 automation control system. | | | | 0.33 Hz (solid borne sound sensors) |
| 8 | Condition diagnostics system | Winergy | Denmark | Up to six inputs per module. Advanced signal processing of vibration levels, load and oil to give automated machinery health diagnostics, forecasts and recommendations for corrective action. Automatic fault identification is provided. Relevant information provided in an automated format to the operations and maintenance centre without any experts being involved. Information delivered to the appropriate parties in real time. Pitch, controller, yaw and inverter monitoring can also be included. | Main shaft, gearbox, generator | Vibration (accelerometer)<br>Oil debris particle counter | Time domain<br>FFT frequency domain analysis | 96 kHz per channel |
| 9 | Condition management system | Moventas | Finland | Compact system measuring temperature, vibration, load, pressure, speed, oil aging and oil particle count. Sixteen analogue channels can be extended with adapter. Data acessed remotely via TCP/IP. Mobile interface available. | Gearbox, generator, rotor, turbine controller | Vibration Oil quality/ particles<br>Torque<br>Temperature | Time domain (possible FFT) | – |
| 10 | Distributed condition monitoring system | National Instruments | USA | Up to 32 channels; default configuration: 16 accelerometer/ microphone, 4 proximity probe and 8 tachometer input | Main bearing, gearbox, generator | Vibration<br>Acoustic | Spectral analysis<br>Level measurements<br>Order analysis<br>Waterfall plots | 24-bit res<br>23.04 kHz of bandwidth with |

*(Continues)*

*Table D.1 (Continued)*

| | Product and company information | | | Product details (based on available literature and contact with industry including EWEC 2008, 2009, 2010, 2011) | | | | |
|---|---|---|---|---|---|---|---|---|
| **Ref.** | **Product** | **Supplier or manufacturer (known users)** | **Country of origin** | **Description** | **Main components monitored** | **Monitoring technology** | **Analysis method(s)** | **Data rate or sampling frequency** |
| | | | | channels. Also provided mixed-measurement capability for strain, temperature, acoustics, voltage, current and electrical power. Oil particulate counts and fibre optic sensing can also be added to the system. Possible integration into SCADA systems. | | | Order tracking<br>Shaft centre-line measurements<br>Bode plots | antialiasing filters per accelerometer/ microphone channel |
| 11 | OneProd wind system | Areva (01dB-Metravib) | France | Eight to 32 channels. Instrumentation includes operating condition channels to trigger data acquisitions, measurement channels for surveillance and diagnosis. Data set comparison when relating to similar operating conditions; data alarm systems warn on the repetitive and abnormal shocks enabling the detection of failure modes; built-in diagnostic tool. Optional additional sensors for shaft displacement, for permanent oil quality monitoring, low frequency sensors on the structure and current and voltage sensors for generator monitoring. | Main bearing on LSS,<br>Bearing on gearbox LSS,<br>Bearing on intermediate gearbox shaft, on gearbox high-speed shaft, on generator<br>Oil debris, structure, shaft displacement, electrical signals | Vibration<br>Electrical signature analysis<br>Thermography<br>Oil debris particle counter | Time domain<br>FFT frequency analysis | – |
| 12 | SMP-8C | Gamesa Eolica | Spain | Continuous on-line vibration measurement of main shaft, gearbox and generator. Comparison of spectra trends. Warnings and alarm transmission | Main shaft, gearbox, generator | Vibration | FFT frequency domain | – |

| | | | | | | | | |
|---|---|---|---|---|---|---|---|---|
| | | | | connected to wind farm management system. | | | | |
| 13 | System 1 | Bently Nevada (GE) | USA | Monitoring and diagnostics of drive train parameters such as vibration and temperature. Correlate machine information with operational information such as machine speed, electrical load and wind speed. | Main bearing, gearbox, generator, nacelle<br>Optional bearing and oil temperature | Vibration (accelerometer) | FFT frequency domain<br>Acceleration enveloping | – |
| 14 | TCM (turbine condition monitoring)<br>Enterprise V6 Solution with SCADA integration | Gram & Juhl A/S | Denmark | Advanced signal analysis and process signals combined with automation rules and algorithms for generating references and alarms. M-System hardware features up to 12/24 synchronous channels, interface for structural vibration monitoring and RPM sensors, external process parameters and analog outputs. TCM server stores data and does the post data processing. Control room with web based operator interface. | Tower, blades, shaft and nacelle<br>Main bearing, gearbox and generator | Vibration (accelerometer)<br>Sound analysis<br>Strain analysis<br>Process signals analysis | FFT and Wavelet frequency domain analysis<br>Envelope analysis<br>RMS analysis<br>Order tracking analysis | – |
| 15 | Wind AnalytiX | ICONICS | USA | This software solution uses fault detection and diagnostics technology that identifies equipment and energy inefficiencies and provides possible causes that help in predicting plant operations, resulting in reduced downtime and costs related to diagnostic and repair. | Main WT components | Vibration (accelerometer) | Unknown | – |

*(Continues)*

*Table D.1 (Continued)*

| | Product and company information | | | Product details (based on available literature and contact with industry including EWEC 2008, 2009, 2010, 2011) | | | | |
|---|---|---|---|---|---|---|---|---|
| **Ref.** | **Product** | **Supplier or manufacturer (known users)** | **Country of origin** | **Description** | **Main components monitored** | **Monitoring technology** | **Analysis method(s)** | **Data rate or sampling frequency** |
| 16 | WindCon 3.0 | SKF (REpower) | Sweden | Lubrication, blade and gearbox oil systems can be remotely monitored through SKF ProCon sofetware. WindCon 3.0 collects, analyses and compiles operating data that can be configured to suit management, operators or maintenance engineers. | Blade, main bearing, shaft, gearbox, generator, tower, generator electrical | Vibration (accelerometer, proximity probe)<br>Oil debris particle counter | FFT frequency domain analysis<br>Envelope analysis<br>Time domain analysis | Analogue: DC to 40 kHz (Variable, chan dependent)<br>Digital: 0.1 Hz–20 kHz |
| 17 | Wind Turbine In-Service | ABS Consulting | USA | Data gathered from inspections, vibration sensors and SCADA system. Ekho for WIND software features regular diagnostics, dynamic performance reports, key performance indicators, fleet-wide analysis, forecasts/schedules and asset benchmarking. It generates alarms and notifications or triggers work orders for inspections or repairs. | Main bearing, gearbox and generator<br>Gearbox and gear oil, rotor blades and coatings | Vibration<br>Inspections | FFT frequency domain analysis<br>Time domain analysis | – |
| 18 | WinTControl | Flender Service GmbH | Germany | Vibration measurements are taken when load and speed triggers are realised. Time and frequency domain analysis are possible. | Main bearing, gearbox, generator | Vibration (accelerometer) | FFT frequency domain<br>Time domain analysis | 32.5 kHz |
| 19 | WiPro | FAG Industrial Services GmbH | Germany | Measurement of vibration and other parameters given appropriate sensors. Time and frequency domain analysis carried out during alarm situations. Allows speed-dependent frequency band tracking and speed-variable alarm level. | Main bearing, shaft, gearbox, generator, temperature (Adaptable inputs) | Vibration (accelerometer) | FFT frequency domain<br>Time domain analysis | Variable up to 50 kHz |

| | | | | | | | | |
|---|---|---|---|---|---|---|---|---|
| 20 | WP4086 | Mita-Teknik | Denmark | Up to eight accelerometers for real-time frequency and time domain analysis. Warnings/alarms set for both time and frequency domains based on predefined statistical/thresholds-based vibration limits. Operational parameters recorded alongside with vibration signals/spectra and full integration into gateway SCADA system. Algorithm toolbox for diagnostic analysis. Approximately 5000–8000 variables covering different production classes. | Main bearing, gearbox, generator | Vibration (accelerometer) | FFT amplitude spectra<br>FFT envelope spectra<br>Time domain magnitude<br>Comb filtering, whitening,<br>Kurtogram analysis | 12-bit chan res<br>Variable up to 10 kHz |
| 21 | HYDACLab | HYDAC Filtertechnik GmbH | Germany | Permanent monitoring system to monitor particles (including air bubbles) in hydraulic and lube oil systems. | Lubrication oil and cooling fluid quality | Oil debris particle counter | N/A | – |
| 22 | PCM200 | Pall Industrial Manufacturing (Pall Europe Ltd) | USA (UK) | Fluid cleanliness monitor reports test data in real-time so ongoing assessments can be made. Can be permanently installed or portable. | Lubrication oil cleanliness | Oil cleanliness sensor | N/A | – |
| 23 | TechAlert 10<br>TechAlert 20 | MACOM | UK | TechAlert 10 is an inductive sensor to count and size ferrous and non-ferrous debris in circulating oil systems. TechAlert 20 is a magnetic sensor to count ferrous particles. | Lubrication oil quality | Inductive or magnetic oil debris particle counter | N/A | – |
| 24 | BLADEcontrol | IGUS ITS GmbH | Germany | Accelerometers are bonded directly to the blades and a hub measurement unit transfers data wirelessly to the nacelle. Blades are assessed by comparing spectra with those stored for common conditions.<br>Measurement and analysis data are stored centrally and blade condition displayed using a web browser. | Blades | Accelerometer | FFT frequency domain | $\approx$ 1 kHz |

*(Continues)*

*Table D.1 (Continued)*

| | Product and company information | | | Product details (based on available literature and contact with industry including EWEC 2008, 2009, 2010, 2011) | | | | |
|---|---|---|---|---|---|---|---|---|
| Ref. | Product | Supplier or manufacturer (known users) | Country of origin | Description | Main components monitored | Monitoring technology | Analysis method(s) | Data rate or sampling frequency |
| 25 | FS2500 | FibreSensing | Portugal | BraggSCOPE measurement unit designed for industrial environments to interrogate up to four Fibre Bragg Grating sensors. Acceleration, tilt, displacement, strain, temperature and pressure measurable. | Blades | Fibre optic | Unknown | Up to 2 kHz |
| 26 | RMS (rotor monitoring system) | Moog Insensys Ltd. | UK | Modular blade sensing system consisting of 18 sensors, 6 per blade, installed in the cylindrical root section of each blade to provide edgewise and flapwise bending moment data. Can be designed-in during turbine manufacture or retrofitted. Monitors turbine rotor performance, mass and aerodynamic imbalance, blade bending moments, icing, damage and lightning strikes. Possible integration, as an external input, in commercial available CMSs. | Blades | Fibre optic strain | Time domain strain analysis | 25 Hz/sensor |

The first observation to make is that the CMS nearly all focus on the same wind turbine sub-assemblies as follows:

- blades
- main bearing
- gearbox internals
- gearbox bearings
- generator bearings

Summarising Table D.1 using the numbers there shows that there are

- 20 systems primarily based on drive train vibration analysis (1–20);
- 3 systems solely for oil debris monitoring (21–23);
- 1 system using vibration analysis for wind turbine blade monitoring (24);
- 2 systems based on fibre optic strain measurement in wind turbine blades (25, 26).

The majority of the systems are based on monitoring methods originating from other, traditional rotating machinery industries. Indeed 19 of the 26 systems in the table are based on vibration monitoring using accelerometers typically using a configuration similar to that shown in Figure D.3 for the Mita-Teknik WP4086 CMS (20).

Of these 19 CMS, all have the capability to carry out some form of diagnostic procedure once a fault has been detected. In most cases, this is done through fast Fourier transform (FFT) analysis of high-frequency data in order to detect fault-specific frequencies. In the case of the SKF WindCon 3.0 (16), the Areva OneProd Wind CMS (11) and several others, high data acquisition is triggered by operational parameters. For example, the SKF WindCon 3.0 CMS can be configured to collect a

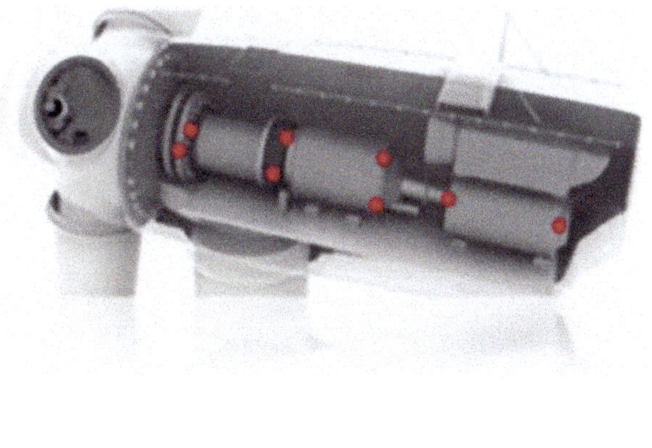

*Figure D.3 Typical CMS accelerometer positions in the nacelle of a wind turbine*

vibration spectrum on either a time basis or when a specific load and speed condition is achieved. The aim of this is to acquire data that are directly comparable between each point and, importantly, to allow spectra to be recorded in apparently stationary conditions. This is important to note when using traditional signal-processing methods such as the FFT which require stationary signals in order to obtain a clear result. The Mita-Teknik WP4086 system (20), however, states that it includes advanced signal-processing techniques such as comb filtering, whitening and Kurtogram analysis which, in combination with re-sampling and order alignment approaches, allow the system to overcome the effects of wind turbine speed variations.

An innovative vibration-based CMS is OrtoSense APPA (2) which is based on auditory perceptual pulse analysis. This patented technology outperforms the human ear by capturing a detailed interference pattern and detecting even the smallest indication of damaged or worn elements within the machine/turbine. OrtoSense states that its product is four to ten times more sensitive compared to prevailing systems.

Five of the vibration-based CMS also state that they are able to monitor the level of debris particles in the wind turbine gearbox lubrication oil system. Further to this, included in the table are three systems which are not in themselves CMS. These three (21–23) are oil quality monitoring systems or transducers rather than full CMS but are included as discussion with industry has suggested that debris in oil plays a significant role in the damage and failure of gearbox components. Systems using these debris in oil transducers are using either cumulative particle counts or particle count rates.

Several of the 20 vibration-based CMS also allow for other parameters to be recorded alongside vibration such as load, wind speed, generator speed and temperatures although the capabilities of some systems are unclear given the information available. There is some interest being shown regarding the importance of operational parameters in wind turbine CM. This arises from the fact that many analysis techniques, for example the FFT, have been developed in constant speed and constant load environments. This can lead to difficulties when moving to the variable speed, variable load wind turbine; however, experienced CM engineers are able to use these techniques and successfully detect faults.

Recent CM solutions, as (1), (10), (14), (20), can be adapted and fully integrated with existing SCADA systems using standard protocols. Thanks to this integration, the analysis of the systems installed on the wind energy plant can also directly consider any other signals or variables of the entire controller network, as for example current performance and operating condition, without requiring a doubling of the sensor system. The database, integrated into a single unified plant operations' view, allows a trend analysis of the condition of the machine.

In some cases, the CMS company also offers custom service solutions from 24/7, 365 days a year, remote monitoring to on-demand technical support, examples are GE Energy ADAPT. Wind (1) and ABS Wind Turbine In-Service (17).

Two in the table (25, 26) are effectively blade monitoring systems (BMS) based on strain measurement using fibre optic transducers. These are aimed at detection of damage to wind turbine blades and, in the case of the Moog Insensys system (26), blade icing, mass unbalance or lightning strikes. Both systems may be fitted to wind turbine blades retrospectively. Compared to vibration-monitoring techniques, these systems can be operated at low sampling rates as they are looking to observe changes in time domain. They are usually integrated in the wind turbine control system but there are also some cases of integration, as an external input, into commercially available conventional vibration-based CMS. In addition to (25) and (26), there is the IGUS BMS (24) using accelerometers to monitor blade damage, icing and lightning strikes. This system compares the blade accelerometer FFT with stored spectra for similar operating conditions and has the power to automatically shut down or restart a wind turbine based on the results. The system appears to be popular within industry.

## D.5 Future of wind turbine condition monitoring

As can be seen from this survey of current CMS, there is a clear trend towards vibration monitoring of wind turbines. This is presumably a result of the wealth of knowledge gained from many years work in other fields. It is likely that this trend will continue; however, it would be reasonable to assume that other CM and diagnostic techniques will be incorporated into existing systems.

Currently, these additions are those such as oil debris monitoring and fibre optic strain measurement. However, it is likely that major innovation will occur in terms of developing signal-processing techniques. In particular, the industry is already noting the importance of operational parameters such as load and speed and so techniques may begin to adapt further to the wind turbine environment leading to more reliable CMS, diagnostics and alarm signals.

Automation of CM and diagnostic systems may also be an important development as wind turbine operators acquire a larger number of turbines and manual inspection of data becomes impractical. Further to this, it is, therefore, essential that methods for reliable, automatic diagnosis are developed with a consideration of multiple signals in order to improve detection and increase operator confidence in alarm signals.

However, it should be noted that a major hindrance to the development of CMS and diagnostic techniques could be data confidentially meaning that few operators are able to divulge or obtain information concerning their own wind turbines. This is an issue which should be addressed if the art of CM is to progress quickly. Confidentiality has also led to a lack of publicly available cost justification of wind turbine CM, which seems likely to provide overwhelming support for wind turbine CM, particularly in the off-shore environment where availability is at a premium.

## D.6 Summary

From this survey we can conclude that

- there is a wide variety of commercially available CMS currently in use on operational wind turbines;

- monitoring technology is currently based on techniques from other, conventional rotating machine industries;
- successful CMS must be able to adapt to the non-stationary, variable speed nature of wind turbines;
- vibration monitoring is currently favoured in commercially available systems using standard time and frequency domain techniques for analysis;
- these traditional techniques can be applied to detect wind turbine faults but require experienced CM engineers for successful data analysis and diagnosis;
- some commercially available CMS are beginning to adapt to the wind turbine environment and to be fully integrated into existing SCADA systems;
- a diverse range of new or developing technologies are moving into the wind turbine CM market.

Finally, it should be noted that there is no current consensus in the wind turbine industry as to the correct route forward for wind turbine CMS. Work in this document and its references suggest that CM of wind turbines will be important for large on-shore wind turbines, essential for all off-shore development and should be considered carefully by the industry as a whole.

## *Appendix E*

# Reliabilities of key off-shore sub-assemblies

Table E.1, from Tavner (2017), summarises the reliabilities of key systems used in off-shore wind farms. Many of the sources were collated in Delorm (2013) for an investigation into wave and tidal device reliability, and these values could prove useful for early predictions of off-shore wind asset reliability, where similar off-shore sub-assemblies are used, for example corrosion protection systems, mooring systems, sub-sea connectors and shore sub-stations.

Lower and upper bound figures are predicted for a ground benign reliability environment.

*Table E.1 Reliabilities of key sub-assemblies*

| **Sub-assemblies** | **Sources** | **Reported failure rate (failures/year), ground benign (GB), environment** | |
|---|---|---|---|
| | | **Lower** | **Upper** |
| Civil engineering steel or concrete structures and foundations for off-shore application | Delorm (2013) | 0.0011 | 0.0500 |
| Monopile or fixed sea-bed location and pinning systems for off-shore application | Delorm (2013) | 0.0011 | 0.0500 |
| Nacelle or housing or duct structure | Delorm (2013) | 0.0011 | 0.0011 |
| Cross-beam structure | Delorm (2013) | 0.0011 | 0.0500 |
| Detachable sea-bed location and pinning systems for off-shore application | Tavner (2016) | 0.0022 | 0.0022 |
| Buoyant, hinged steel structures for application | Tavner (2016) | 0.0022 | 0.0022 |
| Buoyant hull for off-shore application | Delorm (2013) | 0.0011 | 0.0011 |
| Sub-sea hydraulic rams | Tavner (2016) | 0.0090 | 0.0090 |
| Yoke for hydraulic rams | Tavner (2016) | 0.0045 | 0.0045 |
| Retractable turbine legs | Delorm (2013) | 0.0090 | 0.0090 |
| Mooring yoke system | Delorm (2013) | 0.0500 | 0.0500 |
| Connecting links | Delorm (2013) | 0.1000 | 0.1850 |

*(Continues)*

*Table E.1* (*Continued*)

| **Sub-assemblies** | **Sources** | **Reported failure rate (failures/year), ground benign (GB), environment** | |
|---|---|---|---|
| | | **Lower** | **Upper** |
| Catenary mooring, chain | Delorm (2013) | 0.0237 | |
| Catenary mooring, steel wire rope | Delorm (2013) | 0.0000 | |
| Pile anchors | Delorm (2013) | 0.0074 | |
| Corrosion protection | Delorm (2013) | 0.0443 | 0.1168 |
| Ventilation system | Delorm (2013) | 0.1048 | 0.1048 |
| Water-cooled heat exchanger | Delorm (2013) | 0.0936 | 0.0936 |
| Hydraulic power units for off-shore application | Delorm (2013) | 0.0394 | 0.0394 |
| Hydraulic receivers, accumulators and tanks for off-shore application; | Delorm (2013) | 0.0394 | 0.0394 |
| Hub, electric pitch control unit, pitch bearing, per blade | Chung *et al.* (2015) | 0.1950 | 0.1950 |
| Hub, hydraulic pitch control unit, pitch bearing, per blade | Tavner (2012) | 0.1000 | 0.2000 |
| Main shaft, main bearing, couplings | Delorm (2013) | 0.0310 | 0.0550 |
| Rim bearing and seals | Delorm (2013) | 0.0208 | 0.0208 |
| Shaft seals | Delorm (2013) | 0.0613 | 0.0613 |
| Fixed pitch rotor blades for variable speed turbines for off-shore application, per blade | Delorm (2013) | 0.0080 | 0.0800 |
| Pitchable rotor blades for variable or fixed speed turbines for off-shore application, per blade | Tavner (2012) | 0.0267 | 0.0800 |
| Gearbox, lubrication and cooling system | Spinato *et al.* (2009) | 0.2000 | 0.3000 |
| Hydraulically operated brake system | Delorm (2013) | 0.0550 | 0.1350 |
| Direct drive water-cooled permanent magnet synchronous generator | Carroll *et al.* (2014) | 0.0706 | 0.0706 |
| High-speed geared drive water-cooled wound induction synchronous generator | Tavner (2012) | 0.2000 | 0.2500 |
| High-speed geared drive water-cooled permanent magnet synchronous generator | Tavner (2012) | 0.2000 | 0.2500 |
| LV partially rated VSC in the power range 0.6–0.8 MVA to convert generator outputs to fixed LV power frequencies | Chung *et al.* (2015) | 0.1000 | 0.1000 |
| LV fully rated VSC in the power range 2–3 MVA to convert generator outputs to fixed LV power frequencies | Chung *et al.* (2015) | 0.5930 | 0.5930 |
| Flexible LV transmission cable, assume 500-m length | IEEE Std 493 (1997) Gold Book | 0.0012 | 0.0012 |
| MV or LV isolator | IEEE Std 493 (1997) Gold Book | 0.0052 | 0.0052 |
| LV contactor | IEEE Std 493 (1997) Gold Book | 0.0052 | 0.0052 |

(*Continues*)

*Table E.1* *(Continued)*

| Sub-assemblies | Sources | Reported failure rate (failures/year), ground benign (GB), environment | |
|---|---|---|---|
| | | Lower | Upper |
| LV fully rated AC/DC VSC for auxiliary system | Chung *et al.* (2015) | 0.0630 | 0.0630 |
| LV circuit breaker feeding auxiliary supply | IEEE Std 493 (1997) Gold Book | 0.0027 | 0.0027 |
| LV cables for auxiliary system | IEEE Std 493 (1997) Gold Book | 0.0040 | 0.0040 |
| Battery electrical auxiliary system for off-shore | Delorm (2013) | 0.0319 | 0.1472 |
| MV fully rated single channel VSCs in the power range 3–5 MVA to convert generator outputs to fixed LV power frequencies | Carroll *et al.* (2014) | 0.7400 | 0.7400 |
| MV fully rated parallel redundant channel VSCs in the power range 4–6 MVA to convert generator outputs to fixed LV power frequencies | Carroll *et al.* (2014) | 0.4000 | 0.4000 |
| MV contactor | IEEE Std 493 (1997) Gold Book | 0.0153 | 0.0153 |
| VSC controller | Chung *et al.* (2015) | 0.0420 | 0.0420 |
| Device controller and output gain controller for off-shore application | Delorm (2013) | 0.2390 | 0.6300 |
| Process automation and SCADA | Delorm (2013) | 0.2860 | 0.7540 |
| Dry type LV/MV transformers to raise export cable voltage for power output to shore-side substation | IEEE Std 493 (1997) Gold Book | 0.0036 | 0.0036 |
| Dry type MV/LV transformer to step down output voltage for LV auxiliary system | IEEE Std 493 (1997) Gold Book | 0.0018 | 0.0018 |
| Oil-cooled MV/HV transformers to raise export cable voltage for power output to the grid | IEEE Std 493 (1997) Gold Book | 0.0030 | 0.0030 |
| LV circuit breaker feeding the sea-shore export cable | IEEE Std 493 (1997) Gold Book | 0.0044 | 0.0044 |
| MV circuit breakers feeding the shore-side sub-station | IEEE Std 493 (1997) Gold Book | 0.0064 | 0.0064 |
| SCADA monitoring interface, sea-shore or shore-sea for off-shore application and umbilical fibre optic cable | Delorm (2013) | 0.0101 | 0.0159 |
| Cable buoyancy support with swivel | Delorm (2013) | 0.0500 | 0.0500 |
| Buoyant LV or MV cable for off-shore application, assume 100-m length | Delorm (2013) | 0.0422 | 0.1112 |
| Sub-sea connector | Delorm (2013) | 0.0000 | 0.0091 |
| Fixed LV export cable systems, assume 10-km length | IEEE Std 493 (1997) Gold Book | 0.0236 | 0.0236 |

*(Continues)*

*Table E.1* (*Continued*)

| **Sub-assemblies** | **Sources** | **Reported failure rate (failures/year), ground benign (GB), environment** | |
|---|---|---|---|
| | | **Lower** | **Upper** |
| Fixed MV export cable systems, assume 10-km length | IEEE Std 493 (1997) Gold Book | 0.0052 | 0.0052 |
| Shore-side sub-station | IEEE Std 493 (1997) Gold Book | 0.0200 | 0.0200 |
| MV circuit breaker feeding the sea-shore export cable | IEEE Std 493 (1997) Gold Book | 0.0064 | 0.0064 |
| MV circuit breaker feeding off-shore output to power grid | IEEE Std 493 (1997) Gold Book | 0.0064 | 0.0064 |
| Sub-station power flow controller | Chung *et al.* (2015) | 0.0420 | 0.0420 |

*Appendix F*

# Wind power historical timeline

| Year | Originator or origination of development | Technologies including : | Photo, diagram or note |
|---|---|---|---|
| | | • **Turbines**<br>• **Foundations**<br>• **Foundations**<br>• **Electrical machines and electronics**<br>• **Control**<br>• **Reliability** | |
| | Particular thanks to Dr Wenjuan Blackwood, who first compiled this historical table<br>For other historical expositions see: Needham (1994), Hau (2006), Zhang (2009), Burton *et al.* (2011), Gasch *et al.* (2012) and Gipe (2016) | | |
| 200 BC | Persia, vertical axis wind machines were first used | | |
| 70 AD | Greece, Hero's Pneumatica, a reaction steam turbine. There is debate whether Hero invented this device, or it was stimulated from different examples developed by Hero | | |
| 634–44 AD | Persia, Iran<br>Practical wind-mills built in Sistan, Persia–Afghanistan border region by the Rashidun Caliph Umar for grinding grain and pumping water<br>1963: 50 of these types of windmill were still operating in Neh, Iran<br>Hau (2006) | VAWT wind-mill for milling grain, vertical axle, long vertical driven shafts, rectangle blades<br>Enclosed by a 2-storey circular wall, millstones at the top, rotor at the bottom<br>Rotor: spoked with 6–12 upright ribs, each covered with cloth to form separate sails |  |

| | | |
|---|---|---|
| 1119 | The Netherlands | 4-blade HAWT post-mills for draining water, milling grain, sawing wood. Easy to yaw, but support might have been an issue<br>Transition from post-mills to cap-mills  |
| 1191 | England, first VAWT wind-mill in West Suffolk | |
| 1219 | China | Reference to a possible common origin for wind-mills in W China and Persia, see earlier, Needham (1994) and Zhang (2009)<br>Sheng Ruozi refers to wind-mills in the 'Placid Retired Scholar', Yelü Chucai (1190–1244), prominent Jin and Yuan statesman after fall to the Mongols in 1234<br>The passage refers to a journey in 1219 to Turkestan in modern Xinjiang, and of Hechong Fu to Samarkand, in modern Uzbekistan<br>These wind-mills had luffable self-adjusting sails, in response to wind condition as the wind-mill rotated |
| 1200s | Europe | 4-blade, HAWT, squat structure with wooden shutters that were adjustable as luffable blades |
| 1295 | The Netherlands | 4-blade HAWT upwind cap- or tower-mill<br>Post-mills dominated the milling and pumping scene in Europe until the nineteenth century when tower-mills began to replace them<br>The advantage of the cap- or tower-mill over the earlier post-mill is that it is not necessary to turn the whole mill body or buck, with all its machinery into the wind, this allows more space for machinery as well as for storage  |

*(Continues)*

(*Continued*)

| | | | |
|---|---|---|---|
| Early 1500s | The Netherlands | Driving scoop wheels for pumping water, Wipmolen or Kokermolen hollow post-mill | A more compact tower post-mill, which could be described as a cap-mill, where the yawing machinery was concentrated in the cap of the mill |
| 1800s | The Netherlands | Development of precision wooden pin and socket gears for cap-mills | Patent for wooden right-angle gearbox between the horizontal mill rotor axis and the vertical wallower gear axis, transferring horizontal to vertical rotary power<br>Influenced by seventeenth and eighteenth centuries scientific innovations from Leibnitz, Bernoulli and Euler, based on the emerging mathematics of the natural sciences |
| 1866 | USA, pumping water on central plains farms and filling railroad tanks | HAWT, multi-blade upwind turbines<br>Application of US mass-production methods, using iron and steel components, to large, remote, mechanised farms |  |

| | | | |
|---|---|---|---|
| 1854 | USA, Daniel Halladay formed the Wind Engine and Pump Company where it became one of the most successful wind-mill companies, Batavia, Illinois | HAWT, multi-blade upwind turbines for water pumping<br>Automatic yawing, innovation, design and manufacturing excellence. Availability of steel facilitated this rotor technology |  |
| !820–73 | Europe and USA, Oersted, Faraday, Arago, Babbage, Herschel, Pixii, Lenz, Wheatstone, Cooke, Gramme, Pacinotti | Development of the first DC electric generators and motors using current switching via a rotating commutator |  |
| 1882–86 | USA, Edison | First DC electric central power station developed applying large DC generators and underground LV cable for large-scale transmission and usage of DC electric power from Pearl Street Power Station, central New York |  |

*(Continues)*

(*Continued*)

| | | | |
|---|---|---|---|
| 1883 | USA, first automatic wind-mill for electricity production for battery charging by Charles Brush, in Cleveland, Ohio | 12 kW, 144-blade, 4 t, HAWT<br>DC generator only available in the USA and Europe since 1878<br>Innovative combining available US wind turbine manufacturing techniques with new DC electrical generation and utilisation |  |
| 1885 | Italy, Ferraris | Demonstrated a rotating magnetic field by phase-separated single-phase AC flowing into space-quadrature windings, precursor of the induction motor | |
| 1885–90 | UK, Ferranti Subsequently, USA, Tesla and Westinghouse | First AC electric central power station with generators, transformers and underground HV cables for large-scale transmission and usage of AC electric power from Deptford Power Station to Central London |  |
| 1886 | Tesla<br>Dobrowolski | First polyphase AC induction motor/generator developed<br>First three-phase AC induction motor developed, working independently from Tesla<br>Then, developed a three-phase AC induction motor with a polyphase slip-ring rotor connection to resistors for starting control | |

| | | | |
|---|---|---|---|
| 1887 | UK, Scotland, Prof James Blyth of the Royal College of Science, Glasgow, now known as Strathclyde University, for electricity production and battery charging | 10 kW, 7-blade VAWT, driving a DC generator. Believed to have had adjustable blades, which contemporary alternatives did not |  |
| 1891 | Denmark, Poul la Cour developed a wind turbine for electricity production and battery charging<br>la Cour | 10 kW, 4-blade HAWT, fixed pitch wind turbine driving a DC generator<br>Innovative aerodynamic system and electrical energy utilisation for electrolysis and hydrogen storage, mitigating wind variability<br>Note the 20-year life and high reliability of la Cour and Lykkegard wind turbines in Denmark |  |
| 1888–1900 | Denmark, Poul la Cour USA, Halladay | Experimental wind-mills used to generate electricity, based on Danish and American designs. The need for electricity for pumping and light, on large, remote, mechanised farms, in the flat, windy Mid-West USA, stimulated US wind power development | |
| 1900–10 | Denmark, 2,500 electric wind-mills in use, generating up to 30 MW. Flat, windy Danish landscape | | |
| 1908 | Denmark, 72 electric wind turbine generators recorded | 5–25 kW, 4-blade upwind HAWT, *D* 23 m, 24-m high | |
| 1910–30 | USA, produces 100,000 farm HAWT wind-mills/year, for water pumping | Mixture of American and Danish DC and AC designs<br>Proof of high quality of US mass production techniques and the need for power where there was a lack of grid connection<br>Did Danish emigrants to the USA distribute Poul la Cour technology in USA? | |
| 1910–14 | Competition for electric wind-mills, from internal combustion engines following the development of diesel- and then petrol-engine driven generators | | |
| 1904–11 | USA and Germany, Ward Leonard, Kramer, Schrage | Development of variable speed DC and then AC electric generators/motors using switching via rotating commutators | |

*(Continues)*

(*Continued*)

| | | | |
|---|---|---|---|
| 1914–18 | First World War, reduction in oil supplies, 20–35-kW electric wind-mills were being built | | |
| 1918 post-war | Wind-mill development languished | Small wind turbines proving less reliable for electricity production than diesel- or petrol-engine driven generators, but AC electrical grid connection was becoming more widespread | |
| 1920s | Following the First World War, the influence of aerodynamic knowledge from aircraft starts to affect wind turbine design. For example, wing and propeller development | | |
| | Betz (1926) | | |
| 1923–73 | Various inventors and industrial companies | Development of valve (ignitron) and semiconductor (thyristor) switches to convert AC/DC and DC/AC, facilitating reversible power flow through AC generators | |
| 1930s | Denmark, USA, Wind-mills for electricity became common on large farms | High-tensile steel was cheap, and wind-mills were being placed atop prefabricated open steel lattice towers | Decline of the American multi-blade turbine concept in favour of European 4-, 3- and 2-blade concepts |
| 1930s and 1940s | Europe and North America, national AC electrification using fossil-fired power stations feeding national grids, based on Ferranti and Tesla's ideas and Westinghouse's investments | | |
| | Research programmes considered wind power as a supplement to the national grids in Denmark, France, Germany and the United Kingdom | | |
| 1931 | USSR, Yalta, modern wind turbine | 100 kW, 30-m high, 3-blade upwind HAWT, geared-drive, connected to 6.3-kV AC distribution system<br>32% capacity factor. Adjustable blade flaps<br>Post-mill with the whole structure rotating along a track<br>Early large 3-blade machine exhibiting clear signs of growing aeronautical influence<br>Notice similarity to Dutch post-mills |  |

| | | |
|---|---|---|
| 1932–50 | Nyquist, Bode and others | Development of the techniques necessary to control the power flow through reversible AC/DC and DC/AC power systems, essential to the development of variable pitch, variable speed wind turbines |
| 1938–44 | Denmark, FLSmidth | Aeromotor 45-kW range, 2-blade upwind HAWT<br>FLSmidth a former cement production machinery company started the manufacture of wind turbines, based on la Cour's concepts, to challenge the rising cost of petrol- and diesel-powered generation<br>Was this the first off-shore wind turbine?  |

*(Continues)*

(*Continued*)

| | | | |
|---|---|---|---|
| 1940 | Germany, Ventimotor company formed with a test centre near Weimar, Germany to develop wind turbines for the war effort<br>Included Ulrich Hutter among its key personnel<br>Hutter (1954) | W34 2-blade downwind HAWT with excellent aerodynamics<br><br>Light and cost-effective, with guyed tower |  |
| 1940–70 | USA, the United Kingdom, Germany, Result of Second World War experience | Development of statistically based reliability theories to improve the reliability of engineering developments<br>Particularly in the military, aero-space, space and engineering industries. This knowledge subsequently improved the reliability and development of variable speed electrical wind turbines | |

| | | | |
|---|---|---|---|
| 1941 | USA, operation of Smith-Putnam design, world's first megawatt-size wind turbine connected to the local electrical distribution system on Grandpa's Knob, Castleton, Vermont<br><br>Considered to be the grandfather of the modern electric wind turbine<br><br>Putnam (1947) | 1.25 MW, *D* 57 m, 40-m high, 2-blade downwind HAWT, geared-drive, constant speed, full-span pitch control, stall-regulated with synchronous generator. Sophisticated, modern, first AC grid-connected wind turbine<br>Rugged design by Palmer Cosslett Putnam, manufactured by S. Morgan Smith Company, included collaboration by MIT aerodynamicist von Karman |  The turbine completed on Grandpa's Knob. |
| 1939–45 | As a result of the Second World War, another reduction in available oil supplies accelerates wind power development | | |
| 1945–70 | Western Europe and particularly Denmark, new growth in wind power taking place, principally under the direction of those trained by Poul la Cour | | |
| 1954 | UK, Scotland, Costa Head in Orkney. First experimental grid-connected wind turbine by John Brown Engineering Company | 100 kW, *D* 18 m, 3-blade downwind HAWT, geared-drive, pitch-regulated<br>Grid-connected slip ring AC induction generator, which facilitated a variable speed turbine with a huge top-head mass<br>Lack of demand investigation, marketing or mass-production techniques |  |

*(Continues)*

(*Continued*)

| | | |
|---|---|---|
| 1950–70 | Various inventors and industrial companies | Development of new types of power semi-conductor switches, the thyristor, transistor and IGBT |
| 1956 | UK, St. Albans, Enfield-Andreau turbine | 100 kW, *D* 25 m, 2-blade downwind HAWT<br>French Andreau design built by the Enfield Cable Co.<br>Unusual vertical axis air turbine-generator at ground level, driven by high-speed air flow, exhausted from the main turbine blade tips<br>An interesting idea but not subsequently exploited<br>Note the aeronautical influence and guyed tower to reduce structural mass<br> |

| | | | |
|---|---|---|---|
| | Denmark, Gedser, Juul develops the modern wind turbine, forerunner of the Danish Concept. Considered to be the mother of the modern electric wind turbine | 200 kW, $D$ 24 m, 3-blade upwind HAWT, geared-drive, stall-regulated, with aerodynamic tip brakes on rotor blades, released automatically in over-speed<br>Blade tip brakes were a good innovation |  |
| 1958 | France, Station d'Etude de l'Energie du Vent, Nogent-le-Roi | 800 kVA, 3-blade downwind HAWT, geared-drive, pitch-regulated, experimental AC wind turbine<br>Interesting design but no subsequent development, because of French national decision to concentrate on nuclear power<br>Note the large top head mass and therefore need for large structural steel support |  |

(Continues)

(*Continued*)

| | | | |
|---|---|---|---|
| 1957 onwards | Number of inventors | Back-to-back reversible a DC and AC drives using thyristor or other semiconductor bridges with forced static commutation, switched twice per cycle<br>Static commutator eventually renders rotating commutators obsolete<br>These developments were essential for variable speed electric wind turbines using AC generators and converters and variable speed generators with electric pitch motors | |
| 1972 | International oil crisis, triggered by the Yom Kippur war, a renaissance of wind power<br>See Gipe (2016), considering both small and large wind turbines | | |
| 1975 | USA | General Electric Invention of new fast-acting semiconductor devices | |
| 1975–85 | USA and Europe | Aerodynamic and material development of blade aero-foils, airbrakes, pitch mechanisms and blade materials | Evolving from steel and aluminium to composite glass-fibre and innovative balsa and bamboo wood structures |
| 1976–81 | USA and Western Europe, modern small-scale wind turbines | 1–10 kW, VAWT and HAWT | Small-scale inheritors of the 1930s small USA and Danish turbines. The market was still uncertain because the technology is still unresolved |
| 1979 | The United Kingdom, Carmarthen Bay | 130 kW, VAWT450 prototype with furling blades, encouraged by the UK CEGB, very unusual but did not work well |  |

| | | |
|---|---|---|
| Denmark, Nibe, two experimental machines were erected<br><br>One with pitch control and one without | 200 kW, $D$ 24 m, 3-blade upwind HAWT, geared-drive, fixed speed, stall-regulated with aerodynamic rotor blade tip brakes |  |

*(Continues)*

(*Continued*)

| | | | |
|---|---|---|---|
| | If Gedser was the mother of the modern electric wind turbine, these two were her strongest children | 630 kW, *D* 40 m, 3-blade upwind HAWT, geared-drive, full-span pitch control, fixed speed, stall-regulated |  |

| | | | |
|---|---|---|---|
| 1980 | Germany, Wilhelmshaven, MBB $30 million 1-blade Monopteros wind turbine program with three 600-kW prototypes Program featured a line of wind turbines from *D* 15 to 56 m | 600 kW, *D* 15–56 m, 1-blade upwind HAWT, geared-drive<br>Very novel, high performance, light weight with a guyed tower to reduce structural mass<br>Some are still operating, but the concept is unpopular with customers<br>German legacy of Hutter |  |
| | USA, Western Europe and Japan | Replacement of drive bridge thyristors by faster, cheaper, silicon switches, including insulated gate bipolar transistors (IGBT), offering faster switching speeds<br>Development of PWM IGBT voltage source converters (VSC) drives and early controllers, static commutators enabling simple 1-, 2- and 4-quadrant control of wind turbine AC generators, rendering earlier rotating commutators obsolete | |
| | The Netherlands | Development of modern 300-kW geared-drive, 3-blade, stall-regulated, fixed speed wind turbines | |

*(Continues)*

(*Continued*)

| | | | |
|---|---|---|---|
| | USA, great California wind rush | Large numbers of wind turbines $\leq 100$ kW, mostly HAWT but some VAWT<br>Very poor reliability from many different designs but some did survive |  |
| | USA, MOD 0 | 200 kW, 2-blade downwind HAWT, geared-drive, full-span pitch control<br>Note the large top head mass and lattice structural tower |  |

| | | | |
|---|---|---|---|
| 1981 | USA, MOD 1 | 2 MW, 2-blade downwind HAWT, geared-drive, full-span pitch control<br>Overweight and unreliable<br>Note the large top head mass and lattice structural tower |  |
| | USA, MOD 2 | 2.5 MW, *D* 91 m, 2-blade upwind HAWT, geared-drive, full-span pitch control<br>Sophisticated light-weight. No teeter hub so excessive stresses at blade centre |  |

*(Continues)*

(*Continued*)

| | | |
|---|---|---|
| 1982 | Sweden, wind turbine S75-3<br>USA, wind turbine S4 | 2 MW, 3-blade upwind HAWT, geared-drive, full-span pitch control<br>4 MW, 2-blade downwind HAWT, geared-drive, full-span pitch control<br>Sophisticated design. Huge and complex<br> |

| | | | |
|---|---|---|---|
| 1983 | Germany, \$55 million invested in this wind turbine, which operated for only 420 h before experiencing hub fatigue failure | 3 MW, *D* 100 m, 100-m tall, 2-blade downwind HAWT, geared-drive, full-span pitch control, with fully rated cyclo-converter supplying AC power<br>Note trademark German guyed tower to reduce structural mass<br>Very unusual, large, risky and unreliable<br>The large array upwind of the turbine measured local wind-speeds and turbulence incident on the turbine<br>Modern legacy of Hutter and the Ventimotor |  |

*(Continues)*

*(Continued)*

| | | | |
|---|---|---|---|
| 1987 | The United Kingdom, LS1, Burgar Hill, Orkney<br><br>Wind Energy Group prototype large grid-connected wind turbine designed and constructed by Wind Energy Group | 3 MW, *D* 60 m, 2-blade upwind HAWT, geared-drive, full-span pitch control, variable speed<br>Note large top-head mass and consequent strong concrete tower<br><br>Very unusual, big, risky and unreliable but a prototype for WEG's later highly successful MS 2-blade upwind HAWTs |  |

| | | | |
|---|---|---|---|
| | The United Kingdom, HWP 55 Richborough | Howden 1 MW, 3-blade upwind HAWT, geared-drive, stall-regulated, fixed speed, with rotor aerodynamic tip brakes<br>Simple, rugged, grid-connected, reliable, but lack of market penetration<br>Note smaller top head mass and steel tube tower |  |
| 2000 | Denmark's first off-shore wind farm, Middelgrunden | 20 Siemens 2 MW 3-blade upwind HAWT, geared-drive variable speed turbines mounted on concrete gravity caissons |  |

*(Continues)*

(*Continued*)

| | | | |
|---|---|---|---|
| 2002 | UK's first off-shore wind farm, Blyth | 2 Vestas 2-MW 3-blade upwind HAWT, geared-drive, variable speed turbines mounted on steel monopiles<br>Developed by E.ON, Shell Renewables, NUON and Border Wind consortium, operator E.ON<br>UK's New & Renewable Energy Centre (NaREC) established at Blyth, now ORE Catapult, close to this wind farm. The farm was decommissioned in 2019 |  |
| | Germany, Enercon E-112 | 6 MW, *D* 114 m, 3-blade upwind HAWT, direct drive, full-span pitch control installed on-shore<br>Note large top-head mass and consequently strong concrete tower<br>Son of Growian but with good reliability<br>An all-electric turbine with fully-rated grid-connected IGBT variable speed converter<br>Enercon's owner, Aloys Wobben, was a power electronic engineer<br>Although this turbine was installed on-shore it is the forerunner of the huge turbines to be installed in large off-shore wind farms |  |
| 2007 | The Netherlands | Blue H Technologies, installed the world's first floating off-shore downwind 2-blade HAWT |  |

| | | |
|---|---|---|
| 2008–20 | World-wide | Wind turbine improvements, Manwell *et al.* (2009):<br>• Blade manufacture and life improvements by combinations of materials, resins and fibres for strength, including use of carbon fibre, different woods, polyester and epoxy resins. Improved blade coatings to withstand off-shore weather conditions<br>• Direct-drive versus geared-drive wind turbine generators and use of permanent magnet rather than wound rotor wind turbine generators, Polinder *et al.* (2013)<br>• Wider use of static frequency change converters to allow wind turbine variable speed, strengthening grid connection voltage and frequency control but also ensuring intermittent grid fault ride-through capability during lightning storms or grid-switching<br>• Off-shore wind turbine fixed foundations including the following:<br>– cast concrete gravity caissons for shallow water<br>– steel monopiles for deeper water<br>– steel 3-leg and twisted jackets for deep water<br>• Off-shore wind considering floating versus fixed mounting of turbines |
| 2008 | China, Dong Hai Da Qiao (东海大桥) near Shanghai | Installation of a 3-MW offshore 3-blade upwind HAWT, Sinovel SL3000/90, geared drive, fully-rated grid-connected IGBT variable speed converter, pitch-regulated<br>Wind turbines mounted on concrete gravity caissons<br>Note the low top-head mass and slender steel tower<br> |

*(Continues)*

(*Continued*)

| | | | |
|---|---|---|---|
| 2009 | Norway, Statoil's Hywind demonstrator project | Siemens SWT, 2.3 MW, 3-blade, upwind HAWT, geared drive, fully-rated grid-connected IGBT variable speed converter, pitch-regulated, mounted on a floating, moored spar-buoy<br>Note the low top-head mass, slender steel tower and spar-buoy |  |
| 2011 | Portugal, Aguacadoura, WindFloat | MHI-Vestas, 2 MW, 3-blade, upwind HAWT, geared drive, fully-rated grid-connected IGBT converter, pitch-regulated, mounted on a floating, moored three-chamber caisson. Note low top-head mass, slender steel tower but large floating caisson |  |

| | | | |
|---|---|---|---|
| 2013 | UK, London Array Off-shore Wind Project | 630-MW off-shore wind farm, 175 Siemens SWT, 3.6 MW 3-blade, upwind HAWT, geared drive, synchronous permanent magnet generator, fully rated grid-connected IGBT variable speed converter, pitch-regulated wind turbines mounted on steel mono-piles, in relatively shallow water |  |
| | | Wind turbines with step-up transformers connected to two off-shore sub-stations by 30-kV AC collector cables. Substations inter-connected to shore by 2 parallel 53-km 150-kV AC cable links, using AC sub-stations |  |

(Continues)

(*Continued*)

Enabling lower capital cost effective medium-distance electrical power transfer

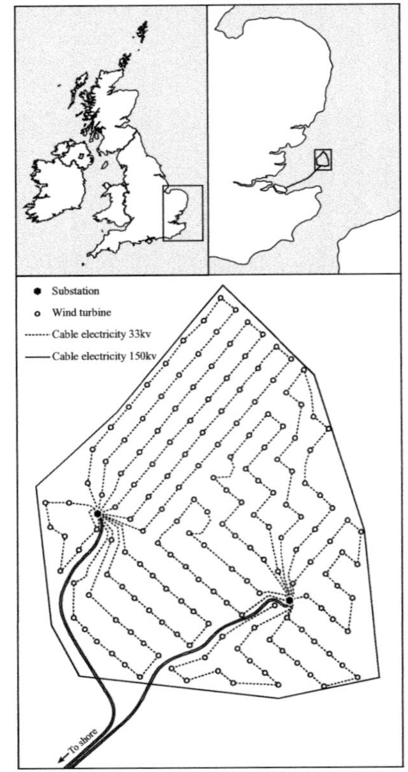

| | | | |
|---|---|---|---|
| 2013 | Germany, BARD Off-shore Wind Project | 400-MW off-shore wind farm, 80 5-MW 3-blade, upwind HAWT, geared drive, doubly fed wound asynchronous generator, partially-rated grid-connected IGBT variable speed converter, mounted on 3-leg steel jackets, for deep water, rather than earlier, more common steel mono-piles |  |
| | | Wind turbines with step-up transformers connected by 170- or 380-kV AC collector cables to off-shore sub-stations and IGBT HVDC converter-stations |  |

*(Continues)*

(*Continued*)

| | | | |
|---|---|---|---|
| | | Inter-connected to shore from HVDC converter-stations by 100-km, ±150-kV DC cable links, enabling optimal economic long-distance electrical power transfer | 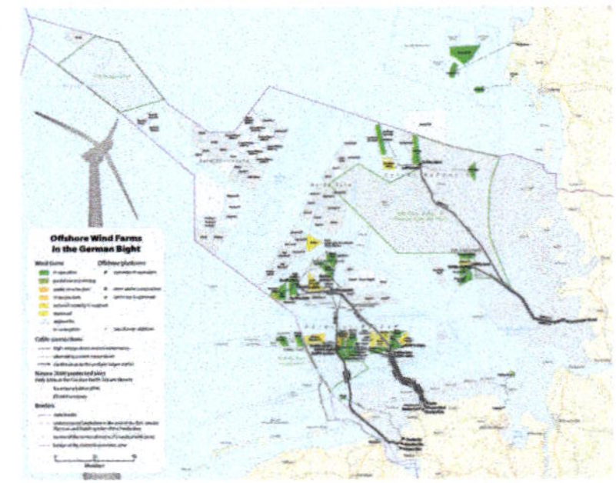 |
| 2019 | UK, Scotland Hywind Pilot Park | 50-MW floating off-shore wind farm, 5 Siemens SWT, 6 MW 3-blade, upwind HAWT, direct drive, synchronous permanent magnet generator, fully rated grid-connected IGBT variable speed converter, pitch-regulated mounted on floating, moored spar-buoys Inter-connected to shore by a 35-km 33-kV AC export cables with battery support |  |

# Index